Mechanisms and Machine Science

Volume 31

Series editor

Marco Ceccarelli, Cassino, Italy

More information about this series at http://www.springer.com/series/8779

Burkhard Corves · Erwin-Christian Lovasz
Mathias Hüsing
Editors

Mechanisms, Transmissions and Applications

Proceedings of the Third MeTrApp Conference 2015

Springer

Editors
Burkhard Corves
Department of Mechanism Theory
and Dynamics of Machines
RWTH Aachen
Aachen
Germany

Mathias Hüsing
Department of Mechanism Theory
and Dynamics of Machines
RWTH Aachen
Aachen
Germany

Erwin-Christian Lovasz
Faculty of Mechanical Engineering
Universitatea Politehnica Timişoara
Timişoara
Romania

ISSN 2211-0984 ISSN 2211-0992 (electronic)
Mechanisms and Machine Science
ISBN 978-3-319-17066-4 ISBN 978-3-319-17067-1 (eBook)
DOI 10.1007/978-3-319-17067-1

Library of Congress Control Number: 2015934916

Springer Cham Heidelberg New York Dordrecht London

Printed on acid-free paper

Springer International Publishing AG Switzerland is part of Springer Science+Business Media (www.springer.com)

Preface

MeTrApp 2015 is already the third edition of a conference that started in 2011 as a workshop organized by the University of Timisoara in Romania. The second edition was organized as a conference in Bilbao, Spain, by the University of the Basque Country. Now this book is already the third of its kind presenting the collection of scientific papers that were presented on the occasion of the Third Conference on Mechanisms, Transmissions and Applications organized by RWTH Aachen University in Aachen, Germany.

The driving force behind this now well-established conference series is the International Federation for the Promotion of Mechanism and Machine Science namely its two Technical Committees "Linkages and Mechanical Controls" and "Gearing and Transmissions".

The aim of this Third Conference on Mechanisms, Transmissions and Applications is to offer a stage for original research presentations for researchers, scientists, industry experts, and students in the fields of mechanisms and transmissions with special emphasis on industrial applications in order to stimulate the exchange of new and innovative ideas. By collecting the peer-reviewed papers that were funnelled through a rigorous two-stage review process, within the Springer Mechanism and Machine Science Series, we take the chance to present and share the outcome of this conference with interested professionals and scientists who are at the front line of mechanism and machine theory. Thus, the content of this book is subdivided into different sections that cover the topics Mechanism and Machine Design, Mechanical Transmissions, Industrial Applications, VDI-Guidelines, Biomechanics and Medical Engineering, Robotics, Mechatronics, and Dynamics of Mechanisms and Machines.

In total, we received 41 papers, which were carefully reviewed by three reviewers per paper in a double-review process. Finally, 35 papers were accepted for presentation during the conference and for publication in this book. Thus, we want to express our thanks to the reviewers who contributed to this process with their experience and scientific background. Only through their effort was it possible to organize a thorough yet speedy review process.

Many thanks also go to the authors for their enthusiasm about this conference and to all who helped in organizing this publication as well as the conference itself. We thank the German Research Foundation for awarding a grant in order to host this conference, as well as RWTH Aachen University for supporting this conference. We also thank the staff at Springer for their support through all stages of preparing this book.

We very much hope that this book also inspires those who could not attend the conference to contribute or attend one of the next issues of MeTrApp.

February 2015

Burkhard Corves
Erwin-Christian Lovasz
Mathias Hüsing

Conference Committees

Conference Chairman

Burkhard Corves, RWTH Aachen University, Germany

Conference Co-Chairmen

Erwin-Christian Lovasz, Polytechnical University of Timisoara, Romania
Daizhong Su, Nottingham Trent University, Great Britain

Program Committee

Oscar Altuzarra, University of the Basque Country, Bilbao, Spain
Juan Antonio Carretero, University of New Brunswick, Canada
Marco Ceccarelli, University of Cassino and South Latium, Italy
Antoni Gronowicz, Wroclaw University of Technology, Poland
Alfonso Hernandez, University of the Basque Country, Bilbao, Spain
Chin-Hsing Kuo, National Taiwan University of Science and Technology, Taiwan
Heikki Martikka, Lappeenranta University of Technology, Finland
Doina Pisla, Technical University of Cluj-Napoca, Romania
Anupam Saxena, IIT Kanpur, India
Eres Soylemez, Middle East Technical University, Ankara, Turkey
Miroslav Vaclavik, VUTS Liberec, Czech Republic
Fernando Viadero, University of Cantabria, Spain
Yao Yan-an, Beijing Jiaotong University, China

Local Organizing Committee at RWTH Aachen

Mathias Hüsing, RWTH Aachen University, Germany
Stefan Kurtenbach, RWTH Aachen University, Germany
Claudia Cornely, RWTH Aachen University, Germany

Under the Patronage of IFToMM

Teresa Zielinska, Secretary-General
Erwin-Christian Lovasz, Chair of the Technical Committee for Linkages and Mechanical Controls
Daizhong Su, Chair of the Technical Committee for Gearings and Transmissions

Contents

Part I
Mechanism and Machine Design

The Infinitesimal Burmester Lines in Spatial Movement

Delun Wang, Wei Wang, Huimin Dong and Son Lin

Abstract The paper studies the infinitesimal Burmester lines of the moving body at any instant, according to the invariance of the constraint ruled surface of the binary link C-C. A line-trajectory is expressed by the invariants of axodes of a rigid body in spatial motion. The Euler-Savary analogue of a line-trajectory in spatial movement is described in the Frenet frame of axodes. Both the stationary line congruence of constant axis curvature and the Ball line are revealed. The degenerated cases of the infinitesimal Burmester lines, the characteristic lines L_{HC} and L_{RC}, are discussed according to the H-C curvature and R-C curvature. An numerical example of a spatial linkage RCCC is given to show some of above results.

Keywords Curvature theory · Ruled surface · Invariant · Axode · Kinematics

1 Introduction

The kinematic synthesis of linkages is essentially to locate the special points or lines in the moving body, which trace constraint curves or ruled surfaces of the binary links with specific curvature properties, such as the Burmester points in planar motion. The curvature theory provides the clues to locate the special lines in the moving body by estimating the approximation of a general ruled surface and a

D. Wang (✉) · W. Wang (✉) · H. Dong
Dalian University of Technology, Dalian, China
e-mail: dlunwang@dlut.edu.cn

W. Wang
e-mail: wangweidlut@mail.dlut.edu.cn

H. Dong
e-mail: donghm@dlut.edu.cn

S. Lin
Tongji University, Shanghai, China
e-mail: slin@tongji.edu.cn

B. Corves et al. (eds.), *Mechanisms, Transmissions and Applications*,
Mechanisms and Machine Science 31, DOI 10.1007/978-3-319-17067-1_1

constraint ruled surface. The infinitesimal Burmester lines in spatial motion are not completely defined and readily located yet in the literature. The constraint curves and ruled surfaces in spatial linkages have been introduced in Ref. [1]. Distelli [2] previously presented the Euler-Savary equation of the straight line in spatial motion. The dual numbers, dual vectors and dual matrices are adopted to study the trajectories of line in space. Yang [3] defined characteristic scalars to reflect the infinitesimal properties of ruled surface. Kose [4] adopted dual vector calculus to study the invariants of a line trajectory in spatial motion. The canonical coordinate system was adopted by Veldkamp [5], Kirson [6], Roth [7] to study instantaneous spatial motion. Ting [8] presented a unified algebraic approach for the modeling of the instantaneous motion of all linear elements. McCarthy [9] derived three curvature parameters, invariants of a ruled surface, to study their local shape by the moving frame.

For the complexity of line-trajectories, the kinematic invariants of a ruled surface, three construction parameters [10], are naturally preferred to describe the local geometrical properties of a ruled surface. The construction parameters of both the moving and fixed axodes are used to reveal the intrinsic property of a rigid body in spatial motion, which occupy a similar important fundamental roles in spatial kinematics [11, 12] both in the expressional forms and the contents as the centrodes do in planar motion. Hence, this paper studies the infinitesimal Burmester line of the moving body in spatial motion at an instant, corresponding to binary link C-C, in terms of the invariance of the constraint ruled surfaces [1].

2 Constraint Ruled Surfaces of Spatial Linkages

In linkages, a binary link with two kinematic pairs kinematically connects a moving body to a base link, or fixed frame. The line of the moving body, or the axis/guidance line of a moving joint, is constrained and defined as a ***characteristic line***, and its trajectory in the base link is called a ***constraint ruled surface***.

For a binary link R-C, the moving joint C is a cylindrical pair and the fixed joint R is a revolute pair. Its constraint ruled surface, a hyperboloid of one sheet, is Σ_{RC} with L_{RC} as the characteristic line. The binary links with fixed joint R, such as R-C, R-R, R-H and R-P, have the same constraint ruled surface, or correspond to the same characteristic line on the moving body. Therefore, a cylindrical pair C can be taken as a typical moving joint for the binary links with moving joints C, R, H and P. Hereinafter, all constraint ruled surfaces are represented by a symbol Σ with subscripts. With C-pair as the moving joint of the binary links, three current kinematic joints, C, R, H, of spatial linkages are listed in Table 1.

Table 1 Binary (generalized) links and their constraint surfaces

Binary links	Constraint ruled surfaces	Symbols of lines	Symbols of surfaces
C-C	Constant axis ruled surface	L_{CC}	Σ_{CC}
H-C	Helicoid	L_{HC}	Σ_{HC}
R-C	Hyperboloid	L_{RC}	Σ_{RC}

3 Infinitesimal Burmester Lines

In order to describe the spatial movement of a rigid body relative to the fixed body, the moving Cartesian coordinate system $\{\mathbf{O}_m; \mathbf{i}_m, \mathbf{j}_m, \mathbf{k}_m\}$ is established on the moving body and the fixed system $\{\mathbf{O}_f; \mathbf{i}_f, \mathbf{j}_f, \mathbf{k}_f\}$ is on the fixed body. As we known, there is an instantaneous screw axis (ISA) for the spatial motion of the moving body relative to the fixed body. For all instants, the ISA traces a moving axode Σ_m in $\{\mathbf{O}_m; \mathbf{i}_m, \mathbf{j}_m, \mathbf{k}_m\}$ and fixed axode Σ_f in $\{\mathbf{O}_f; \mathbf{i}_f, \mathbf{j}_f, \mathbf{k}_f\}$.

3.1 The Axodes of a Rigid Body in Spatial Movement

The fixed axode Σ_f is a ruled surface, whose vector expression can be written by an adjoint approach [11] in $\{\mathbf{O}_f; \mathbf{i}_f, \mathbf{j}_f, \mathbf{k}_f\}$ as $\mathbf{R}_f = \boldsymbol{\rho}_f + \mu \mathbf{s}_f$, where $\boldsymbol{\rho}_f$ is the vector of the striction curve of Σ_f, $\mathbf{s}_f$ is the unit vector of the generator of Σ_f. The construction parameters α_f, β_f, γ_f of Σ_f can be derived by Frenet formulas [10].

On the other hand, the moving axode Σ_m is also a ruled surface, whose vector expression can be written in $\{\mathbf{O}_m; \mathbf{i}_m, \mathbf{j}_m, \mathbf{k}_m\}$ as $\mathbf{R}_m = \boldsymbol{\rho}_m + \mu \mathbf{s}_m$, where $\boldsymbol{\rho}_m$ is the vector of the striction curve of Σ_m, $\mathbf{s}_m$ is the unit vector of the generator of Σ_m. The construction parameters α_m, β_m, γ_m of Σ_m can be derived by Frenet formulas. The ***induced construction parameters*** α^*, β^*, γ^*, as key kinematic invariants for the spatial movement, can be obtained as $\alpha^* = \alpha_f - \alpha_m$, $\beta^* = \beta_f - \beta_m$, $\gamma^* = \gamma_f - \gamma_m = 0$. The arc lengths of the spherical image curves of $\mathbf{s}_f$ and $\mathbf{s}_m$ are respectively σ_f and σ_m.

We designate σ to represent σ_f and σ_m for short since they are equal to each other. The properties of axodes are presented in Ref. [11], which lay the groundwork for the following spatial kinematics in this paper.

3.2 A Line-Trajectory and Its Frenet Frame

For a moving body in spatial motion, a point P with Cartesian coordinates (x_{Pm}, y_{Pm}, z_{Pm}) in $\{\mathbf{O}_m; \mathbf{i}_m, \mathbf{j}_m, \mathbf{k}_m\}$ traces a spatial trajectory Γ_P in $\{\mathbf{O}_f; \mathbf{i}_f, \mathbf{j}_f, \mathbf{k}_f\}$. A line L with unit directional vector $\boldsymbol{l}(\delta_l, \theta_l)$ in Frenet frame $\{\boldsymbol{\rho}_m; \mathbf{E}_{1m}, \mathbf{E}_{2m}, \mathbf{E}_{3m}\}$ of Σ_m passes through P. δ_l is the inclined angle between $\mathbf{l}$ and $\mathbf{E}_{1m}$, and θ_l is the

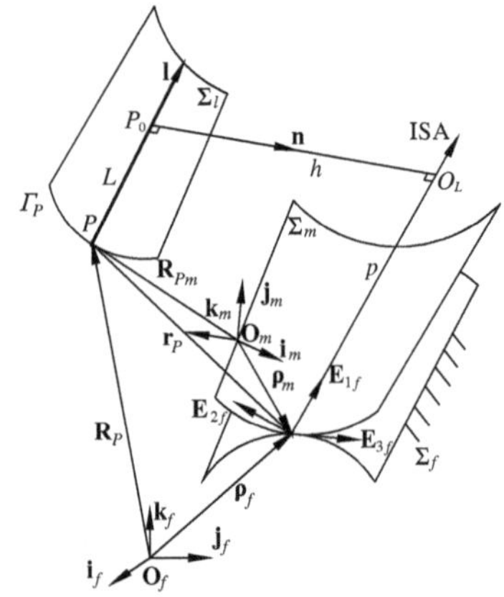

Fig. 1 A line-trajectory adjoint to ISA

directional angle of the projection vector of $\mathbf{l}$ on the plane $\boldsymbol{\rho}_m - \mathbf{E}_{2m}\mathbf{E}_{3m}$. The position of L is located by parameters (p, h) in $\{\boldsymbol{\rho}_m; \mathbf{E}_{1m}, \mathbf{E}_{2m}, \mathbf{E}_{3m}\}$, as shown in Fig. 1.

At an instant, L corresponds to ISA and traces a trajectory-ruled surface Σ_l in $\{\mathbf{O}_f; \mathbf{i}_f, \mathbf{j}_f, \mathbf{k}_f\}$, which can be expressed by

$$\begin{cases} \Sigma_l : \mathbf{R}_l = \boldsymbol{\rho}_l + \mu\mathbf{l} = \boldsymbol{\rho}_f + p\mathbf{E}_{1f} + H\mathbf{E}_{1f} \times \mathbf{E}_1 + \mu\mathbf{l}, \\ H = h/\sqrt{1 - l_1^2}; \quad \mathbf{l} = (l_1, l_2, l_3) = (\cos\delta_l, \sin\delta_l\cos\theta_l, \sin\delta_l\sin\theta_l) \end{cases} \tag{1}$$

The line L is described by the four parameters (l_1, l_2, p, h) in Frenet frame $\{\boldsymbol{\rho}_f; \mathbf{E}_{1f}, \mathbf{E}_{2f}, \mathbf{E}_{3f}\}$ of Σ_f. The differential of the arc length σ_l of the spherical image curve of $\mathbf{l}$ is $d\sigma_l = |d\mathbf{l}/d\sigma|\, d\sigma = \beta^*\sqrt{1 - l_1^2}d\sigma$. The vector equation of striction curve of Σ_l is $\boldsymbol{\rho}_l = \boldsymbol{\rho}_f + p\mathbf{E}_{1f} + h\mathbf{E}_{1f} \times \mathbf{E}_1/|\mathbf{E}_{1f} \times \mathbf{E}_1|$ and the derivative of the striction directrix distance b_l is $db_l/d\sigma = (l_2p - l_1l_3H + l_3\gamma)/(1 - l_1^2)$. The Frenet frame $\{\boldsymbol{\rho}_l; \mathbf{E}_1, \mathbf{E}_2, \mathbf{E}_3\}$ of Σ_l can be established, and the three construction parameters can be derived by the Frenet formulas of a ruled surface as

$$\alpha_l = \frac{l_1\alpha^* - \beta^*(1 - l_1^2)H + db_l/d\sigma}{\beta^*(1 - l_1^2)^{1/2}}, \quad \beta_l = \frac{l_3 + \beta^* l_1(1 - l_1^2)}{\beta^*(1 - l_1^2)^{3/2}}, \quad \gamma_l = \frac{\alpha^*}{\beta^*} + l_1H \tag{2}$$

3.3 *The Infinitesimal Burmester Lines*

3.3.1 Constant Axis Curvature

Based on the properties of the constraint ruled surface, we can discuss the infinitesimal Burmester lines of the rigid body in spatial movement. As presented in Ref. [1], the ***constant axis curvature*** or the C-C ***curvature*** includes β_l = constant and

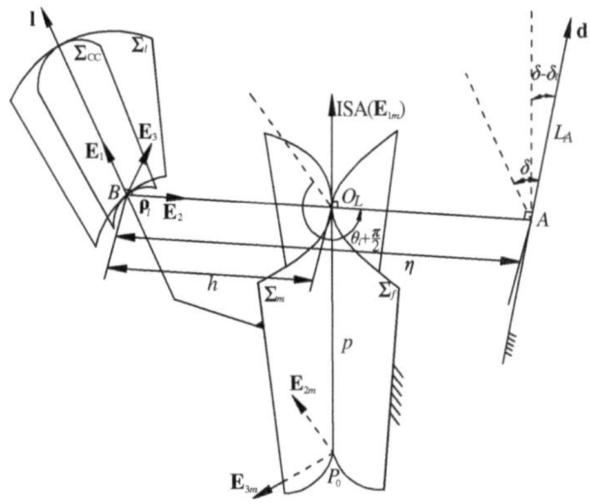

Fig. 2 Euler-Savary analogue of a line-trajectory

$\alpha_l - \beta_l \cdot \gamma_l$ = constant. This implies the spherical image curve of Σ_l is a circle, and the striction curve of Σ_l is a cylindrical curve.

For the constant axis curvature surface Σ_{CC} of Σ_l, the position can be located by the normal of Σ_l at the striction point, or the fixed point A on the fixed axis L_A of Σ_{CC}. The position vector of the fixed point A is

$$\mathbf{R}_A = \boldsymbol{\rho}_l + \eta \mathbf{E}_2, \quad \eta = (\alpha_l - \beta_l \gamma_l)/(1 + \beta_l^2) \tag{3}$$

where η is the distance between the fixed point A and the striction point B along the common normal, as shown in Fig. 2. Based on the Eq. (3), both the direction and the position of Σ_{CC} can be expressed as

$$\begin{cases} \cot(\delta - \delta_l) + \cot \delta_l = -\beta^* / \sin \theta_l \\ \dfrac{\eta + h}{\sin^2(\delta - \delta_l)} - \dfrac{h}{\sin^2 \delta_l} = -\dfrac{\alpha^* - \beta^* \gamma}{\sin \theta_l} + \dfrac{\cos \theta_l}{\sin^2 \theta_l} \beta^* p \end{cases} \tag{4}$$

At an instant, the Eq. (4) is called the ***Euler-Savary analogue of a line-trajectory in spatial movement***.

3.3.2 Stationary Line Congruence

If a line-trajectory Σ_l contacts Σ_{CC} in the third order, the condition equations $d\beta_l/d\sigma = 0$ and $d\eta/d\sigma = 0$ should be satisfied, which are

$$\begin{cases} \cot \delta_l = 1/(M \sin \theta_l) + 1/(N \cos \theta_l) \\ a_{11} p + a_{12} H + a_{13} = 0 \end{cases} \tag{5}$$

where $1/M = (\beta_m - \beta^*)/3$, $1/N = (d\beta^*/d\sigma)/3\beta^*$. The coefficients a_{11}, a_{12}, a_{13} are all functions of (l_1, l_2, l_3) and the induced construction parameters of axodes. All such lines of the moving body constitute ***stationary line congruence of constant axis***

curvature. Any line of the line congruence has a constant inclination angle and the equivalent distance with a fixed line at four infinitesimal positions.

In particular, if the line-trajectory Σ_l has the properties that $\beta_l = 0$, $\eta = 0$, $d\beta_l/d\sigma = 0$ and $d\eta/d\sigma = 0$, the line L is regarded as ***Ball line***. It intersects with a fixed line orthogonally at four infinitesimal positions. The directions of the Ball lines are determined by an eight-degree algebraic equation of $\cos\delta_l$, which can be solved to get the eight sets of directional angles. Actually, there may exist four directions of Ball lines due to the symmetrical spherical image of unit direction vector of lines.

3.3.3 Infinitesimal Burmester Lines

A line L of a rigid body in spatial motion, whose trajectory Σ_l has the properties as $d\beta_l/d\sigma = 0$, $d\eta/d\sigma = 0$, $d^2\beta_l/d\sigma^2 = 0$ and $d^2\eta/d\sigma^2 = 0$, is called ***constant axis line***, or ***infinitesimal Burmester line***. The condition equations are

$$\begin{cases} \cot\delta_l = 1/(M\sin\theta_l) + 1/(N\cos\theta_l) \\ (1+\tan^2\theta_l)[\frac{2-\beta_m M}{M^2} + \frac{dM/d\sigma + 3M/N}{M^2}\tan\theta_l + \frac{1+dN/d\sigma}{N^2}\tan^2\theta_l \\ + \frac{\beta_m - 1/M}{N}\tan^3\theta_l - \frac{1}{N^2}\tan^4\theta_l] + \tan^2\theta_l = 0 \\ a_{11}p + a_{12}H + a_{13} = 0 \\ a_{21}p + a_{22}H + a_{23} = 0 \end{cases} \tag{6}$$

where a_{21}, a_{22}, a_{23} are also functions of (l_1, l_2, l_3, α^*, β^*, γ^*). At any instant, there exist at most six directions of the Burmester lines in the moving body, whose five infinitesimal successive positions locate on a constant axis ruled surface Σ_{CC}.

If the fixed joint C is replaced by H-pair or R-pair, the constant axis ruled surface Σ_{CC} degenerates to be a constant parameter surface in geometrical shape, such as a helicoid and a hyperboloid of one sheet, denoted by Σ_{HC} and Σ_{RC} respectively, and the characteristic line L_{CC} becomes L_{HC} and L_{RC}.

Based on the geometrical properties [1] of Σ_{HC}, α_l = const, β_l = const and γ_l = const are referred to as the ***constant curvature*** or **H-C *curvature***. If a line-trajectory Σ_l contacts Σ_{HC} in second order, the condition equation $d\gamma_l/d\sigma = 0$ should be satisfied. There exist infinite (∞^3) such lines in the moving body at an instant. If a line-trajectory Σ_l contacts Σ_{HC} in third order, the contact conditions are $d\alpha_l/d\sigma = d\beta_l/d\sigma = d\gamma_l/d\sigma = 0$ and $d^2\gamma_l/d\sigma^2 = 0$. There exist finite such lines in the moving body at an instant.

Based on the geometrical properties [1] of Σ_{RC}, α_l = const, β_l = const, γ_l = const and $\alpha_l\beta_l + \gamma_l = 0$ are defined as the ***hyperbolic curvature*** or **R-C *curvature***. If a line-trajectory Σ_l contacts Σ_{RC} in second order, the condition equations $\alpha_l\beta_l + \gamma_l = 0$ and $d\gamma_l/d\sigma = 0$ should be satisfied. There exist infinite such lines in the moving body at an instant. If a line-trajectory Σ_l contacts Σ_{RC} in third order, the contact conditions are $\alpha_l\beta_l + \gamma_l = 0$, $d\alpha_l/d\sigma = d\beta_l/d\sigma = d\gamma_l/d\sigma = 0$ and $d^2\gamma_l/d\sigma^2 = 0$. The line may not exist because four parameters of the line should meet five constraint equations.

4 Numerical Example

A spatial RCCC linkage with parameters $\alpha_{01} = 30°$, $\alpha_{12} = 55°$, $\alpha_{23} = 45°$, $\alpha_{30} = 60°$, $a_0 = 5$, $a_1 = 2$, $a_2 = 4$, $a_3 = 3$, $h_0 = 0$ is given in Ref. [13]. The Cartesian coordinate systems are respectively built up on different links by the Denavit-Hartenberg appointment. With the motion of the linkage determined by the input angle θ_1, the ISA traces the moving axode Σ_m and fixed axode Σ_f in the coupler link and frame link. At the instant $\theta_1 = 1.00$, the construction parameters of Σ_f and Σ_m are $\alpha_f = -3.4818$, $\beta_f = 0.4993$, $\gamma_f = 0.8023$ and $\alpha_m = -6.0744$, $\beta_m = -0.0498$, $\gamma_m = 0.8023$, so the induced construction parameters are $\alpha^* = 2.5926$ and $\beta^* = 0.5490$.

For the coupler link 2 of the RCCC linkage, there are at most six infinitesimal Burmester lines at any instant, whose positions and orientations can be calculated through Eq. (6). At the instant $\theta_1 = 1.00$, we can locate four infinitesimal Burmester lines in the coupler link, whose parameters are listed in Table 2.

In order to show the positions of the Burmester lines, we chose the coordinate plane $\mathbf{R}_C - \mathbf{x}_2\mathbf{y}_2$ of the coupler link as a reference plane, as shown in Fig. 3a. The four Burmester lines intersect the reference plane at four reference points, as shown in Fig. 3b. Obviously, two of the four Burmester lines are just the two axes $\mathbf{z}_1$ and $\mathbf{z}_2$ of the C-pairs of the coupler link with points B' and C as the reference points.

Table 2 Parameters of the Burmester lines at instant $\theta_1 = 1.00$

Burmester lines	Parameters (δ_l, θ_l, p, H) in Frenet frame of Σ_m			
	δ_l	θ_l,	p	H
L_{B1}	1.1305	4.5703	5.2179	2.9518
L_{B2}	0.7009	6.1193	4.5109	11.7035
L_{B3}	1.3182	5.5787	0.0982	2.2858
L_{B4}	1.3262	5.5571	13.8430	7.2734

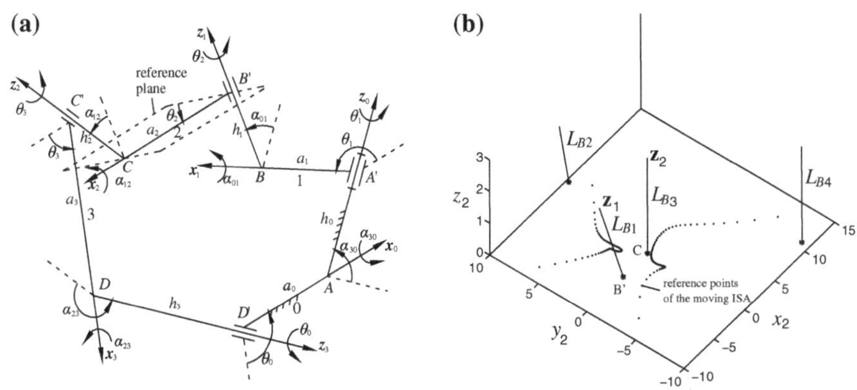

Fig. 3 The orientation and position of the infinitesimal Burmester lines. **a** RCCC linkage, **b** Instantaneous Burmester lines

5 Conclusions

Based on the constraint ruled surfaces of the binary link C-C in spatial linkages and their generalized curvatures, the Burmester lines of a rigid body at five infinitesimal successive spatial positions are defined for the binary link C-C and concisely located in the Frenet frame of axodes. There are at most six infinitesimal Burmester lines at an instant for a spatial motion. The curvatures of a line trajectory in spatial movement are developed to the generalized curvatures for the constraint ruled surfaces of the binary links in spatial linkages, which provide a solid ground for the curvature theory of the line trajectory in spatial movement.

Acknowledgments The authors want to acknowledge the support from the National Natural Science Foundation of China (Grant No. 51275067).

References

1. Wang, D.L., et al.: Geometrical characteristics of some typical spatial constraints. Mech. Mach. Theory **35**, 1413–1430 (2000)
2. Disteli, M.: Uber des analogon der savaryschen formel und konstruktion in der kinematischen geometrie des raumes. Zeitschrift fur Mathematik und Physik **62**, 261–309 (1914)
3. Yang, A.T., et al.: On a kinematic curvature theory for ruled surfaces. In: Proceedings of Fourth World Congress on the Theory of Machines and Mechanisms, pp. 737–742 (1975)
4. Köse, O.: Kinematic differential geometry of a rigid body in spatial motion using dual vector calculus: part-I. Appl. Math. Comput. **183**, 17–29 (2006)
5. Veldkamp, G.R.: Canonical systems and instantaneous invariants in spatial kinematics. J. Mech. **3**, 329–388 (1967)
6. Kirson, Y., Yang, A.T.: Instantaneous invariants in 3D kinematics. ASME J. Appl. Mech. **45**, 409–414 (1978)
7. Roth, B., Yang, A.T.: Applications of instantaneous invariants to the analysis and synthesis of mechanisms. ASME J. Eng. Ind. **99**, 97–103 (1977)
8. Ting, K.L., Zhang, Y.: Rigid body motion characteristics and unified instantaneous motion representation of points, lines and planes. ASME J. Mech. Des. **126**, 593–601 (2004)
9. McCarthy, J.M., Roth, B.: The curvature theory of line trajectories in spatial kinematics. ASME J. Mech. Des. **103**, 718–724 (1981)
10. Sasaki, S.: Differential Geometry (in Japanese). Kyolitsu Press, Tokyo (1956)
11. Wang, D.L., et al.: Kinematic differential geometry of a rigid body in spatial motion-I: a new adjoint approach and instantaneous properties of a point trajectory in spatial kinematics. Mech. Mach. Theory **32**, 419–432 (1997)
12. Wang, D.L., et al.: Kinematic differential geometry of a rigid body in spatial motion-II: a new adjoint approach and instantaneous properties of a line trajectory in spatial kinematics. Mech. Mach. Theory **32**, 433–444 (1997)
13. Yang, A.T., Freudenstein, F.: Application of dual-number quaternion algebra to the analysis of spatial mechanism. ASME J. Appl. Mech. **31**, 300–308 (1964)

Educational and Research Kinematic Capabilities of GIM Software

Erik Macho, Victor Petuya, Mónica Urízar, Mikel Diez and Alfonso Hernández

Abstract In this paper an educational and research software named GIM is presented. This software has been developed with the aim of approaching the difficulties students usually encounter when facing up to kinematic analysis of mechanisms. A deep understanding of the kinematic analysis is necessary to go a step further into design and synthesis of mechanisms. In order to support and complement the theoretical lectures, GIM software is used during the practical exercises, serving as an educational complementary tool reinforcing the knowledge acquired by the students.

Keywords Motion simulation · Computational kinematics · Mechanism synthesis · General purpose software

1 Introduction

In the teaching of subjects related to Machine Theory, supporting and complementing theoretical lectures with a simulation and analysis software, helps the students to understand deeply and visually the theoretical bases of the Mechanisms

E. Macho (✉) · V. Petuya · M. Urízar · M. Diez · A. Hernández
University of the Basque Country, Bilbao, Spain
e-mail: erik.macho@ehu.es

V. Petuya
e-mail: victor.petuya@ehu.es

M. Urízar
e-mail: monica.urizar@ehu.es

M. Diez
e-mail: mikel.diez@ehu.es

A. Hernández
e-mail: a.hernandez@ehu.es

B. Corves et al. (eds.), *Mechanisms, Transmissions and Applications*,
Mechanisms and Machine Science 31, DOI 10.1007/978-3-319-17067-1_2

Science. In the Department of Mechanical Engineering of the University of the Basque Country (UPV/EHU) two main Bachelor subjects can be highlighted in this field: Applied Mechanics [1, 2] and Kinematics of Mechanisms [3]. In these subjects, GIM software is used.

GIM is a registered software created by the COMPMECH Research Group (www.ehu.es/compmech). The software has been developed focusing, not only on educational purposes but also on research in the field of computational kinematics and mechanism design applications. The software presented in this article also has potential to be used by students of Master Degrees in Mechanical Engineering and other subjects related to Robotics, Mechanism Design, etc.

GIM has been developed in a modular structure. After defining the kinematic structure of a linkage in the Geometry module, the user can perform the motion simulation in the Motion module. GIM is mainly oriented to the field of kinematic analysis, motion simulation and dimensional synthesis of planar mechanisms. In any case, it also includes other modules for workspace and singularity evaluation [4] and static analysis of mechanical structures.

Currently other Universities are using different kinematic softwares during their lessons. In RWTH Aachen, IGM students use the interactive geometry software Cinderella (freeware tool provided by Springer) during mechanism lectures [5]. Other Spanish Universities use well known commercial softwares as GeoGebra [6] or ADAMS [7]. In this paper, we will present the main GIM capabilities used mainly by Bachelor and Master student but also by some of our PhD students.

2 Kinematic Analysis

In this section are presented and briefly described the main capabilities of the so-called *Motion* module. This module is able to simulate the motion of any n-dof planar mechanism with any kind of revolute and prismatic joints. Also the disk-disk and disk-line rolling and cam contacts can be modelled.

2.1 Position, Velocity and Acceleration Problems

Figure 1 shows an example of the most elemental results that can be depicted when the motion of a mechanism is obtained and simulated. Obviously, once the geometry has been defined, the first step to compute the motion is to define as many actuators in the mechanism as degrees of freedom. There are different types of actuators, i.e. rotary ones, linear ones, as well as several input function types, i.e. polynomial or sinusoidal, available to control the position, velocity and acceleration of each actuator.

The trajectories, velocities (v) and accelerations (a) of any point, and the angular velocities (ω) and accelerations (α) of all elements can be drawn. Also the center of

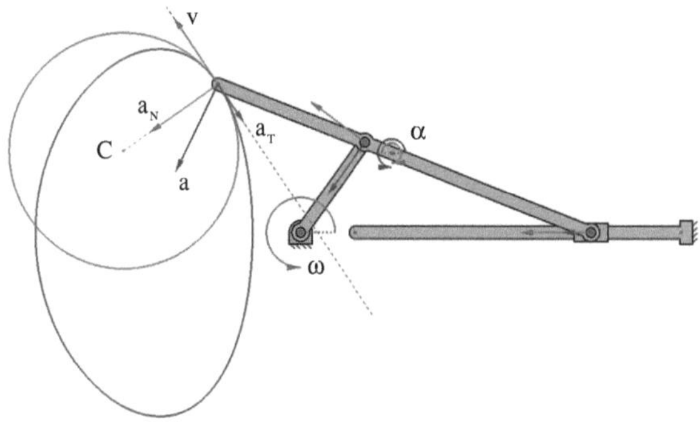

Fig. 1 General motion simulation results

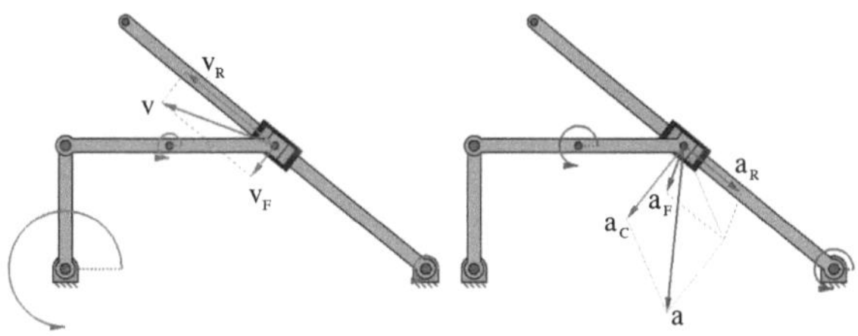

Fig. 2 Relative motion composition in velocities and accelerations

curvature (C) of the trajectory and the intrinsic components of the acceleration (a_T, a_N) can be represented. From an academic point of view, the compliance of all the well-known properties of these magnitudes can be directly observed and understood along the whole motion, e.g. the velocity is always tangent to the trajectory, or the normal acceleration points always towards center of curvature.

All motion results can be obtained not only in the fixed frame, but also with respect to any relative reference. Such a capability, as shown in Fig. 2, provides the best support for explaining (or understanding) the frame (v_F, a_F) and relative (v_R, a_R) motion compositions (a_C is Coriolis term).

In the same way, the user can check how the specific properties of the rolling motion, with or without sliding, are verified, e.g., as shown in Fig. 3, when a rolling motion with no sliding component is performed, contact points in both elements have the same velocity and the same projection of the acceleration over the contact tangent line. Those properties are not satisfied when exist sliding in the rolling.

2.2 Kinematic Geometry

GIM software is able to compute and represent the main kinematic geometrical entities of any element of the mechanism, such as the instantaneous center of rotation (P), the pole of accelerations (Q), the fixed and moving centrodes (c.f. and

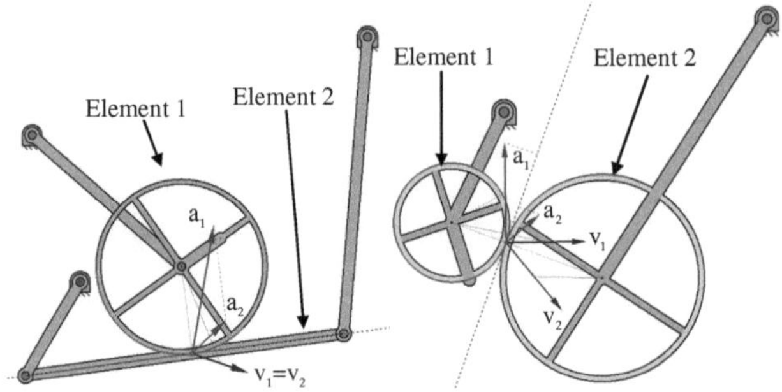

Fig. 3 Velocities and accelerations of contact points in a pure rolling and in a cam joint

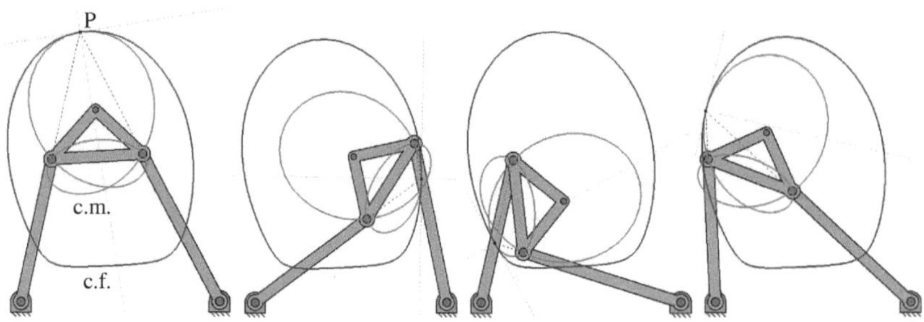

Fig. 4 Rolling motion sequence between the fixed and moving centrodes

c.m.) and the inflection and Bresse circles (c.i. and c.b.). When studying these entities, students have an interactive tool at hand which facilitates a deeper understanding.

For example, as shown in Fig. 4, during the motion simulation, it is checked that the fixed and moving centrodes are always tangent at the instantaneous center of rotation, and the moving centrode moves welded the element, rolling over the fixed one. In the same way, as this tool enables to drag the coupler point of an element to any position in the moving plane, it can be observed that the pole of velocity has acceleration and the pole of acceleration has velocity (Fig. 5, up). Also, any point on the inflection circle is passing through an inflection point on its trajectory and lacks of normal acceleration, as well as any point on the Bresse circle lacks of tangent acceleration (Fig. 5, down).

2.3 Advanced Computations

Apart from the basic kinematic geometry, some advances features can be issued. Here are found some examples. Figure 6 shows the envelope of a line that moves fixed to the coupler element of a mechanism. As it is known, in positions where

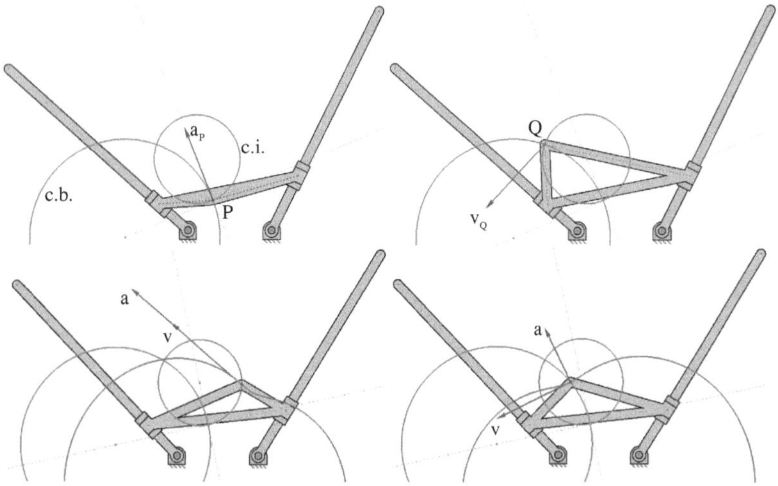

Fig. 5 Instantaneous centers of rotation and acceleration and inflection and Bresse circles

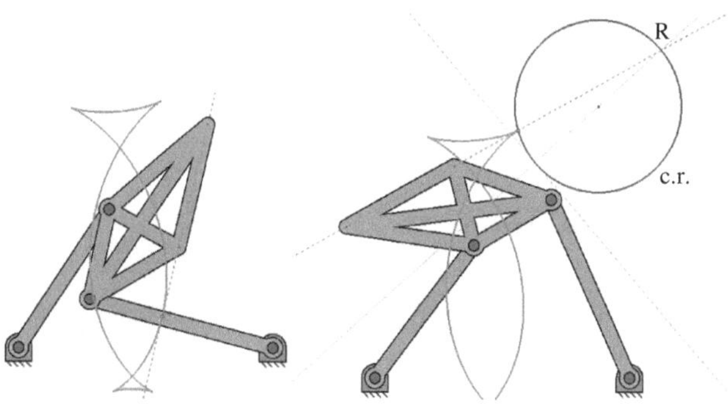

Fig. 6 Moving line envelope and return circle

such a line passes through the return pole (R), the envelope curve has a cusp point (c.r. is the return circle).

For some simple mechanisms, as the 4-bar linkage, the cubic of stationary curvature (c.s.c.) and the pivot point curve (p.p.c.) can be traced. When the coupler point is located at the intersection between the cubic of stationary curvature and the inflection circle (Ball point, B) it is achieved a quasi-straight-line trajectory in the proximity of such a point (Fig. 7).

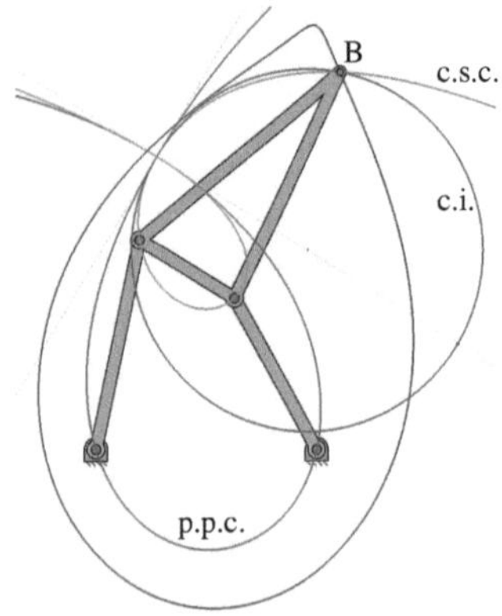

Fig. 7 Cubic of stationary curvature, pivot point curve and ball point

3 Dimensional Synthesis

The Motion module presented in the previous section has a general purpose and can solve a large amount of planar mechanisms. Apart from this module, the program provides another one that deals with the dimensional synthesis of one specific mechanism, the four-bar linkage. Three traditional synthesis problem types have been addressed, i.e. path generation, rigid body guiding and function generation.

3.1 Path Generation Synthesis

This synthesis is based on the use of precision points and consists in determining a mechanism whose coupler point trajectory passes exactly through some specific positions. With this software tool, the user can drag the precision points to change their position and the computation of the mechanism that fulfills all conditions is done in real time. This non-linear problem admits many different solutions for the same input values. The designer has the full control to visualize each of them, as depicted in Fig. 8.

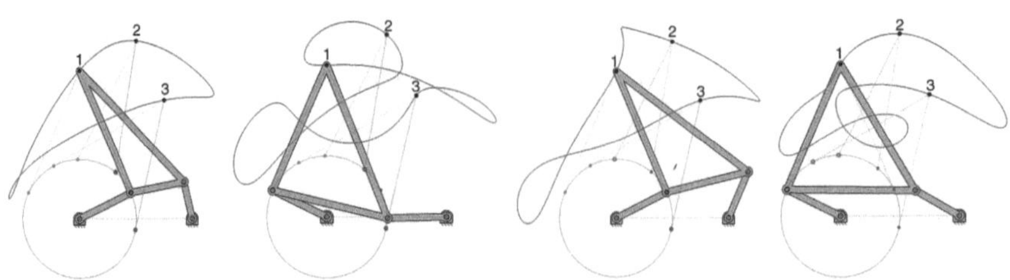

Fig. 8 Synthesis with three precision points. Different solutions for the same data

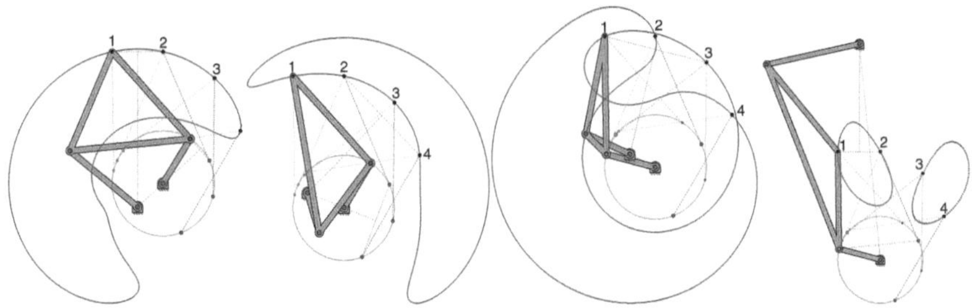

Fig. 9 Synthesis with four precision points. Order error and loop error

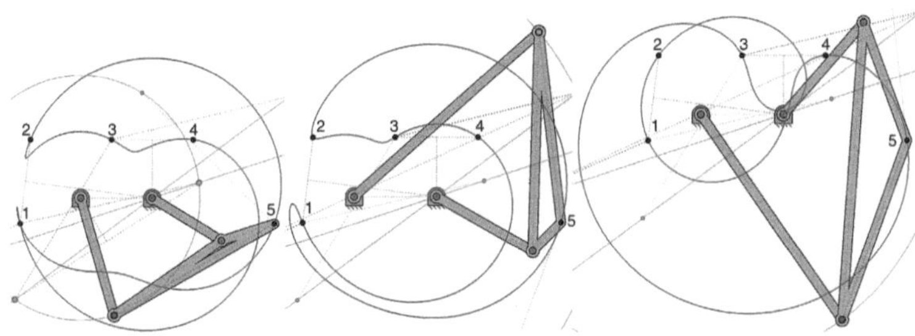

Fig. 10 Synthesis with five precision points. Different solutions for the same data

The program offers the option for making the mechanism synthesis being defined three, four, or five precision points of the trajectory. This ability is shown in Figs. 8, 9 and 10. From the academic point of view, one of main advantages is the possibility of checking that some solutions do not accomplish the right order in the sequence of precision points, order error, and also, in the case of Grashof's four bar linkages, could appear the so-called branch error, that means that some points belong to the disconnected path corresponding to the crossed quadrilateral.

Also, the software allows the visualization of the classical geometrical constructions for making the synthesis. User can choose between displaying or not such auxiliary constructions. This feature is valuable for an academic purpose.

3.2 Rigid Body Guiding

The solid element guiding synthesis computes the mechanism that is able to fully locate an element in a set of desired postures (position and orientation). Figure 11 shows the three alternatives given in the program. The first one allows the user to specify three postures and the relative positions of the floating joints (the shape of the coupler). Using the second option, apart from three desired postures, positions of fixed joints can be specified. Finally, solid element guiding for four target postures can be done.

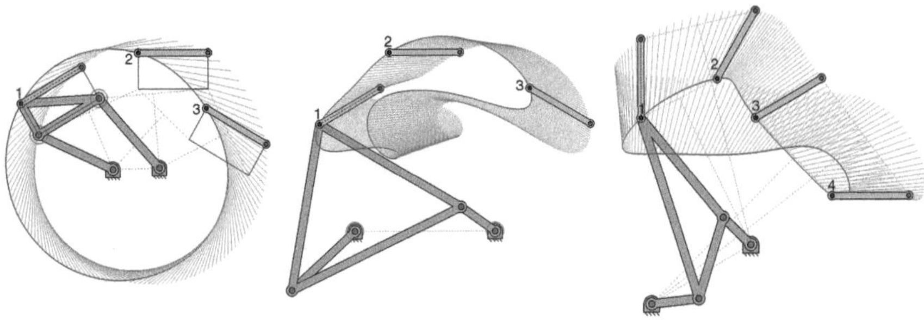

Fig. 11 Synthesis for solid element guiding

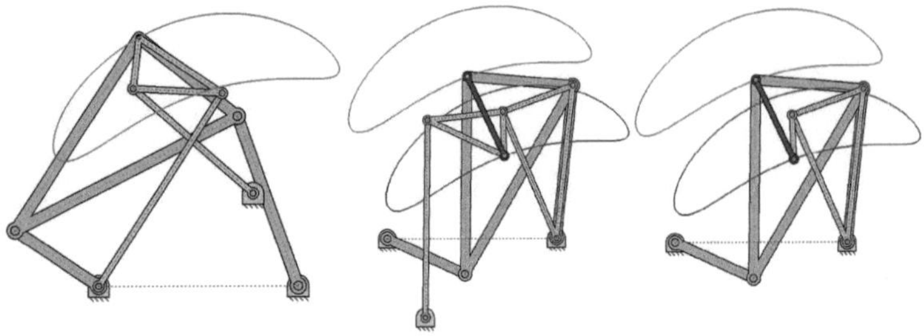

Fig. 12 Four-bar mechanism cognate and translational mechanisms

3.3 Additional Functionalities

In the Synthesis module the user can find some extra features related with the design alternatives based in the four-bar linkage. Some of them are presented in Fig. 12. Apart from the crossed mechanism, cognates, which trace the same coupler trajectory, are obtained. Based on them, the so-called 1-dof translational mechanisms are obtained. The two known versions can be represented: the redundant one and the non-redundant one.

4 Conclusions

The software presented in this paper allows students and researchers to model and analyse in a quick and simple way n-dof planar linkages. Using the software capabilities, the user is able to carry out a deep kinematic performance analysis of the whole mechanism. GIM software has proven to be a very effective tool to complement and reinforce the theoretical concepts explained during the lectures of subject related to Mechanism and Machine Science.

GIM software can be freely downloaded from the COMPMECH web site in the following link: www.ehu.es/compmech/software.

Acknowledgments The authors wish to acknowledge the financial support received from the Spanish Government through the Ministerio de Economía y Competitividad (Project DPI2011-22955) and the Regional Government of the Basque Country through the Departamento de Educación, Universidades e Investigación (Project IT445-10) and UPV/EHU under program UFI 11/29.

References

1. Bilbao, A., Amezua, E.: Mecánica Aplicada: Estática y Cinemática. Editorial Síntesis, Madrid (2008)
2. Bilbao, A., Amezua, E., Altuzarra, O.: Mecánica Aplicada: Dinámica. Editorial Síntesis, Madrid (2008)
3. Hernández, A.: Cinemática de mecanismos: Análisis y Diseño. Editorial Síntesis, Madrid (2008)
4. Petuya, V., Macho, E., Altuzarra, O., Pinto, Ch., Hernández, A.: Educational software tools for the kinematic analysis of mechanisms. Comput. Appl. Eng. Educ. **22**, 72–86 (2014) (Wiley Periodicals Inc.)
5. Corves, B., Riedel, M., Hüsing, M.: Descriptive and intuitive mechanism design and synthesis using geometry-based computer-aided methods. In: 13th World Congress in Mechanism and Machine Science, Guanajuato, Mexico: International Federation for the Promotion of Mechanism and Machine Science, 19–25 June 2011
6. Iriarte, X., Aginaga, J., Ros, J.: Teaching Mechanism and Machine Theory with GeoGebra. New Trends in Educational Activity in the Field of Mechanism and Machine Theory. MMS Series, vol. 19, pp. 211–219. Springer, Berlin (2014)
7. Suñer, J.L., Mata, V.: Utilización de programas de simulación en la docencia de la ingeniería mecánica. Actas del X Congreso de Innovación Educativa en las Enseñanzas Técnicas, Valencia, Spain (2002)

Geared Linkages with Linear Actuation Used as Kinematic Chains of a Planar Parallel Manipulator

Erwin-Christian Lovasz, Sanda Margareta Grigorescu, Dan Teodor Mărgineanu, Corina Mihaela Gruescu, Cristian Pop, Valentin Ciupe and Inocentiu Maniu

Abstract The paper deals with a novel kinematic chain, which uses a geared linkage with linear actuation. The structure is used to actuate the mobile platform of a parallel planar manipulator (3-R$\underline{R}$R). The main characteristics of the geared linkage with linear actuation are the large rotation angle with proper transmission angle and the approximately linear transmission function in a large range. These properties of the geared linkages with linear actuation as kinematic chain allow the avoiding of the first type singularities and an easier control of the mobile platform movement.

Keywords Parallel planar manipulator · Geared linkage with linear actuator · Kinematics · Singularity

E.-C. Lovasz (✉) · S.M. Grigorescu · D.T. Mărgineanu · C.M. Gruescu · C. Pop · V. Ciupe · I. Maniu
Politehnica University, Timişoara, Romania
e-mail: erwin.lovasz@upt.ro

S.M. Grigorescu
e-mail: sanda.grigorescu@upt.ro

D.T. Mărgineanu
e-mail: dan.margineanu@upt.ro

C.M. Gruescu
e-mail: corina.gruescu@upt.ro

C. Pop
e-mail: cristian.pop@upt.ro

V. Ciupe
e-mail: valentin.ciupe@upt.ro

I. Maniu
e-mail: inocentiu.maniu@upt.ro

B. Corves et al. (eds.), *Mechanisms, Transmissions and Applications*, Mechanisms and Machine Science 31, DOI 10.1007/978-3-319-17067-1_3

1 Introduction

Parallel manipulators due to their major advantages became a stand-alone branch of robotics science. There was a continuous and valuable gain of knowledge during the last decades, which now allow thorough analysis and practical creation. Nowadays, parallel manipulators science enriches with new structures fit to dedicated tasks.

The large number of possible architectures inspired approaches in generalization or classification [1, 2]. Different solutions of structure and kinematics were proposed and analyzed aiming to increase or optimize characteristics such as workspace, precision, dexterity and others [3–7].

With parallel structures an important issue to study and solve is singularity-free workspace. Each new structure ought to be analyzed from this point of view. Various approach methods and their results on this matter are presented for different specific schemes in [3, 8–15].

Each application needs a certain structure in accordance with a given task. Many authors seek for solution of reconfigurable robots, dedicated to multitasking [16, 17].

Some studies go to subtle problems such as dynamic control in the presence of parameter uncertainties and external disturbances [18].

The present paper contributes to the development of the specific class of planar parallel manipulators. It describes a new kinematic chain and focuses mainly on the problem of singularity avoiding.

2 Structural Analysis of the Novel Parallel Manipulator

Many studies of the authors illustrated that geared linkages with linear actuation allow a large rotation angle with proper transmission angle for the output element and an approximately linear transmission function in a large range (approximately constant transmission ratio) [19, 20]. This property recommends our mechanism to be used as part of the kinematic chain for the parallel manipulators and it is expected that this structure avoids the first type singularities in a wider range. This mechanism type is very compact; the self-locking condition is fulfilled by using a screw-nut for converting the movement and also for ensuring a high transmission ratio without gear box. Additionally, while the geared linkages, as actuation chains, are jointed with the frame; it is expected that the dynamic behavior does not worsen. Figure 1 shows the kinematic schema of the planar parallel manipulator using geared linkages with linear actuation.

The proposed notation of this type of manipulator chain should be 3-R(R$\underline{\text{P}}$RGR) RR, where R—revolute joint, P—prismatic joint and G—gear pair. In brackets ( ) is indicated the parallel connected actuating kinematic chain and underlined (ex. $\underline{\text{P}}$) the active joint. The considered geared linkage chains use 3 linear actuators $n_{act} = 3$, which actuate the output elements 2, 2′ and 2″ connected to the mobile platform. The number of the elements of each kinematic chain of the planar parallel manipulator increases to 5 elements $n_{chain} = 5$. The proposed planar parallel

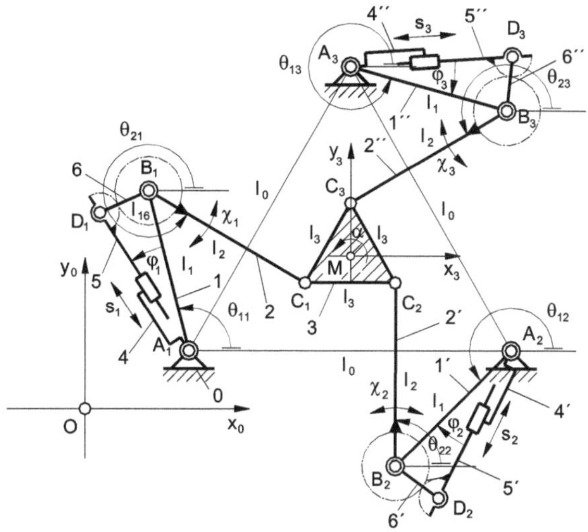

Fig. 1 Kinematic schema of the planar parallel manipulator 3-R(RPRGR)RR, using geared linkages with linear actuation

manipulator allows 3 DOF for the mobile platform (3): 2 translations along the *Ox* and *Oy* axis and the rotation around the *Oz* axis, which means the mobility of the whole structure is computed with the relationship:

$$M = 3 \cdot (n - 1) - 2 \cdot e_1 - e_2 = 3 \cdot (17 - 1) - 2 \cdot 21 - 3 = 3, \qquad (1)$$

where

n is the number of the elements,
e_1 number of kinematic pair with $f = 1$,
e_2 number of kinematic pair with $f = 2$

Considering the relationship (1), the mechanism structure of the planar parallel manipulator follows a constrained motion expressed through the condition:

$$M = n_{act}. \qquad (2)$$

3 Kinematic Analysis of the Novel Parallel Manipulator

A simplified computing model, illustrated in Fig. 2, is considered for the kinematic analysis of the novel planar parallel manipulator 3-R(RPRGR)RR.

Considering Chasles vector equation [23], starting from the reference system of the fixed platform (OX_0Y_0), there is [21, 22, 24]:

$$\overrightarrow{OA_i} + \overrightarrow{A_iC_i} = OM + \underline{R} \cdot \overrightarrow{MC_i^{(3)}}, \quad i = \overline{1,3}, \qquad (3)$$

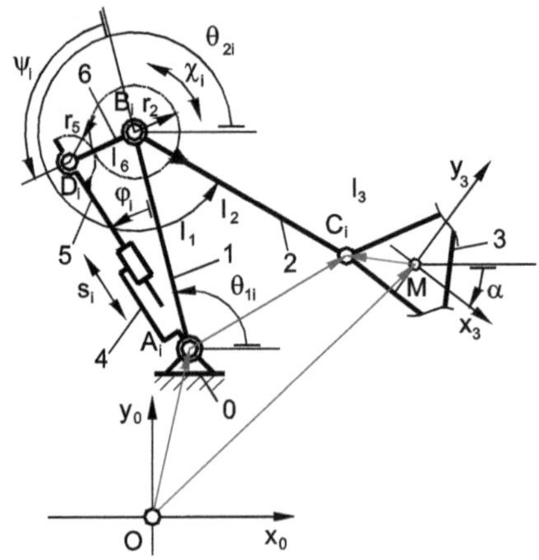

Fig. 2 Simplified computing model for the kinematic analysis 3-R(RPRGR)RR

where

$$\overrightarrow{OA_i} = \begin{bmatrix} x_{A_i} \\ y_{A_i} \end{bmatrix}, \quad \overrightarrow{OM} = \begin{bmatrix} x_M \\ y_M \end{bmatrix}, \quad \underline{R} = \begin{bmatrix} \cos\alpha & -\sin\alpha \\ \sin\alpha & \cos\alpha \end{bmatrix}, \quad \overrightarrow{MC_i}^{(3)} = \begin{bmatrix} x_{C_i}^{(3)} \\ y_{C_i}^{(3)} \end{bmatrix}, \tag{4}$$

$$\overrightarrow{A_iC_i} = \begin{bmatrix} l_1 \cos\theta_{1i}(s_i) + l_2 \cos\theta_{2i}(\chi_i(s_i)) \\ l_1 \sin\theta_{1i}(s_i) + l_2 \sin\theta_{2i}(\chi_i(s_i)) \end{bmatrix}. \tag{5}$$

$$\theta_{2i}(\chi_i(s_i)) = \chi_i(s_i) + \Delta\chi_i. \tag{6}$$

with $\Delta\chi_i$ constant values imposed by the planar parallel manipulator designing conditions.

The geared linkage is composed of an inverted slider-crank, as basic structure, connected in parallel with a gear train. The planetary gear is fixed with the slider of the basic structure and performs a planetary motion (Fig. 2). The drive is the slider, which performs a stroke s_i. The driven gear 2 performing the rotation angle χ_i, is fixed with the binary link B_iC_i, connected to the mobile platform 3.

The transmission function $\chi_i(s_i)$ of the geared linkage is [24]:

$$\chi_i(s_i) = (1-\rho)\cdot\psi_i(s_i) + \rho\cdot\varphi_i(s_i), \tag{7}$$

where

$$\rho = \pm r_5/r_2, \tag{8}$$

$$\varphi_i(s_i) = \arccos\Big(\big(l_6^2 - l_1^2 - (s_0+s_i)^2\big)\Big/(2\cdot l_1\cdot(s_0+s_i))\Big), \tag{9}$$

and

$$\psi_i(s_i) = \arccos\left(\left((s_0 + s_i)^2 - l_1^2 - l_6^2\right) / (2 \cdot l_1 \cdot l_6)\right). \tag{10}$$

The instantaneous transmission ratio correlated with (7)–(10) gives:

$$\chi_i'(s_i) = (-(s_0 + s_i) + \rho \cdot l_1 \sin \varphi_i)/(l_1 l_6 \sin \psi_i). \tag{11}$$

For solving the forward kinematics problem of the planar parallel manipulator 3-R(R$\underline{P}$RGR)RR the following matrix equation is used:

$$\begin{bmatrix} x_M \\ y_M \end{bmatrix} = \begin{bmatrix} x_{A_i} \\ y_{A_i} \end{bmatrix} + \begin{bmatrix} l_1 \cos \theta_{1i}(s_i) + l_2 \cos \theta_{2i}(\chi_i(s_i)) \\ l_1 \sin \theta_{1i}(s_i) + l_2 \sin \theta_{2i}(\chi_i(s_i)) \end{bmatrix} - \begin{bmatrix} x_{C_i}^{(3)} \cos \alpha - y_{C_i}^{(3)} \sin \alpha \\ x_{C_i}^{(3)} \sin \alpha + y_{C_i}^{(3)} \cos \alpha \end{bmatrix}, \quad i = \overline{1,3}, \tag{12}$$

where the known parameters are the drive manipulator parameters ($\theta_{1i}(s_i)$ and $\theta_{2i}(\chi_i(s_i))$), the geometry of the fixed platform (x_{A_i}, y_{A_i}) and mobile platform ($x_{C_i}^{(3)}, y_{C_i}^{(3)}$), the length of the links l_1, l_2, l_6 and the gear ratio ρ.

For the inverse kinematics problem the matrix Eq. (12) will be transformed in the form:

$$\begin{bmatrix} l_1 \cos \theta_{1i}(s_i) + l_2 \cos \theta_{2i}(\chi_i(s_i)) \\ l_1 \sin \theta_{1i}(s_i) + l_2 \sin \theta_{2i}(\chi_i(s_i)) \end{bmatrix} = \begin{bmatrix} x_{A_i} \\ y_{A_i} \end{bmatrix} + \begin{bmatrix} x_{C_i}^{(3)} \cos \alpha - y_{C_i}^{(3)} \sin \alpha \\ x_{C_i}^{(3)} \sin \alpha + y_{C_i}^{(3)} \cos \alpha \end{bmatrix} - \begin{bmatrix} x_M \\ y_M \end{bmatrix}, \quad i = \overline{1,3}, \tag{13}$$

where the known parameters are the coordinates of the characteristic point M (x_M, y_M), the rotation angle of the mobile platform (α), the geometry of the fixed platform (x_{A_i}, y_{A_i}) and mobile platform ($x_{C_i}^{(3)}, y_{C_i}^{(3)}$), the length of the links l_1, l_2, l_6 and the gear ratio ρ.

4 Singularity Calculus of the 3-R(R$\underline{P}$RGR)RR Planar Parallel Manipulator

Taking into account the invers kinematic problem Eq. (13), by separating the sub matrix containing terms θ_{1i}, one can eliminate the θ_{1i}, $i = \overline{1,3}$, reaching the expression:

$$
\begin{aligned}
l_1^2 = {} & \left(x_M + x_{C_i}^{(3)} \cos\alpha - y_{C_i}^{(3)} \sin\alpha - x_{A_i} - l_2 \cos\theta_{2i}(\chi_i(s_i))\right)^2 \\
& + \left(y_M + x_{C_i}^{(3)} \sin\alpha + y_{C_i}^{(3)} \cos\alpha - y_{A_i} - l_2 \sin\theta_{2i}(\chi_i(s_i))\right)^2.
\end{aligned} \tag{14}
$$

Let's consider the Eq. (14) as an implicit function $F_i(x_M, y_M, \alpha)$, where

$$
\begin{aligned}
F_i(x_M, y_M, \alpha) = {} & \left(x_M + x_{C_i}^{(3)} \cos\alpha - y_{C_i}^{(3)} \sin\alpha - x_{A_i} - l_2 \cos\theta_{2i}(\chi_i(s_i))\right)^2 \\
& + \left(y_M + x_{C_i}^{(3)} \sin\alpha + y_{C_i}^{(3)} \cos\alpha - y_{A_i} - l_2 \sin\theta_{2i}(s_i)\right)^2 - l_1^2 = 0
\end{aligned} \tag{15}
$$

and $\theta_{2i}(\chi_i(s_i))$ depending on $\chi_i(s_i)$, i = 1, 2, 3 is determined by the drive stroke s_i.

Three dimensional functions $F_i(x_M, y_M, \alpha)$ are functions of driven elements 2, 2′ and 2″, expressed by $\chi_i(s_i)$, so it can be written:

$$
F_i(X, Q) = 0, \tag{16}
$$

with: $X = [s_1, s_2, s_3]$ and $Q = [x_M, y_M, \alpha]$.

The function differentiation with respect to time is leading to the known relation:

$$
J_s \cdot \dot{X} + J_q \cdot \dot{Q} = 0, \tag{17}
$$

where J_s and J_q are Jacobian matrices that show the mechanism singularities of second type and first type, respectively.

The first type singularities express the cases in which the actuation of the drive elements does not move the mobile platform. The computation of this type of singularity implies to satisfy the condition to have the Jacobian matrix determinant J_s equal to zero [23]:

$$
\det(J_s) = \det \begin{bmatrix}
\frac{\partial F_1(X,Q)}{\partial s_1} & \frac{\partial F_1(X,Q)}{\partial s_2} & \frac{\partial F_1(X,Q)}{\partial s_3} \\
\frac{\partial F_2(X,Q)}{\partial s_1} & \frac{\partial F_2(X,Q)}{\partial s_2} & \frac{\partial F_2(X,Q)}{\partial s_3} \\
\frac{\partial F_3(X,Q)}{\partial s_1} & \frac{\partial F_3(X,Q)}{\partial s_2} & \frac{\partial F_3(X,Q)}{\partial s_3}
\end{bmatrix} = 0, \tag{18}
$$

where

$$
\begin{aligned}
\frac{\partial F_i(X,Q)}{\partial s_i} = 2l_2\Big\{ & \Big[x_M + x_{C_i}^{(3)} \cos\alpha - y_{C_i}^{(3)} \sin\alpha \\
& - x_{A_i} - l_2 \cos\theta_{2i}(\chi_i(s_i))\Big] \sin\theta_{2i}(\chi_i(s_i)) \\
& - \Big[y_M + x_{C_i}^{(3)} \sin\alpha + y_{C_i}^{(3)} \cos\alpha - y_{A_i} \\
& - l_2 \sin\theta_{2i}(\chi_i(s_i))\Big] \cos\theta_{2i}(\chi_i(s_i))\Big\} \cdot \chi_i'(s_i),
\end{aligned} \tag{19}
$$

$$\frac{\partial F_i(X,Q)}{\partial s_j} = 0, \quad i \neq j, \quad i = \overline{1,3}, \quad j = \overline{1,3}. \tag{20}$$

Through the substitution of (19) and (20) in the relationship (18) result the singularities of first type:

$$0 = \frac{\partial F_1(X,Q)}{\partial s_1} \cdot \frac{\partial F_2(X,Q)}{\partial s_2} \cdot \frac{\partial F_3(X,Q)}{\partial s_3}. \tag{21}$$

Because the partial derivative in (21) are similar and if any of them is zero, than the condition (18) is satisfied and the study of avoiding the singularities of first type can be reduced on study of one partial derivative of one actuating kinematic chain:

$$0 = \frac{\partial F_i(X,Q)}{\partial s_i}, \quad i = \overline{1,3}. \tag{22}$$

The second type singularities correspond to the positions when non-actuated drives allow an infinitesimal motion of the mobile platform inside the workspace. These can be determined when the Jacobian matrix determinant J_q is zero. In addition J_q depicts the mechanism workspace boundaries [23, 24].

$$\det(J_q) = \det\begin{bmatrix} \frac{\partial F_1(X,Q)}{\partial x_M} & \frac{\partial F_1(X,Q)}{\partial y_M} & \frac{\partial F_1(X,Q)}{\partial \alpha} \\ \frac{\partial F_2(X,Q)}{\partial x_M} & \frac{\partial F_2(X,Q)}{\partial y_M} & \frac{\partial F_2(X,Q)}{\partial \alpha} \\ \frac{\partial F_3(X,Q)}{\partial x_M} & \frac{\partial F_3(X,Q)}{\partial y_M} & \frac{\partial F_3(X,Q)}{\partial \alpha} \end{bmatrix} = 0 \tag{23}$$

where:

$$\frac{\partial F_i(X,Q)}{\partial x_M} = 2 \cdot \left[x_M + x_{C_i}^{(3)} \cos\alpha - y_{C_i}^{(3)} \sin\alpha - x_{A_i} - l_2 \cos\theta_{2i}(\chi_i(s_i))\right], \tag{24}$$

$$\frac{\partial F_i(X,Q)}{\partial y_M} = 2 \cdot \left[y_M + x_{C_i}^{(3)} \sin\alpha + y_{C_i}^{(3)} \cos\alpha - y_{A_i} - l_2 \sin\theta_{2i}(\chi_i(s_i))\right], \tag{25}$$

$$\begin{aligned} \frac{\partial F_i(X,Q)}{\partial \alpha} = 2 \cdot \Big\{ &\left[y_M + x_{C_i}^{(3)} \sin\alpha + y_{C_i}^{(3)} \cos\alpha - y_{A_i} - l_2 \sin\theta_{2i}(\chi_i(s_i))\right] \\ &\cdot \left(x_{C_i}^{(3)} \cos\alpha - y_{C_i}^{(3)} \sin\alpha\right) \\ &- \left[x_M + x_{C_i}^{(3)} \cos\alpha - y_{C_i}^{(3)} \sin\alpha - x_{A_i} - l_2 \cos\theta_{2i}(\chi_i(s_i))\right] \\ &\cdot \left(x_{C_i}^{(3)} \sin\alpha + y_{C_i}^{(3)} \cos\alpha\right) \Big\} \end{aligned} \tag{26}$$

The substitution of partial derivatives (24), (25) and (26) in the relationship (23) allows the computing of second type singularities. The singularities of second type use identical terms in J_q with the planar manipulator 3-RRR [24]. Because the explicit form of the equations expressing the singularities of first and second type is too complicated, it is not indicated in the paper.

5 Example Problem

Taking the dimensions of the planar parallel manipulator 3-R(RPRGR)RR given in Table 1 and exploring the plane for the characteristic coordinates $x_M \in [0, 110]$ and $y_M \in [-10, 100]$, with a constant rotation angle $\alpha = 30°$, we obtained the following graphics for singularities analysis (Figs. 3 and 4). The link lengths of the parallel manipulator l_0, l_1, l_2 and l_3 are chosen in order to reduce the singularities of second order and the optimal link lengths of the geared linkages were dimensioned on the criterion of maximizing the angle of driven elements 2, 2′, 2″ with proper values of the transmission angle ($\mu > 30°$) by the inverted slider-crank mechanism.

The absence of zero values for the partial derivatives (22) of each parallel connected actuating kinematic chain (Fig. 3) confirm the supposition that the 3-R (RPRGR)RR structure avoids the singularities of first type in the considered range.

Figure 4b shows the second type singularities of the 3-R(RPRGR)RR manipulator in the workspace, which does not depend on the parallel connected actuating kinematic chains. The results are similar with other studies about planar 3-RRR manipulator [24].

Table 1 Geometrical parameters of the planar 3-R(RPRGR)RR manipulator

Frame platform length (0)	l_0 = 120 mm	Carrier length (6)	l_6 = 10 mm
Mobile platform length (3)	l_3 = 25 mm	Gear ratio	ρ = 1.5
Chain link length (1,1′,1″)	l_1 = 50 mm	Initial stroke	s_0 = 42.5 mm
Chain link length (2,2′,2″)	l_2 = 50 mm	Stroke	h = 15 mm

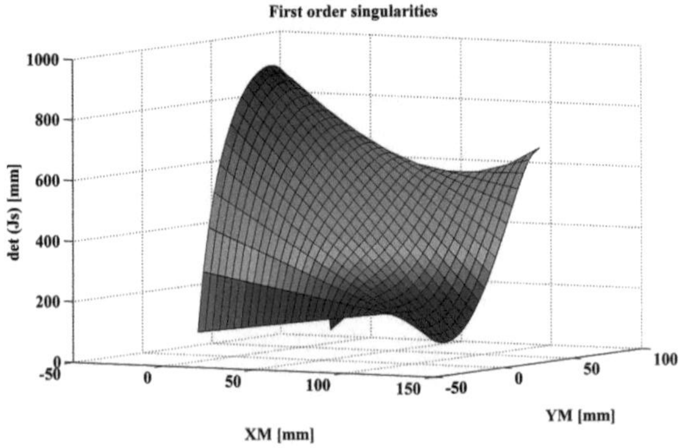

Fig. 3 Analysis of the singularities of first type for the planar 3-R(RPRGR)RR manipulator

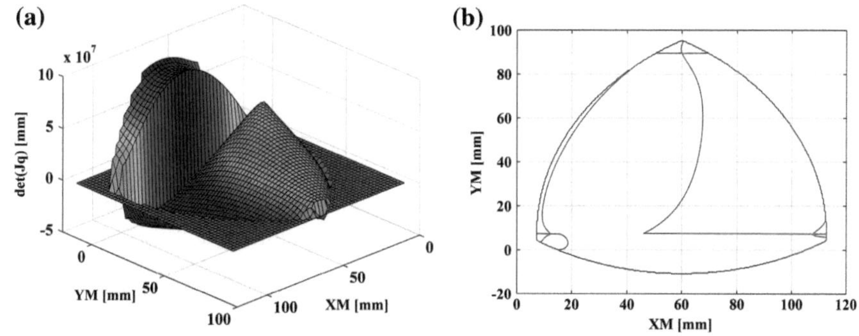

Fig. 4 Analysis of the singularities of second type for the planar 3-R(RPRGR)RR manipulator in 3D (**a**) and 2D (**b**) representation

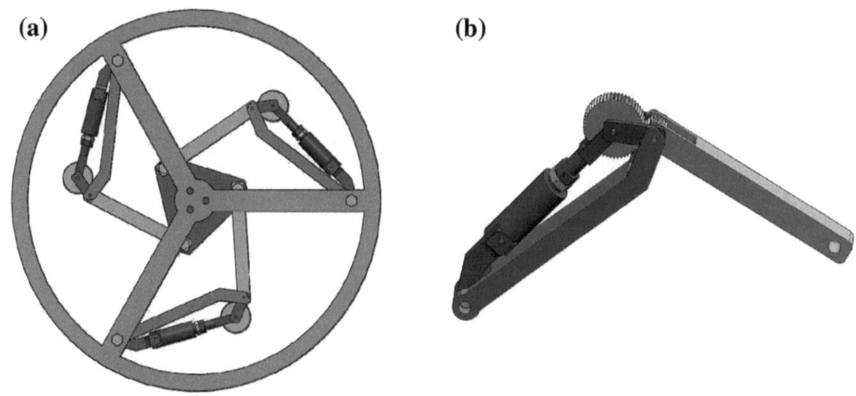

Fig. 5 Design of the planar parallel manipulator 3-R(RPRGR)RR (**a**) and one kinematic chain with geared linkage and linear actuation (**b**)

A proposed design of the planar parallel manipulator 3-R(RPRGR)RR and a view of the kinematic chain using parallel connected geared linkage with inverted slider crank basic structure and linear actuation are shown in Fig. 5a, b.

6 Conclusions

The paper proposes an original solution of a planar parallel manipulator 3-R (RPRGR)RR, which uses a parallel connected geared linkage with linear actuation.

The kinematic chain implements geared linkages actuated with an electrical linear actuator beside binary elements and rotation joints. The extension of the rotation and its forward/backward sense is controlled by means of a geared linkage with inverted slider crank basic structure.

The analysis of the singularities of this manipulator shows that the geared linkages with linear actuation allow the avoiding of singularities of the first type by choosing of optimal initial stroke and working stroke.

Some major advantages of this novel type of planar parallel manipulator are indicated. So, the used mechanism allows a large rotation angle, is light weighted for a simple and sturdy construction. The linear transmission function can be achieved by means of a fairly simple and reliable system. The using of linear actuators with screw-nut fulfills the self-locking condition and ensures a high transmission ratio without gear box.

References

1. Ibarreche, J.I., Altuzarra, O., Petuya, V., Hernandez, A., Pinto, C.: Structural synthesis of the families of parallel manipulators with 3 degrees of freedom. In: Proceedings of 19th CSIM IFToMM Symposium, pp. 35–42. Springer, Berlin (2013)
2. Hernandez, A.: Ibarreche, J.I., Petuya, V., Altuzarra, O.: Structural synthesis of 3-DoF spatial fully parallel manipulators. Int. J. Adv. Rob. Syst. **11**(101), 1–8 (2014)
3. Kotlarski, J. Abdellatif, H., Ortmaier, T., Heimann, B.: Enlarging the useable workspace of planar parallel robots using mechanisms of variable geometry. In: Proceedings of ReMAR 2009, pp. 63–72. London (2009)
4. Joubair, A., Slamani, M., Bonev, I.A.: Kinematic calibration of a 3-DOF planar parallel robot. Ind. Robot Int. J. **39**(4), 392–400 (2012)
5. Huang, M.Z.: Design of a planar parallel robot for optimal workspace and dexterity. Int. J. Adv. Robot. Syst. **8**(4), 176–183 (2011)
6. Williams, R.L., Joshi, A.R.: Planar parallel 3-RPR manipulator. In: Proceedings of the Sixth Conference on Applied Mechanisms and Robotics. Cincinnati, pp. 1–8. OH (1999)
7. Buium, F., Leonchi, D., Doroftei, I.: A Workspace characterization of the 0/3/3R parallel mechanism. Appl. Mech. Mater. **658**, 563–568 (2014)
8. Huang, M.Z., Thebert, L. L.: A study of workspace and singularity characteristics for design of 3-DOF planar parallel robots. Int. J. Adv. Manuf. Technol. **51**(5–8), 789–797 (2010)
9. Kotlarski, J., Heimann, B., Ortmaier, T.: Influence of kinematic redundancy on the singularity-free workspace of parallel kinematic machines. Front. Mech. Eng. **7**(2), 120–134 (2012)
10. Kotlarski, J., de Nijs, R., Abdellatif, H., Heimann, B.: New interval-based approach to determine the guaranteed singularity-free workspace of parallel robots. In: ICRA 2009 Proceedings of IEEE International Conference on Robotics and Automation, pp. 1256–1261. Kobe, Japan (2009)
11. Chablat, D., Staicu, S.: Kinematics of A 3-PRP planar parallel robot. U.P.B. Sci. Bull. D **71**(2), 3–15 (2009)
12. Chablat, D., Wenger, P.: Self Motions of a Special 3-RPR Planar Parallel Robot. Advances in Robot Kinematics, pp. 221–228. Springer, Berlin (2006)
13. Briot, S., Bonev, I., Chablat, D., Wenger, P., Vigen, A.: Self-motions of general 3-RPR planar parallel robots. Int. J. Robot. Res. **27**(7), 855–866 (2008)
14. Duca, C., Buium, F.: Transmission indices adoption at 0/3/0 structural group. Appl. Mech. Mater. **658**, 55–58 (2014)
15. Duca, C., Buium, F.: Singularities classification for structural groups of dyad type. Appl. Mech. Mater. **658**, 47–54 (2014)
16. Pisla, D., et al.: Kinematics and design of two variants of a reconfigurable parallel robot. In: Proceedings of ReMAR, pp. 624–631. London (2009)

17. Plitea, N., Lese, D., Pisla, D., Vaida, C.: Structural design and kinematics of a new parallel reconfigurable robot. Robot. Comput. Integr. Manuf. **29**(1), 219–235 (2013)
18. Singh, Y., et al.: Dynamic modelling and control of a 3-DOF planar parallel robotic (XYθZ motion) platfom. Procedia ICAMME 2014. Mater. Sci. **5**, 1528–1539 (2014)
19. Modler K.-H., Hollmann C., Lovasz E.-C., Perju D.: Geared linkages with linear displacement actuator used as function generating mechanisms. In: Proceedings of the 11-th World Congress on TMM, pp. 1254–1259. Tian Jin, 1–5 April 2004 (2005)
20. Lovasz E.-C., Ciupe V, Modler K-H, Gruescu C.M., Hanke U, Maniu I, Mărgineanu D.: Experimental Design and Control Approach of an Active Knee Prosthesis with Geared Linkage, New Advances in Mechanisms, Transmissions and Applications, vol. 17, pp. 149–156. Springer (Series: Mechanisms and Machine Science), Berlin (2013)
21. Zlatanov D.: Generalized singularity analysis of mechanisms. PhD thesis, Department of Mechanical and Industrial Engineering, University of Toronto (1988)
22. Bonev I.: Geometric analysis of planar mechanisms. PhD thesis, Department de Genie Mecanique, Faculte des Sciences et de Genie, Universite Laval, Quebec (2002)
23. Merlet J-P: Parallel Robots. Kluwer Academic Publishers (2006)
24. Buium, F., Duca, C., Leonchi, D.: Problems regarding singularities analysis of a 0/3/3 parallel mechanism. Appl. Mech. Mater. **658**, 569–574 (2014)

Kinematic Analysis of 3 RSS+CP Parallel Mechanisms

Tarcisio Antonio Hess Coelho and Fernando Malvezzi

Abstract The paper presents an asymmetric three-degree-of-freedom parallel mechanism purposely conceived as a robotic device for tool positioning applications. The kinematic structure of the mechanism contains one central passive limb that constrains the end-effector motion to two translations and one rotation. This work deals with important issues related to the parallel mechanism, such as the position kinematics analysis and the singular configurations prediction.

Keywords Parallel mechanism · Parallel robot · Kinematics · Singularity

1 Introduction

Parallel mechanisms have been studied during the last two decades, due to the fact that they present some potential advantages in a comparison with serial structures, namely, the ability to perform fast motions, modular construction, lightness and high load capacity [1–3]. In fact, this superiority has been observed, for instance, in pick-and-place parallel robots, capable of reaching very high speeds and accelerations [4].

Regarding the utilization of parallel mechanisms on milling applications, a great number of architectures have been investigated. The first parallel milling machines, employed for this purpose, like the Variax and Hexaglide [1, 5], had six degrees of freedom. However, such structures present some drawbacks, namely, the small ratio of workspace volume to machine size and the need to control six axes to position the cutting tool. On the other hand, the Neos Tricept [6] presents a hybrid architecture, where the parallel portion is a tetrapod with one central passive limb to

T.A.H. Coelho
University of São Paulo, São Paulo, Brazil
e-mail: tarchess@usp.br

F. Malvezzi (✉)
Maua Institute of Technology, São Paulo, Brazil
e-mail: fernando.malvezzi@maua.br; fermaldo@ig.com.br

B. Corves et al. (eds.), *Mechanisms, Transmissions and Applications*,
Mechanisms and Machine Science 31, DOI 10.1007/978-3-319-17067-1_4

constrain the motion of the moving platform. Differently of its predecessors, this machine has been used in automotive and aerospace industry. Recently, there is a trend for using parallel mechanisms with simpler structures, such as the 3-degree-of-freedom Universal Cartesian Robot [7], Tripteron [8] and Orthoglide [9].

Some architectures, as those analyzed in [7, 8, 10], have reduced modeling complexity since their kinematic equations are linear and fully decoupled. However, due to the fact that these robots are over constrained mechanisms, they require a very special care on manufacturing and assembly of their parts. Such requirements often demand more tight dimensional and geometrical tolerances, increasing their costs.

Furthermore, most of the proposed parallel robot architectures present symmetric kinematic chains, while there are only few works dealing with asymmetric architectures [11–13].

This work deals with a three-degree-of-freedom parallel mechanism purposely conceived as a robotic device for tool positioning applications. Initially, the mechanism topology is briefly described in Sect. 2. Then, the position kinematics analysis is presented in Sect. 3. Finally, Sect. 4 focuses in details on the singular configurations.

2 Topology Description

The mechanism analyzed here was proposed in [14, 15] and performs three independent motions. Topologically, the mechanism is parallel and asymmetric. The kinematic structure of the mechanism contains one central passive limb composed by two joints: a cylindrical one (C) and a prismatic one (P) (Fig. 1b).

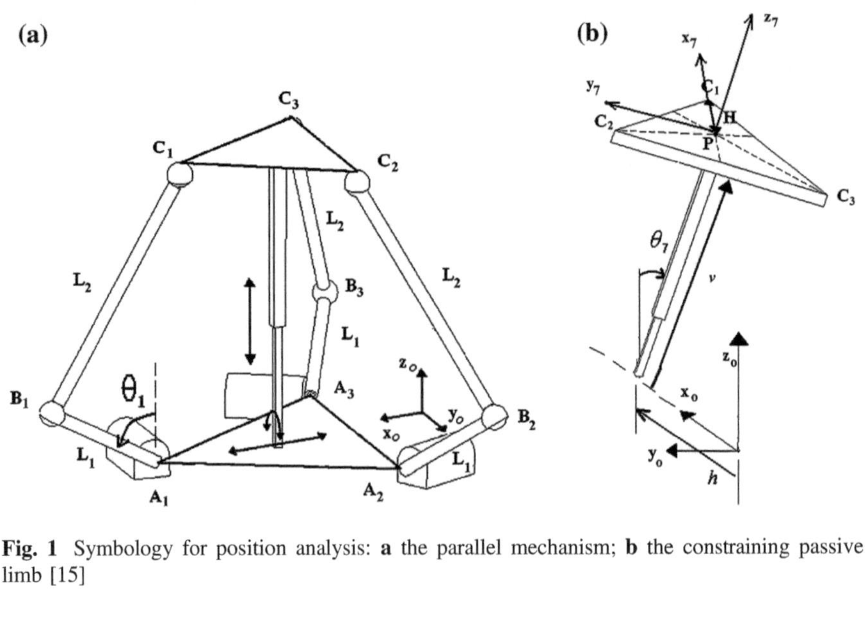

Fig. 1 Symbology for position analysis: **a** the parallel mechanism; **b** the constraining passive limb [15]

As a consequence, the moving platform has got three degrees of freedom and is constrained to perform two translations and one rotation. The connectivity of the other three active limbs, 3RSS, is equal to six (Fig. 1a). An underlined letter represents an active joint, which states the presence of an actuator.

3 Inverse Kinematic Model

The goal of the position kinematic is to obtain the mathematical transformation between the actuators displacements, θ_1, θ_2, and θ_3, and the coordinates h, θ_7 and v of a point P that belongs to the moving platform (Fig. 1b). In the inverse kinematics analysis, the variables h, θ_7 and v of point P are known, while the actuator coordinates are unknown. The mentioned coordinates of point P correspond to the displacement along X_0-axis, Z_7-axis and rotation around X_0-axis, respectively. Applying the homogeneous transformation [16] of frame 7 with respect to the fixed frame 0 ${}^0_7T$, the location of the point P can be determined. The approach followed here to develop the mathematical model is similar to that one applied to develop the inverse kinematic model showed in [15].

$$\begin{bmatrix} {}^0P \\ 1 \end{bmatrix} = {}^0_7T \cdot \begin{bmatrix} {}^7P \\ 1 \end{bmatrix} \tag{1}$$

where

$${}^0_7T = \text{Trans}(X_0, h)\ \text{Rot}(X_0, \theta_7)\ \text{Trans}(Z_7, v) \tag{2}$$

After describing the location of point P, the center of each spherical joints C_j can be also determined. For the sake of simplicity, α_1 is chosen zero (Fig. 2).

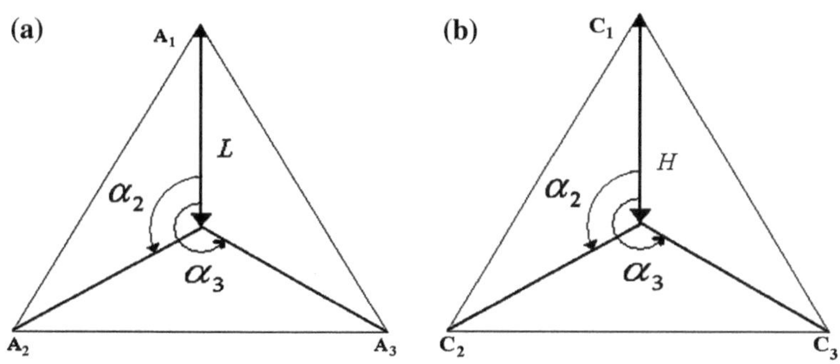

Fig. 2 Mechanism parameters in: **a** the base platform and **b** the moving platform [15]

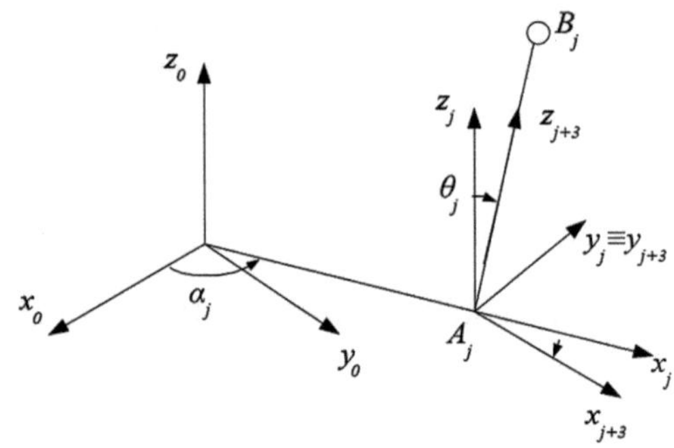

Fig. 3 Sketch of limb j indicating used reference frames [15]

$$\begin{bmatrix} {}^0C_j \\ 1 \end{bmatrix} = {}^0_7T \cdot \begin{bmatrix} H \cdot c\alpha_j \\ H \cdot s\alpha_j \\ 0 \\ 1 \end{bmatrix} = \begin{bmatrix} H \cdot c\alpha_j + h \\ (H \cdot s\alpha_j) \cdot c\theta_7 - v \cdot s\theta_7 \\ (H \cdot s\alpha_j) \cdot s\theta_7 + v \cdot c\theta_7 \\ 1 \end{bmatrix} \quad j = 1, 2, 3 \tag{3}$$

The point B_j, which represents the center of the other spherical joints in each limb j, can be obtained applying Eq. (4) (Fig. 3).

$$\begin{bmatrix} {}^0B_j \\ 1 \end{bmatrix} = {}^0_jT \cdot {}^{j}_{j+3}T \cdot \begin{bmatrix} {}^{j+3}B_j \\ 1 \end{bmatrix} \quad j = 1, 2, 3 \tag{4}$$

where

$${}^0_jT = \begin{bmatrix} c\alpha_j & -s\alpha_j & 0 & Lc\alpha_j \\ s\alpha_j & c\alpha_j & 0 & Ls\alpha_j \\ 0 & 0 & 1 & 0 \\ 0 & 0 & 0 & 1 \end{bmatrix}, \quad {}^{j}_{j+3}T = \begin{bmatrix} c\theta_j & 0 & s\theta_j & 0 \\ 0 & 1 & 0 & 0 \\ -s\theta_j & 0 & c\theta_j & 0 \\ 0 & 0 & 0 & 1 \end{bmatrix}, \quad {}^{j+3}B_j = \begin{bmatrix} 0 \\ 0 \\ L_1 \end{bmatrix}$$

Therefore,

$${}^0B_j = \begin{bmatrix} (L_1 \cdot c\alpha_j) \cdot s\theta_j + Lc\alpha_j \\ (L_1 \cdot s\alpha_j) \cdot s\theta_j + Ls\alpha_j \\ L_1 c\theta_j \end{bmatrix} \quad j = 1, 2, 3 \tag{5}$$

The distance between points Bj and Cj is equal to L_2. Therefore,

$$({}^0B_j - {}^0C_j)^T ({}^0B_j - {}^0C_j) - L_2^2 = 0 \quad j = 1, 2, 3 \tag{6}$$

Then,

$$K_j c\theta_j + M_j s\theta_j + N_j = 0 \quad j = 1, 2, 3 \tag{7}$$

where

$$\begin{aligned}K_j &= -2vc\theta_7 L_1 - 2Hs\alpha_j s\theta_7 L_1\\ M_j &= (-H + 2L - 2hc\alpha_j - Hc2\alpha_j - 2Hc\theta_7 s^2\alpha_j + 2vs\alpha_j s\theta_7)L_1\\ N_j &= h^2 + H^2 - HL + L^2 + v^2 + 2h(H - L)c\alpha_j\\ &\quad - HL(c2\alpha_j + 2c\theta_7 s^2\alpha_j) + 2Lvs\alpha_j s\theta_7 + L_1^2 - L_2^2\end{aligned}$$

Equation (7) can be modified into a 2th-order polynomial equation

$$\left(N_j - K_j\right) \cdot u_j^2 + \left(2M_j\right) \cdot u_j + K_j + N_j = 0 \quad j = 1, 2, 3 \tag{8}$$

where $u_j = \tan\,(\theta_j/2)$. Equation (8) may have up to two different solutions and, as a consequence the mechanism itself may have up to eight assembly modes. These assembly modes are only theoretically possible and, consequently, due to constructive reasons, one among the others is preferable. The chosen assembly mode represented in Fig. 1a is adequate because it avoids interference between each peripheral active limb with the central passive one.

4 Singularity Analysis

In singular configurations, a parallel mechanism can either gain or lose degrees of freedom [2]. The conditions for the occurrence of singularities can be investigated by the inspection of the determinants of Jacobian matrices, J_q and J_x [2, 17]. These Jacobian matrices are obtained by deriving the position equations—Eq. (7) with respect to time, which leads to Eq. (9). Figure 4a shows an example of such

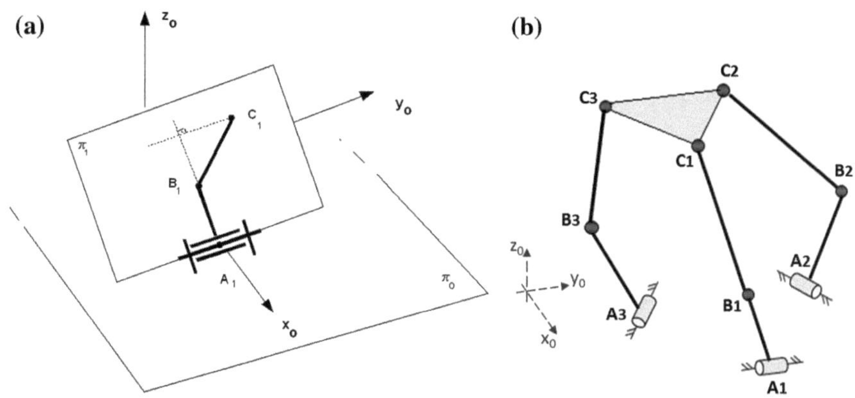

Fig. 4 Examples of singular configurations for limb 1. The singularity occurs when the points A_1, B_1 and C_1 belong: **a** to the same plane π_1; **b** in the same line

configurations. By setting the variable $J_{q(1,1)}$ to zero, for instance, the determinant of J_q becomes null. Figure 4b shows a geometric interpretation of this kind of singularity. This occurs not only when points A_1, B_1 and C_1 are in the same line, but also when these points belong to the same plane π_1.

$$J_x\dot{x} = J_q\dot{q} \tag{9}$$

where

$$\dot{x} = \begin{bmatrix} \dot{h} & \dot{v} & \dot{\theta}_7 \end{bmatrix}^T \qquad \dot{q} = \begin{bmatrix} \dot{\theta}_1 & \dot{\theta}_2 & \dot{\theta}_3 \end{bmatrix}^T$$

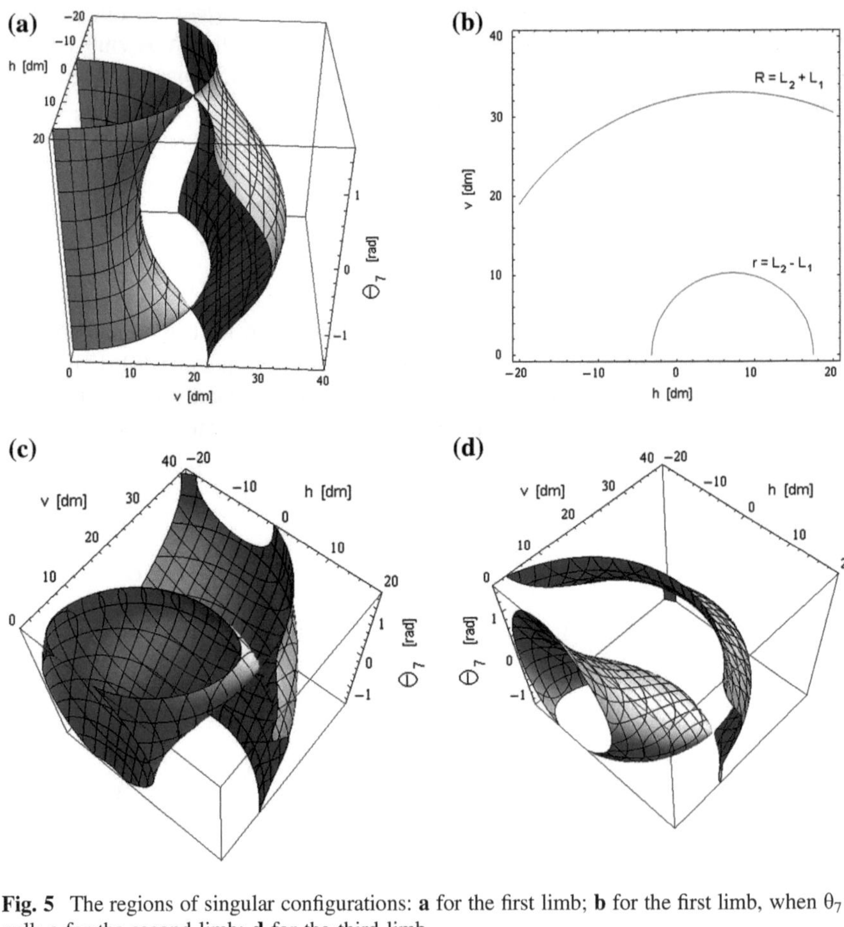

Fig. 5 The regions of singular configurations: **a** for the first limb; **b** for the first limb, when θ_7 is null; **c** for the second limb; **d** for the third limb

Table 1 Parameters of parallel mechanism

H (dm)	L (dm)	L1 (dm)	L2 (dm)	α_1 (deg.)	α_2 (deg.)	α_3 (deg.)
4.8	11.9	11.4	18.5	0	120	240

$$J_q = \begin{bmatrix} J_{q(1,1)} & 0 & 0 \\ 0 & J_{q(2,2)} & 0 \\ 0 & 0 & J_{q(3,3)} \end{bmatrix} \qquad J_x = \begin{bmatrix} J_{x(1,1)} & J_{x(1,2)} & J_{x(1,3)} \\ J_{x(2,1)} & J_{x(2,2)} & J_{x(2,3)} \\ J_{x(3,1)} & J_{x(3,2)} & J_{x(3,3)} \end{bmatrix}$$

$$\begin{aligned} J_{x(j,1)} &= 2[h + (H - L)c\alpha_j - c\alpha_j s\theta_j + L_1] \\ J_{x(j,2)} &= 2[v + Ls\alpha_j s\theta_7 + (-c\theta_j c\theta_7 + s\alpha_j s\theta_j s\theta_7)L_1] \\ J_{x(j,3)} &= 2\{Ls\alpha_j(vc\theta_7 + Hs\alpha_j s\theta_7) + [c\theta_j(-Hc\theta_7 s\alpha_j + vs\theta_7) \\ &\quad + s\alpha_j s\theta_j(vc\theta_7 + Hs\alpha_j s\theta_7)]L_1\} \\ J_{q(j,j)} &= [-2s\theta_j(vc\theta_7 + Hs\alpha_j s\theta_7) + c\theta_j(H - 2L + 2hc\alpha_j + Hc2\alpha_j \\ &\quad + 2Hc\theta_7 + s^2\alpha_j - 2vs\alpha_j s\theta_7)]L_1 \end{aligned}$$

Figure 5 presents the regions of singular configurations for the parallel mechanism with the mechanism parameters presented in Table 1.

These surfaces are generated by annulling the elements $J_{q(j,j)}$ in Eq. (9). Figure 5b shows a 2D-plot of variable v with respect to h, when θ_7 is zero. This represents the loci of point C_1: two circles of radii, $L_2 - L_1$ and $L_2 + L_1$, with center in A_1. Notice that the surfaces shown in Fig. 5c, d are symmetric with respect to the angle θ_7.

5 Conclusions

This work presented the kinematic analysis of 3 RSS+CP parallel mechanism to be applied as a robotic device for tool positioning applications. The main contribution of this paper is a singularity analysis, which shows the regions where the singular configurations can occur. The Jacobian analysis is applied to identify these singularities.

In a comparison with the previous proposed concepts, this asymmetric and parallel kinematic structure has got some advantages. The architecture is simpler than Tricept because the 3 *RSS+CP* does not need a serial *RR* wrist to guide the cutting tool in order to perform three independent translations. Other proposed concepts, for instance, Orthoglide and Tripteron, need to satisfy special assembly conditions: the orthogonality and the parallelism between the axes of passive and active joints in their three limbs. For the 3 *RSS+CP*, such special conditions refer only to the central limb, which might reduce its manufacturing and assembly costs.

References

1. Merlet, J.P.: Parallel Robots, 2nd edn. Springer, Berlin (2006)
2. Tsai, L.W.: Robot Analysis: The mechanics of Serial and Parallel Robots. Wiley, New York (1999)
3. Ceccarelli, M.: Fundamentals of Mechanics of Robotic Manipulation. Kluwer, The Netherlands (2004)
4. Adept Technology, Inc.: Adept Quattro s 650H, Livermore, CA, USA. April 30 (2010). http://www.adept.com (2010)
5. Khol, R.A.: Machine tool built from mathematics. Am. Machinist **138**, 53–55 (1994)
6. Neumann, K.E.: Next generation tricept—a true revolution in parallel kinematics. In: Proceedings of the 4th Chemnitz Parallel Kinematics Seminar, PKS2004, Verlag Wissenschaftliche Scripten. Reports from IWU, vol. 24, pp. 591–594 (2004)
7. Kim, H.S., Tsai, L.-W.: Design Optimization of a Cartesian Parallel Manipulator. ASME DETC Confer, Montreal (2002)
8. Gosselin, C.M., Kong, X., Foucault, S., Bonev, I.A.: A fully-decoupled 3-dof translational parallel mechanism. In: Proceedings of 4th Chemnitz Parallel Kinematics Seminar, PKS2004, Verlag Wissenschaftliche Scripten. Reports from IWU, vol. 24, pp. 595–610 (2004)
9. Chablat, D., Wenger, P., Staicu, S.: Dynamics of the orthoglide parallel robot. UPB Sci. Bull. Ser. D Mech. Eng. **71**(3), 3–16 (2009)
10. Di Gregorio, R., Parenti-Castelli, V.: Design of 3-dof parallel manipulators based on dynamic performances. In: Proceedings of the 4th Chemnitz Parallel Kinematics Seminar, PKS2004, Verlag Wissenschaftliche Scripten. Reports from IWU, vol. 24, pp. 385–397 (2004)
11. Li, Q., Huang, Z.: Type synthesis of 5-DOF parallel manipulators. In: Proceedings of the 2003 IEEE International Conference on Robotics and Automation Taipei, Taiwan, pp. 1203–1208 (2003)
12. Huang, T., Li, M., Zhao, X.M., Mei, J.P., Chetwynd, D.G., Hu, S.J.: Conceptual design and dimensional synthesis for a 3-DOF module of the TriVariant—a novel 5-DOF reconfigurable hybrid robot. IEEE Trans. Rob. **21**(3), 449–456 (2005)
13. Kumazawa, V.D., Hess-Coelho, T.A., Rinaudi, D., Carbone, G., Ceccarelli, M.: Kinematic analysis and operation feasibility of a 3-dof asymmetric parallel mechanism. In: Proceedings of the 20th COBEM, Gramado (2009)
14. HESS-Coelho, T.A., Branchini, D.M., Malvezzi, F.: A new family of 3-DOF parallel robot manipulators for pick-and-place operations. In: 18th COBEM 2005, Ouro Preto (2005)
15. HESS-Coelho, T.A.: An alternative procedure for type synthesis of parallel mechanisms. In: 12th IFTOMM World Congress, Besançon (2007)
16. Craig, J.J.: Introduction to Robotics: Mechanics and Control, 2nd edn. Pearson Education, New Jersey (1989)
17. Azulay, H., et al.: Comparative analysis of a new 3_PPRS parallel kinematic mechanism. Robot. Comput.-Integr. Manuf. **30**, 369–378 (2014)

On the Accuracy Analyses of a Class of 2-DOF Planar Parallel Manipulators

Mümin Özsipahi and Eres Söylemez

Abstract This study is about the accuracy analysis of two degrees of freedom planar parallel manipulators consisting of five links (one of which is fixed) that are connected to each other with five revolute (R) or prismatic (P) joints. Among these manipulators, topologically RRRRR, PRRRP, RPRPR planar parallel manipulators are selected. Accuracy analysis is a common method to evaluate the performance of parallel manipulators. In this study, accuracy analysis of these planar parallel manipulators is presented by using local conditioning index and maximum positional error methods. For each of the manipulators, mathematical expressions for the local conditioning index and maximum positional errors are derived. These expressions are implemented using Matlab© environment and examples are presented to show the accuracy distributions over the workspace. The consistency between local conditioning index and maximum positional error is discussed.

Keywords Planar parallel manipulators · Conditioning index · Maximum positional error

1 Introduction

Parallel kinematic machines have been gaining more importance in the industry. Over the last decade, there has been several research papers published in this area [1]. Two degrees of freedom mechanisms consisting of five links (one of which is fixed) that are connected to each other with five revolute (R) or prismatic (P) joints are called two degrees of freedom Planar Parallel Manipulators (PPM). Topologically, these manipulators are the simplest form of parallel manipulators and various

M. Özsipahi (✉) · E. Söylemez
Middle East Technical University, Ankara, Turkey
e-mail: ozsipahi@metu.edu.tr

E. Söylemez
e-mail: eres@metu.edu.tr

B. Corves et al. (eds.), *Mechanisms, Transmissions and Applications*,
Mechanisms and Machine Science 31, DOI 10.1007/978-3-319-17067-1_5

studies are presented about the kinematic, singularity and workspace analysis of such systems [2–7]. Accuracy is a commonly used performance measure for parallel manipulators. Researchers generally prefer using conditioning index to evaluate the accuracy performance of two degrees of freedom planar parallel manipulators [8–11]. Another accuracy evaluation method used for different type of manipulators is maximum position or orientation error [12–14].

In this study, accuracy analysis of three different planar parallel manipulators is studied. For each of the manipulators, accuracy analysis is performed and mathematical expressions for the conditioning index and maximum positional error are derived. These expressions are implemented using Matlab© environment and examples are presented to show the accuracy distributions over the workspace. The consistency between local conditioning index and maximum positional error is investigated.

2 Selected Class of Manipulators

Topologically, thirty-two different two-degree of freedom planar parallel manipulators can be formed with prismatic (P) or revolute (R) joints. Cervantes-Sanchez and Rendon-Sanchez reduced this number to six, considering some physical

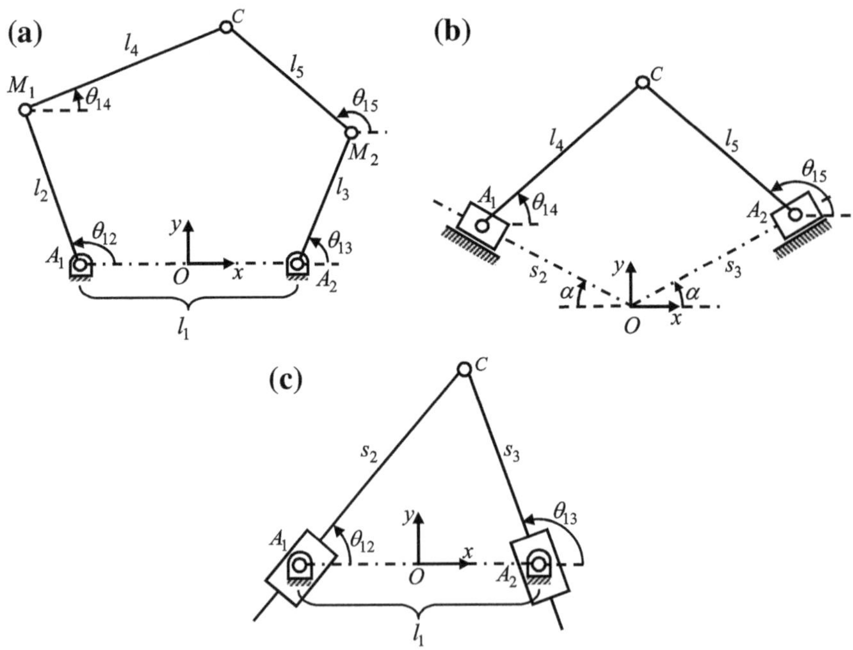

Fig. 1 Schematic view of **a** RRRRR, **b** PRRRP and **c** RPRPR manipulators

constraints [4]. Three different manipulators remain if topological symmetry condition is added. These are RRRRR, PRRRP and RPRPR planar parallel manipulators. The schematic view and geometric parameters of these manipulators are given in Fig. 1. The input parameters are $[\theta_{12}, \theta_{13}]$ for RRRRR manipulator and $[s_2, s_3]$ for PRRRP and RPRPR manipulators.

3 Accuracy Analyses

In the literature, accuracy analysis is used to evaluate dexterities of parallel manipulators. Conditioning index is mostly utilized method to obtain dexterity of parallel robots. Other accuracy analysis method that is presented in this study is maximum positional error.

In both aforementioned methods, Jacobian matrix of the parallel robot is required. In addition, inverse or forward kinematic analyses should be performed to obtain joint variables. Kinematic analysis of the three manipulators shown in Fig. 1 can be found in the literature [8–11]. Using the time derivatives of loop closure equations, Jacobian matrices of RRRRR, PRRRP and RPRPR manipulator can be found as follows,

$$\bar{J}_{RRRRR} = \frac{1}{\sin(\theta_{15} - \theta_{14})} \begin{bmatrix} -l_2 \sin\theta_{15} \sin(\theta_{12} - \theta_{14}) & l_3 \sin\theta_{14} \sin(\theta_{13} - \theta_{15}) \\ l_2 \cos\theta_{15} \sin(\theta_{12} - \theta_{14}) & -l_3 \cos\theta_{14} \sin(\theta_{13} - \theta_{15}) \end{bmatrix} \tag{1}$$

$$\bar{J}_{PRRRP} = \frac{1}{\sin(\theta_{14} - \theta_{15})} \begin{bmatrix} \sin\theta_{15} \cos(\theta_{14} + \alpha) & \sin\theta_{14} \cos(\theta_{15} - \alpha) \\ -\cos\theta_{15} \cos(\theta_{14} + \alpha) & -\cos\theta_{14} \cos(\theta_{15} - \alpha) \end{bmatrix} \tag{2}$$

$$\bar{J}_{RPRPR} = \frac{1}{\sin(\theta_{12} - \theta_{13})} \begin{bmatrix} -\sin\theta_{13} & \sin\theta_{12} \\ \cos\theta_{13} & -\cos\theta_{12} \end{bmatrix} \tag{3}$$

3.1 Conditioning Index

The condition number, K is defined in the literature as [15],

$$\mathrm{K} = \left\|\bar{J}\right\| \left\|\bar{J}^{-1}\right\| \tag{4}$$

Here, $\|.\|$ is the matrix norm, i.e.,

$$\|\bar{J}\| = \sqrt{tr\left(\bar{J}\,\overline{W}\,\bar{J}^T\right)} \quad \text{and} \quad \overline{W} = \frac{1}{n}\bar{I} \tag{5}$$

where n is the dimension of the square matrix and $\bar{I}$ is the $n \times n$ identity matrix. For a 2×2 Jacobian matrix, let

$$\bar{J} = \begin{bmatrix} J_{11} & J_{12} \\ J_{21} & J_{22} \end{bmatrix} \tag{6}$$

Then, using Eqs. (5) and (6), it can be shown that,

$$\|\bar{J}\| = \sqrt{\frac{1}{2}\left(J_{11}^2 + J_{12}^2 + J_{21}^2 + J_{22}^2\right)} \tag{7}$$

The condition number will be between zero and infinity [15]. Therefore, local conditioning index, *LCI* is defined as,

$$LCI = 1/\mathrm{K} = 1\Big/\left(\|\bar{J}\|\left\|\bar{J}^{-1}\right\|\right) \tag{8}$$

Using the Jacobian matrices from Eqs. (1)–(3) with Eqs. (6)–(8), the expressions for the *LCI* can be derived for RRRRR, PRRRP and RPRPR as follows,

$$LCI_{RRRRR} = \frac{2l_2 l_3 |\sin(\theta_{15} - \theta_{14})| \cdot |\sin(\theta_{12} - \theta_{14})| \cdot |\sin(\theta_{13} - \theta_{15})|}{l_2^2 \sin^2(\theta_{12} - \theta_{14}) + l_3^2 \sin^2(\theta_{13} - \theta_{15})} \tag{9}$$

$$LCI_{PRRRP} = \frac{2|\cos(\theta_{14} + \alpha)| \cdot |\cos(\theta_{15} - \alpha)| \cdot |\sin(\theta_{14} - \theta_{15})|}{\cos^2(\theta_{14} + \alpha) + \cos^2(\theta_{15} - \alpha)} \tag{10}$$

$$LCI_{RPRPR} = |\sin(\theta_{12} - \theta_{13})| \tag{11}$$

Note that, *LCI* can only have values between zero and one. *LCI* value close to one means a manipulator with a good accuracy and close to zero refers to manipulators with a bad accuracy [15]. *LCI* can have different values for each position of the end-effector. So, global conditioning index, *GCI* is defined as,

$$GCI = \left(\int_W \frac{1}{\mathrm{K}} dW\right) \Big/ \left(\int_W dW\right) \tag{12}$$

In this study, *W* represents all the reachable workspace bounded by singularities. As the value of the *GCI* gets closer to one, the accuracy performance of the manipulator gets better.

3.2 Maximum Positional Error

Merlet noted that the most contribution to the errors of the end-effector is from the resolution of the sensors at the input motors [16]. Therefore, errors from other sources (assembly, manufacturing, etc.) are neglected in this study. In addition, if the errors of the end-effector are small, x and y components of the positioning errors due to the resolution of actuators, Δx_c and Δy_c can directly be obtained using,

$$\begin{bmatrix} \Delta x_c \\ \Delta y_c \end{bmatrix} = \bar{J} \begin{bmatrix} \Delta q_{12} \\ \Delta q_{13} \end{bmatrix} \quad (13)$$

where Δq_{12} and Δq_{13} are the resolution of the actuators. The maximum positional error, MPE can be defined as,

$$MPE = \max\left(\sqrt{\Delta x_c^2 + \Delta y_c^2}\right) \quad (14)$$

Since the manipulator is symmetric, the resolution of the inputs can be selected as equal, i.e., $\Delta q_{12} = \pm\Delta q$ and $\Delta q_{13} = \pm\Delta q$. Then, using Eqs. (6) and (13) with Eq. (14), it can be shown that,

$$MPE = |\Delta q|\sqrt{J_{11}^2 + J_{12}^2 + J_{21}^2 + J_{22}^2 + 2|J_{11}J_{12} + J_{21}J_{22}|} \quad (15)$$

Note that, $\Delta q = \Delta\theta$ for RRRRR and $\Delta q = \Delta s$ for PRRRP and RPRPR manipulators. Using the Jacobian matrices from Eqs. (1)–(3) with Eq. (15), the expression for the MPE can be derived for RRRRR, PRRRP and RPRPR as follows,

$$MPE_{RRRRR} = |\Delta\theta| \cdot \left(\sqrt{G+H}/|\sin(\theta_{14} - \theta_{15})|\right)$$
$$\text{where} \begin{cases} G = l_2^2 \sin^2(\theta_{12} - \theta_{14}) + l_3^2 \sin^2(\theta_{13} - \theta_{15}) \\ H = 2l_2l_3|\cos(\theta_{14} - \theta_{15}) \sin(\theta_{12} - \theta_{14}) \sin(\theta_{13} - \theta_{15})| \end{cases} \quad (16)$$

$$MPE_{PRRRP} = |\Delta s| \cdot \left(\sqrt{G+H}/|\sin(\theta_{14} - \theta_{15})|\right)$$
$$\text{where} \begin{cases} G = \cos^2(\theta_{14}+\alpha) + \cos^2(\theta_{15} - \alpha) \\ H = 2|\cos(\theta_{14} - \theta_{15}) \cos(\theta_{14}+\alpha) \cos(\theta_{15} - \alpha)| \end{cases} \quad (17)$$

$$MPE_{RPRPR} = |\Delta s| \cdot \left(\sqrt{2 + 2|\cos(\theta_{12} - \theta_{13})|}\Big/|\sin(\theta_{12} - \theta_{13})|\right) \quad (18)$$

4 Examples

The equations obtained in this study are implemented in Matlab© environment. The workspaces of the manipulators are scanned to calculate LCI and MPE. Their distributions within the workspace are determined. Note that, *MPE* values are calculated taking the actuator resolution values as one unit, i.e., $\Delta\theta = 1$ and $\Delta s = 1$.

In this study, examples are presented using arbitrary link length dimensions for each of the manipulator.

For RRRRR manipulator, link lengths are chosen as $l_1 = 1$, $l_2 = 0.8$, $l_3 = 0.8$, $l_4 = 1.2$, $l_5 = 1.2$ units and distribution of the *LCI* and *MPE* are presented in Fig. 2. Note that, "±" working mode of the RRRRR manipulator is used in this study (shown in Fig. 1). This is the most commonly used working mode of RRRRR manipulator.

GCI value that is calculated for all the reachable workspace of the manipulator is shown. Maximum and minimum values for *MPE* is infinity and zero respectively. Maximum value for *MPE* is encountered at the singularity positions of the manipulator. *MPE* values larger than 10 units are not shown for the sake of figure readability.

When the distribution of *LCI* and *MPE* are compared for RRRRR manipulator, these two methods give similar results in general. In some regions, the change of *LCI* and *MPE* contradict with each other. Merlet also mentioned about this problem in his studies [17]. On the other hand, it can be claimed that *MPE* method gives more realistic results in terms of accuracy performance since it has a direct physical meaning.

For the PRRRP manipulator with $0.0 \le s_2 \le 4.0$, $0.0 \le s_3 \le 4.0$, $l_4 = 2$, $l_5 = 2$ units and $\alpha = 15°$ distributions of the LCI and MPE are presented in Fig. 3.

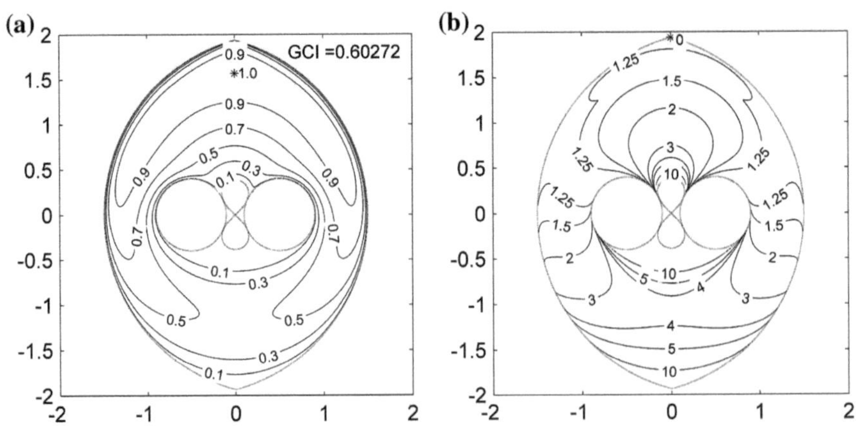

Fig. 2 **a** *LCI*, **b** *MPE* distribution of RRRRR manipulator

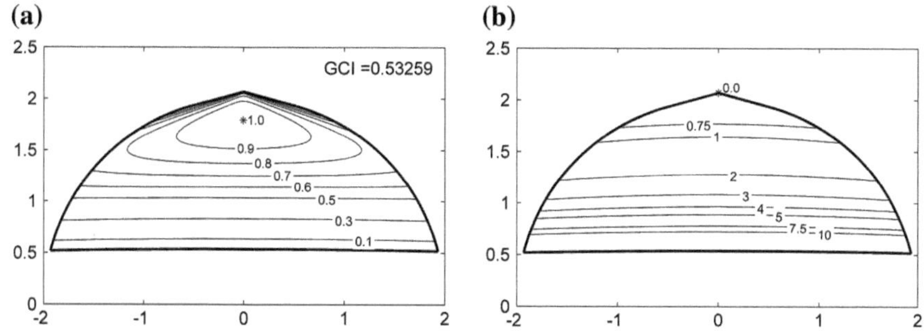

Fig. 3 **a** *LCI*, **b** *MPE* distribution of PRRRP manipulator

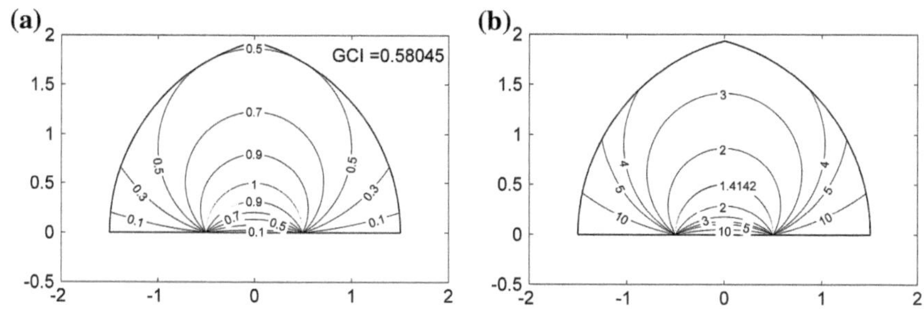

Fig. 4 **a** *LCI*, **b** *MPE* distribution of RPRPR manipulator

For PRRRP manipulator, *LCI* and *MPE* have worst values closer to the the x axis. The distributions of *LCI* and *MPE* indicate parallel results but the distributions deviate from each other around the region at the tip of the workspace.

For the RPRPR manipulator with $l_1 = 1$, $0 \leq s_2 \leq 2$, $0 \leq s_3 \leq 2$ distribution of the LCI and MPE is presented in Fig. 4.

For RPRPR manipulator, the *LCI* and *MPE* plots show the same increase-decrease characteristics. Maximum *LCI* value equals to one and minimum *MPE* value equals to 1.4142 ($\sqrt{2}$). These points lay on the same semi-circle whose diameter is the fixed link (l_1). This can also be obtained from Eqs. (11) and (18).

5 Conclusions

In this study, accuracy analyses of three different planar parallel manipulators are studied. In the accuracy analyses, for each of the manipulator, the expressions for local conditioning index and maximum positional error are derived. These expressions can directly be used to determine the accuracy values and performances of the selected manipulators. Then, some examples of accuracy distributions are presented by using Matlab© environment. The consistency between local conditioning index and maximum positional error is discussed using these examples. It is seen that the distributions of *LCI* and *MPE* differ from each other at some workspace regions.

Acknowledgments The financial support supplied by the Scientific and Technological Research Council of Turkey (TUBITAK) to the first author is acknowledged.

References

1. Merlet, J.-P.: Parallel Robots. Solid Mechanics and Its Applications. Springer, Berlin (2006)
2. Alıcı, G.: An inverse position analysis of five-bar planar parallel manipulators. Robotica **20**, 195–201 (2002)
3. Cervantes-Sanchez, J.J., Rendon-Sanchez, J.G.: A simplified approach for obtaining the workspace of a class of 2-dof planar parallel manipulators. Mech. Mach. Theory **34**, 1057–1073 (1999)
4. Cervantes-Sanchez, J.J., Hernandez-Rodriguez, J.C., Rendon-Sanchez, J.G.: On the workspace, assembly configurations and singularity curves of the RRRRR-type planar manipulator. Mech. Mach. Theory **35**, 1117–1139 (2000)
5. Liu, X.-J., Wang, J., Pritschow, G.: Kinematics, singularity and workspace of planar 5R symmetrical parallel mechanisms. Mech. Mach. Theory **41**, 145–169 (2006)
6. Figielski, A., Bonev, I.A., Bigras, P.: Towards development of a 2-DOF planar parallel robot with optimal workspace use. In: Proceedings of IEEE International Conference on Systems, Montreal, pp. 1562–1566 (2007)
7. Macho, E., Altuzarra, O., Pinto, C., Hernandez, A.: Workspaces associated to assembly modes of the 5R planar parallel manipulator. Robotica **26**, 395–403 (2008)
8. Liu, X.-J., Wang, J., Zheng, H.-J.: Optimum design of the 5R symmetrical parallel manipulator with a surrounded and good-condition workspace. Robot. Auton. Syst. **54**, 221–233 (2005)
9. Liu, X.-J., Wang, J., Pritschow, G.: On the optimal kinematic design of the PRRRP 2-DOF parallel mechanism. Mech. Mach. Theory **41**, 1111–1130 (2005)
10. Liu, X.-J., Guan, L., Wang, J.: Kinematics and closed optimal design of a kind of PRRRP parallel manipulator. J. Mech. Des. **129**(5), 558–563 (2006)
11. Szep, C., Stan, S., Csibi, V., Manic, M., Balan, R.: Kinematics, workspace, design and accuracy analysis of RPRPR medical parallel robot. In: 2nd Conference on Human System Interactions, Catania, pp. 75–80 (2009)
12. Merlet, J-P., Daney, D.: Dimensional synthesis of parallel robots with a guaranteed given accuracy over a specific workspace. In: Proceedings of IEEE International Conference on Robotics and Automation, Barcelona, pp. 942–947 (2005)
13. Briot, S., Bonev, I.A.: Accuracy analysis of 3-DOF planar parallel robots. Mech. Mach. Theory **43**, 445–458 (2007)
14. Yu, A., Bonev, I.A., Zsombor-Murray, P.: Geometric approach to the accuracy analysis of a class of 3-DOF planar parallel robots. Mech. Mach. Theory **43**, 364–375 (2007)
15. Gosselin, C., Angeles, J.: A global performance index for the kinematic optimization of robotic manipulators. Trans. ASME **113**, 220–226 (1991)
16. Merlet, J.-P.: Computing the worst case accuracy of a PKM over a workspace or a trajectory. In: The 5th Chemnitz Parallel Kinematics Seminar, Chemnitz, pp. 83–96 (2006)
17. Merlet, J.-P.: Jacobian, manipulability, condition number, and accuracy of parallel robots. J. Mech. Des. **128**, 199–205 (2006)

Determination of Linkage Parameters from Coupler Curve Equations

Shaoping Bai

Abstract A fundamental problem in four-bar linkage synthesis is to determine linkage parameters from a given coupler-curve equation, which is essentially an inverse problem to the linkage analysis. For many decades, this synthesis problem has been considered as overdetermined, which yields only approximate solutions. This paper presents a new result, which shows that a determined system of coupler-curve coefficient equations can be formulated, which leads to exact solutions of the linkage parameters. A method of linkage parameter determination from known coupler-curve equation is further developed, demonstrated with examples.

Keywords Four-bar linkage synthesis · Coupler-curve coefficient (C^3) equation · Determined path-generation synthesis problem · Roberts-Chebyshev theorem

1 Introduction

Path generation synthesis is to find dimensions of a mechanism which is able to trace a path with a number of prescribed locations [1–3]. Traditionally, the path generation is carried out by solving a set of synthesis equations, which consists of a number of multi-variable higher-order polynomial equations, for which it is difficult to obtain solutions. Approach of polynomial continuation was applied to solve 9-point synthesis problem [4]. With the huge amount of possible solutions, the identifying of physically meaningful linkages becomes another problem. An alternative way for the path problem is to search a library that is pre-built based on kinematic analysis to match a curve [5]. This approach avoids the challenging

S. Bai (✉)
Department of Mechanical and Manufacturing Engineering, Aalborg University, Aalborg, Denmark
e-mail: shb@m-tech.aau.dk

B. Corves et al. (eds.), *Mechanisms, Transmissions and Applications*,
Mechanisms and Machine Science 31, DOI 10.1007/978-3-319-17067-1_6

equation solving. However, both the library building and path curve matching are time consuming. Solving the synthesis problem with optimization techniques has been extensively explored [6, 7]. While optimization methods allow us to specify more points (10–20 or more) to describe a desired curve, solution convergence is not always guaranteed.

In addition to these approaches, there is another approach that is rarely considered in reported works. Blechschmidt and Uicker [8] proposed an approach of synthesis from the algebraic curves of the coupler points. Ananthasuresh and Kota [9] further proposed a two-step method: (1) curve generation; and (2) parameter evaluation. However, this method was not really implemented. A major problem is the overdeterminacy related to the parameter determination.

The overdeterminacy in the parameter determination problem refers to the facts that the number of coefficient equations exceeds the number of linkage parameters. Recall that the coupler curve equation of a four-bar linkage is a sextic, namely,

$$f(x,y) = \sum_{i,j=0}^{6} P_{i,j}x^i y^j = 0; \quad i+j \leq 6 \tag{1}$$

The coupler-curve equation of four-bar linkages is, moreover, a special kind of the generic sextic given above, which includes only 15 coefficients. As the coefficient parameters, or design parameters, are only nine, the synthesis problem appears to be overdetermined, and has been considered as such for decades.

In this paper, the problem of determining linkage parameters for four-bar linkages is studied. A new formulation of the coupler-curve synthesis is developed. The formulation shows that a determined system of coefficient equations can be derived, which allows us to find exact solutions. A method of linkage parameter determination is further developed, demonstrated with examples.

2 Problem Formulation

A four-bar linkage is illustrated in Fig. 1, whose coupler curve is Γ that is visited by a landmark point P of the body. The linkage is generally defined with nine parameters, $\mathbf{x} = [m, h, b_1, b_2, d_1, d_2, l_2, l_3, l_4]^T$, where the link lengths $l_2, l_3, l_4 > 0$. The length l_1 of the base link can be calculated from the coordinates of two points B and D.

The problem is to find the nine parameters, which allow the linkage tracing Γ given in the form of Eq. 1. This problem is essentially the inverse problem to the linkage analysis, in which the coupler curve of a known linkage is determined and analyzed.

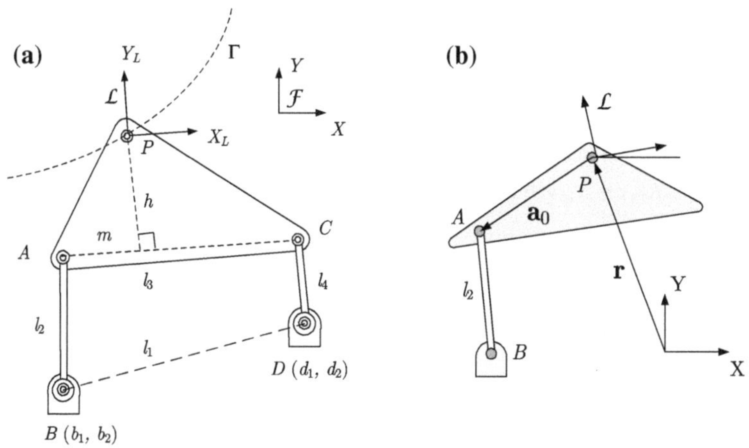

Fig. 1 **a** A generic four-bar linkage with nine design parameters, including link lengths, l_2–l_4, the position of point P in the coupler link located by (m, h), and the positions of two joint centers B (b_1, b_2) and $D(d_1, d_2)$; **b** one Revolute-Revolute (RR) dyad

3 Equations of Coupler-Curve Coefficients (C^3)

We start with the coupler-curve equation of four-bar linkages, which is available from the literature [10]. We include directly the equation, as given below

$$\begin{aligned} & A_1^2C_2^2 - 2A_1C_2A_2C_1 + A_2^2C_1^2 + B_1C_2A_1B_2 - B_1^2C_2A_2 \\ & \quad - C_1B_2^2A_1 + C_1B_2A_2B_1 - A_1^2B_2^2 + 2A_1B_2A_2B_1 - A_2^2B_1^2 = 0 \end{aligned} \tag{2}$$

with

$$A_1 = \mathbf{r}^T\mathbf{a}_0 - \mathbf{b}^T\mathbf{a}_0; \quad A_2 = \mathbf{r}^T\mathbf{c}_0 - \mathbf{d}^T\mathbf{c}_0 \tag{3}$$

$$B_1 = \mathbf{r}^T\mathbf{E}\mathbf{a}_0 - \mathbf{b}^T\mathbf{E}\mathbf{a}_0; \quad B_2 = \mathbf{r}^T\mathbf{E}\mathbf{c}_0 - \mathbf{d}^T\mathbf{E}\mathbf{c}_0 \tag{4}$$

$$C_1 = \frac{1}{2}\left[(\mathbf{r}-\mathbf{b})^T(\mathbf{r}-\mathbf{b}) + \mathbf{a}_0^T\mathbf{a}_0 - l_2^2\right] \tag{5}$$

$$C_2 = \frac{1}{2}\left[(\mathbf{r}-\mathbf{d})^T(\mathbf{r}-\mathbf{d}) + \mathbf{c}_0^T\mathbf{c}_0 - l_4^2\right] \tag{6}$$

where $\mathbf{b}(\mathbf{d})$ is the position vector of point $B(D)$, while $\mathbf{a}_0(\mathbf{c}_0)$ is the position vector of point $A(C)$ expressed in the coupler attached $X_L - Y_L$ frame. $\mathbf{r}$ is the position vector of P, while $\mathbf{E}$ is the matrix of rotation through an angle of 90° CCW.

Equation (2) is the equation of the coupler curve for the four-bar linkage applicable to any point on the coupler link. The equation is a sixth-order bivariate polynomial. Knowing that the circularity of the curve is three, we can write the coupler curve equation as

$$\begin{aligned}f(x,y) &= K_1(x^2+y^2)^3 + (K_2x+K_3y)(x^2+y^2)^2 + (K_4x^2+K_5xy+K_6y^2)(x^2+y^2)\\ &+ K_7x^3 + K_8x^2y + K_9xy^2 + K_{10}y^3 + K_{11}x^2 + K_{12}xy + K_{13}y^2\\ &+ K_{14}x + K_{15}y + K_{16} = 0\end{aligned} \quad (7)$$

This means that the coupler curve of four-bar linkages is a special case of the sextic bivariate polynomial of Eq. (1). Equation (7) can be derived from Eq. (2) using computer algebra, with all coefficients expressed with the nine parameters. Of them, the first coefficient is given as $K_1 = -l_3{}^2/4$. Other coefficients are online available at http://homes.m-tech.aau.dk/shb/four-bar/4barsyn_coeffs.pdf.

By introducing $k_i = K_{i+1}/K_1$, $i = 1,\ldots,15$, Eq. (7) becomes

$$\begin{aligned}f(x,y) &= (x^2+y^2)^3 + (k_1x+k_2y)(x^2+y^2)^2 + (k_3x^2+k_4xy+k_5y^2)(x^2+y^2)\\ &+ k_6x^3 + k_7x^2y + k_8xy^2 + k_9y^3 + k_{10}x^2 + k_{11}xy + k_{12}y^2\\ &+ k_{13}x + k_{14}y + k_{15} = 0\end{aligned} \quad (8)$$

which yields a polynomial equation of exact degree 6. In total, we have 15 coefficients $\{k_i\}_1^{15}$, all functions of nine parameters.

With any given algebraic equation of a four-bar linkage coupler curve, a system of coupler-curve coefficient (C^3) equations with all nine linkage parameters can be established

$$k_i(\mathbf{x}) - k_i^* = 0; \quad i = 1,\ldots,15 \quad (9)$$

where $\{k_i^*\}_1^{15}$ are the known coefficients of the given coupler curve.

Equation (9), with more equations than variables, stands for an overdetermined system, which admits only approximate solutions.

4 A Determined System of C^3 Equations

In this work, we show that a determined system can be established to yield exact solutions, as described presently.

Recall that $K_1 = -l_3{}^2/4$, we introduce a parameter $r = l_3/2$, which implies

$$k_i = -K_{i+1}(\mathbf{x})/r^2, \quad i = 1,\ldots,15 \quad (10)$$

Substituting $l_3 = 2r$ into all coefficients k_i yields new expressions of coefficients in terms of l_2, l_4, m, h and the coordinates (b_1, b_2), (d_1, d_2) and parameter r.

Further analysis on the coefficient functions reveals their properties:

1. Coefficients k_1, k_2 and k_4 are independent of l_2 and l_4
2. Coefficients k_3 and k_5 are linear in l_2^2 and l_4^2; moreover, the coefficient of l_2^2 in k_3 is identical to its coefficient in k_5. The same holds for the term l_4^2
3. Coefficient pairs (k_6, k_8) and (k_7, k_9) have the same features as the pair (k_3, k_5)
4. Coefficients $k_{10}, \ldots, k_{15}$ contain terms of l_2^4, $l_2^2 l_4^2$, l_4^4, l_2^2, l_4^2 only.

These properties allow us to build a system of equations with a reduced number of unknowns through algebraic manipulations. In this vein, a seven-dimensional array of design variables is defined as $\mathbf{y} = [r, m, h, b_1, b_2, d_1, d_2]^T$.

Note that three coefficient functions k_1, k_2 and k_4 are independent of l_2 and l_4. Their corresponding equations in (9) can be expressed in terms of variable $\mathbf{y}$ as

$$k_j(\mathbf{y}) - k_j^* = 0, \quad j = 1, 2, 4 \tag{11a}$$

Parameters l_2 and l_4 can also be eliminated from other coefficient equations. This is done by subtracting one from the other in each pair of the coefficient equations of (k_3, k_5), (k_6, k_8) and (k_7, k_9), which lead to three new equations

$$k_i - k_j - (k_i^* - k_j^*) = 0, \quad (i,j) \in \{(3,5), (6,8), (7,9)\} \tag{11b}$$

all independent of l_2 and l_4.

For the remaining six coefficient equations for $k_{10}, \ldots, k_{15}$, they contain terms including l_2^4, $l_2^2 l_4^2$, l_4^4, l_2^2, l_4^2 only. These equations can be written as:

$$t_{i,1} l_2^4 + t_{i,2} l_2^2 l_4^2 + t_{i,3} l_4^4 + t_{i,4} l_2^2 + t_{i,5} l_4^2 + t_{i,6} = 0; \quad i = 1, \ldots, 6 \tag{11c}$$

which can be cast in a compact form as

$$\mathbf{T}\mathbf{q} = \mathbf{0} \tag{11d}$$

where $\mathbf{q} = [l_2^4, l_2^2 l_4^2, l_4^4, l_2^2, l_4^2, 1]^T$, and $\mathbf{T}$ is a 6×6 matrix whose entries include the terms of Eq. (11c). Equation (11d) implies that

$$g(\mathbf{y}) = \det(\mathbf{T}) = 0 \tag{11e}$$

Equations (11a), (11b) and (11e) comprise a system of seven equations in seven unknowns, thereby establishing a determined system of algebraic equations. Upon solving for $\mathbf{y}$, l_2 and l_4 can be uniquely determined, thereby finding all unknowns.

From the above formulation, we achieve an important result:

If the coupler curve equation of a four-bar linkage is known, a determined system of coefficient equations can be established, whose solutions yield the complete set of the linkage parameters.

Parameter determination The determined system formulated above leads to exact solutions of linkage parameter. A possible method to solve this system is continuation [4]. We propose here an alternative method to solve the synthesis equations.

The new method adopts a trial-and-error approach, combining both graphic and numerical techniques. Let r be the key variable, from Eqs. (11a) and (11b), the solutions of other unknowns in $\mathbf{y}$ are readily solved for each given value of r. These solutions are further substituted back into two coefficient equations, for example, k_5 and k_6, to find corresponding solutions of l_2 and l_4. These solutions are then substituted into the remaining equations, while recording the rms error in meeting these equations. The point with minimum rms error is chosen as a solution.

Note that there are multiple real solutions for each r; we thus take the minimum error, which means, $e = \min\{s_i\}_1^n$, where $\{s_i\}_1^n$ is the set of all rms values for n sets of real solutions.

5 Examples

We include two examples to show the effectiveness of the method. The first example is an asymmetric Watt's curve:

$$\begin{aligned} f = {} & x^6 + 3x^4y^2 + 3x^2y^4 + y^6 - 0.66666667x^5 - 1.33333333x^3y^2 \\ & - 0.66666667xy^4 + 0.11277778x^4 + 0.14222222x^2y^2 + 0.02944444y^4 \\ & + 0.00944444x^3 + 0.00944444xy^2 - 0.00333264x^2 - 0.00277708y^2 \\ & + 0.00000833x + 0.000025 \end{aligned} \tag{12}$$

with the developed method, the errors with different values of r are examined, as shown in Fig. 2. Three minima are identified, which are $r = 0.15$, 0.067, 0.033. Solutions corresponding to the minima are further found, as listed in Table 1. Each minimum yields two solutions, both generating the same linkage. All together, three cognate linkages, anticipated by the Roberts-Chebyshev Theorem, were obtained, as displayed in Fig. 3.

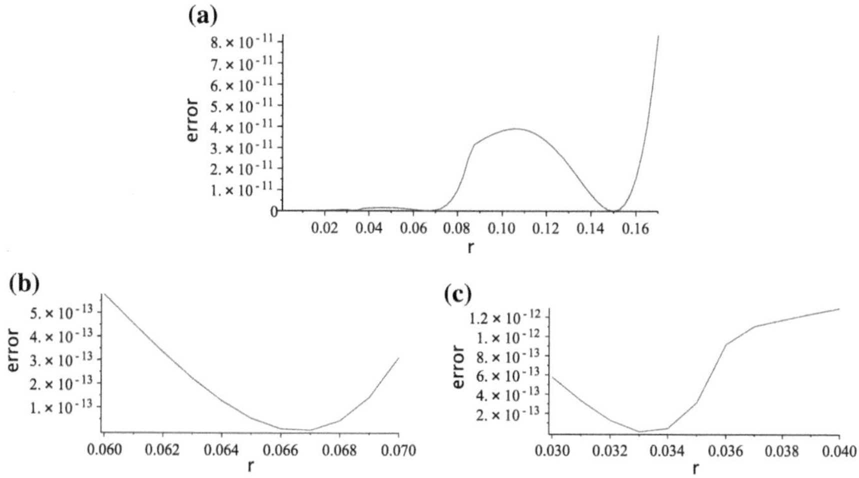

Fig. 2 Error plots. **a** Full view, **b** zoomed view around $r = 0.067$, **c** zoomed view around $r = 0.033$

Table 1 Linkage synthesis results for the example (unit: m)

No.	r	m	h	$[b_1, b_2]$	$[d_1, d_2]$	l_2	l_3	l_4
1	0.15	0.1	0.0	[0, 0]	[0.25, 0]	0.2	0.3	0.2
	0.15	0.2	0.0	[0.25, 0]	[0, 0]	0.2	0.3	0.2
2	0.067	−0.067	0.0	[0.0833, 0]	[0.25, 0]	0.1335	0.134	0.201
	0.067	0.201	0.0	[0.25, 0]	[0.0833, 0]	0.201	0.134	0.1335
3	0.033	0.2004	0.0	[0, 0]	[0.0833, 0]	0.100798	0.0668	0.0672
	0.033	−0.1336	0.0	[0.0833, 0]	[0, 0]	0.0672	0.0668	0.1008

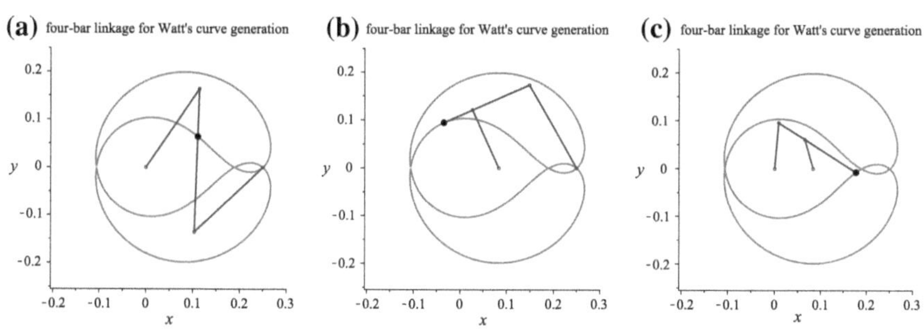

Fig. 3 Three cognate linkages for a Watt's curve

Table 2 Linkage synthesis results for the example (unit: m)

No.	m	h	$[b_1, b_2]$	$[d_1, d_2]$	l_2	l_3	l_4
1	−0.1568	−0.0196	−0.025, 0.1	0.2, −0.2	0.1254	0.294	0.3356
2	0.15	0.1	−0.2, 0	0.2, −0.2	0.15	0.4	0.35
3	−0.1255	0.08369	−0.2, 0	−0.025, 0.1	0.1801	0.068	0.1572

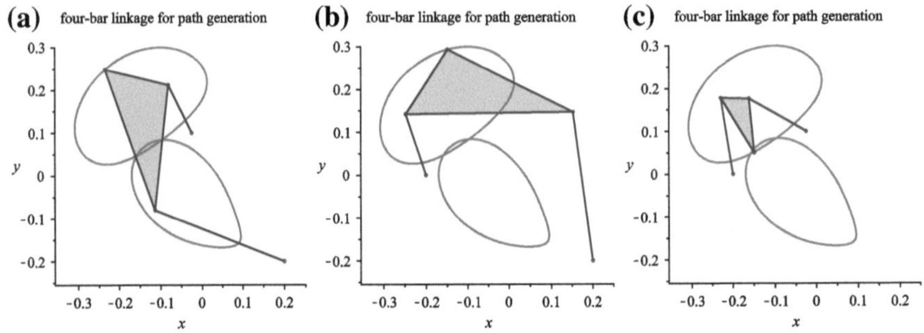

Fig. 4 Three cognate linkages showing together with the coupler curve

In the second example, the coupler-curve equation is given as:

$$\begin{aligned} f = {} & x^6 + 3.0x^4y^2 + 3.0x^2y^4 + y^6 + 0.05x^5 + 0.2x^4y + 0.1x^3y^2 \\ & + 0.4x^2y^3 + 0.05xy^4 + 0.2y^5 - 0.109375x^4 + 0.18x^3y - 0.13875x^2y^2 \\ & + 0.18xy^3 - 0.029375y^4 + 0.00875x^3 - 0.004375x^2y - 0.01525xy^2 \\ & - 0.044375y^3 + 0.0107375x^2 + 0.001425xy + 0.00214375y^2 + 0.0008525x \\ & + 0.00107375y - 0.0000479375025 = 0 \end{aligned} \tag{13}$$

Using the foregoing method, the linkage parameters are then obtained, as listed in Table 2. Figure 4 displays the three linkages, along with the coupler curve. In the figure, we show only the linkages in one branch.

6 Conclusions

The determination of exact linkage parameters from coupler-curve equations is studied. A determined system of coefficient equations was derived. A method combining a numerical method with graphics tools was developed, which allows us to determine linkage parameters directly and simultaneously for the three cognate linkages.

This work solves a fundamental problem in linkage synthesis. The finding of the determinacy of the coefficient equations has several implications on the kinematic synthesis study. First, it implies that exact solutions, rather than approximate solutions, for the synthesis from a coupler curve can be obtained. Moreover, this result could be extended to other types of linkages, including six-bar linkages, spherical and spatial linkages, etc. It can also be used in the synthesis with finitely separated positions, for which how to find a coupler curve from given points remains as an open problem for future study.

Acknowledgments The author acknowledges the discussion with prof. Jorge Angeles, McGill University, and his comments.

References

1. Burmester, L.: Lehrbuch der Kinematik. Arthur Felix Verlag, Leipzig (1888)
2. Bottema, O., Roth, B.: Theoretical Kinematics. North-Holland Pub. Co., New York (1979)
3. McCarthy, J.M., Soh, G.S.: Geometric Design of Linkages. Springer, New York (2011)
4. Wampler, C.W., Morgan, A.P., Sommese, A.J.: Numerical continuation methods for solving polynomial systems arising in kinematics. ASME J. Mech. Des. **112**(1), 59–68 (1990)
5. Kota, S.: Automatic selection of mechanism designs from a three-dimensional design map. J. Mech. Des. **114**(2), 359–367 (1992)
6. Sancibrian, R., Viadero, F., García, P., Fernández, A.: Gradient-based optimization of path synthesis problems in planar mechanisms. Mech. Mach. Theor. **39**(8), 839–856 (2004)
7. Cabrera, J.A., Simon, A., Prado, M.: Optimal synthesis of mechanisms with genetic algorithms. Mech. Mach. Theor. **37**(10), 1165–1177 (2002)
8. Blechschmidt, J.L., Jr. Uicker, J.J.: Linkage synthesis using algebraic curves. ASME J. Mech. Des. **108**(4):543–548 (1986)
9. Ananthasuresh, G.K., Kota, S.: Synthesis of four-bar linkages for path generation via the coupler curve equation. In: Proceedings of Third National Applied Mechanisms and Robotics Conference, Cincinnati, OH, USA, 8–10 Nov 1993, paper 83
10. Bai, S.: Dimensional synthesis of six-bar linkages with incomplete data set. In: Proceedings 5th European Conference on Mechanism Science, pp. 1–8, Guimarães, Portugal, 16–20 Sept 2014

Introduction of a 1-DOF Rolling Contact Element for a Planar Reconfigurable Manipulator

Stefan Kurtenbach, Fritz Ehreiser, Mathias Hüsing and Burkhard Corves

Abstract This paper introduces a rolling contact element and its usage in a planar manipulator. Starting from the state of the art in this field of motion technology, the kinematic structure of the manipulator is introduced. Three different cases for moving an object in a planar workspace are used for the derivation of the kinematics. Superposing these cases leads to the general kinematic equations where finally the input angles can be calculated.

Keywords Rolling contact element · Planar reconfigurable manipulator · Kinematic equations · Workspace

1 Introduction

Current developments in the field of grasping technology are focussing on more tailored solution to adapt to an underlying motion task in an ideal way. Thus, a very wide range of different grippers, grasping units and other contact elements is generated. Depending on material characteristics, surface properties, the shape and some other features of the object to be manipulated these different grippers can be selected to be used within the motion task.

This paper proposes a more general approach where a rolling contact element is used which has left one single degree of freedom to perform a rolling motion

---

S. Kurtenbach (✉) · F. Ehreiser · M. Hüsing · B. Corves
Department of Mechanism Theory and Dynamics of Machines, RWTH Aachen University, Aachen, Germany
e-mail: kurtenbach@igm.rwth-aachen.de

F. Ehreiser
e-mail: ehreiser@igm.rwth-aachen.de

M. Hüsing
e-mail: huesing@igm.rwth-aachen.de

B. Corves
e-mail: corves@igm.rwth-aachen.de

B. Corves et al. (eds.), *Mechanisms, Transmissions and Applications*,
Mechanisms and Machine Science 31, DOI 10.1007/978-3-319-17067-1_7

exactly in the contact point. Thus, one single end-effector becomes widely independent from being affected by object properties. Further, the entire handling system (planar manipulator) becomes flexible, reconfigurable and versatile [1]. These properties enable the manipulation of a wide range of different objects possessing diverse properties using a minimum of end-effectors. The significant difference of this kind of handling systems is the integration of the object into the kinematic structure. The object itself becomes a link of the structure.

To enable a motion, within a planar workspace of the object to be handled, in total three degrees of freedom must be realized. This corresponds with a motion in two perpendicular translational directions and rotation around the third one. The kinematic structure of a manipulator satisfying these requirements is shown in Fig. 1. The object (link 3) in this task is a cylinder with a constant curvature. Three rolling contact joints belonging to three two bar linkages adapt to the cylinder and integrate it into the kinematic structure. The base mounted links 1, 5 and 7 form the input links. The constrained motion in all rolling contact points is one further design task and can be realized through form-closure or force-closure. In this paper the input links only provide the input torque for moving the object with 3 DOF. Since there is no redundant arrangement, the drives cannot provide any retention force.

The solution of the vector contours (Eq. 1) is shown in the following chapters.

$$v_1 = p_{obj} + c_1 \tag{1}$$

2 State of the Art

The application of rolling contact grippers is proposed over the decades. They are integrated in different robotic structures which are planar as well as spatial ones and adapted to object with different shapes.

Cole et al. [2] present in the kinematics and control for multifingered hands manipulating an ellipsoidal object. The object is moved with 6 DOF in a three dimensional workspace. The kinematics, dynamics and control of the introduced system is presented.

Ponce et al. in [3] analysed the influence of varying the object shape. They put focus on n-polygons where the rolling contact element stays on one face of the polygon during the motion task.

The dextrous manipulability for multifingered robots is subject matter in [4]. A general technique was developed for dextrous manipulation of a sphere with a flat finger.

As well as [3] Hasegawa in [5] and Yoshida et al. [6] analysed the stability in grasping processes. Hasegawa implemented a motion planning algorithm for multifingered manipulation with rolling contact elements in 3D. This algorithm automatically suggests the optimal grasping points leading to a stable process. Yoshida analysed the stability of a grasping process using soft fingertip material.

Jeong et al. in [7] presents a planar manipulator with a 2D-Workspace and a redundant actuation with four rolling contact elements. Due to that a release and re-grasp was enabled since just three contact elements need to stay in contact at the same time.

3 Motion Sequence of the Manipulator

In this chapter the basic kinematic equations are derived for the manipulator shown in Fig. 1.

The system contains three manipulators which all have a rolling contact point with the moving object. During a movement of the object the contact point of each manipulator moves on the surface of the object as well as on the surface of the manipulator. Due to that the motion of each manipulator is depending on the geometrical shape of both surfaces. In this application the end-effector has a constant curvature (see Fig. 1) with the radius r as well as the object with the radius R.

Since all three two-bar-linkages have the same structure, also the kinematic is the same. Therefore the investigated system is reduced to the left mechanism with the input motion φ_1 and the object. To derive a mathematical description of the motion, it is separated in the three planar ones: translation in x- and y-direction and rotation around the z-axis.

3.1 Rotation Around the Z-Axis

The motion is separated in three steps which are schematically presented in the following. Starting from the initial position (Step 1 in Fig. 2) the entire system is rotated around the centre of the cylinder with the angle $\psi_{rotation}$ (Step 2).

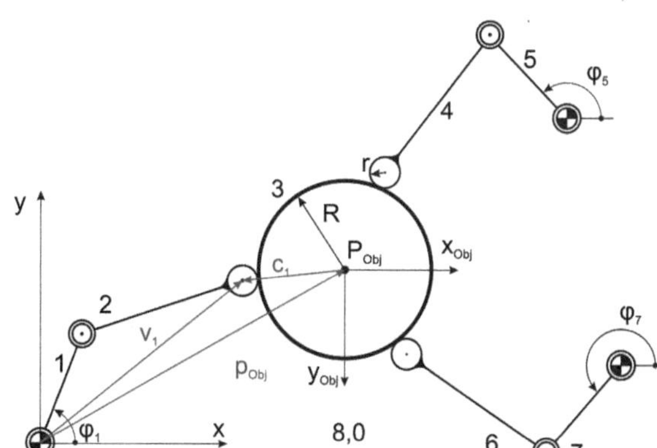

Fig. 1 Kinematic structure of the mechanism

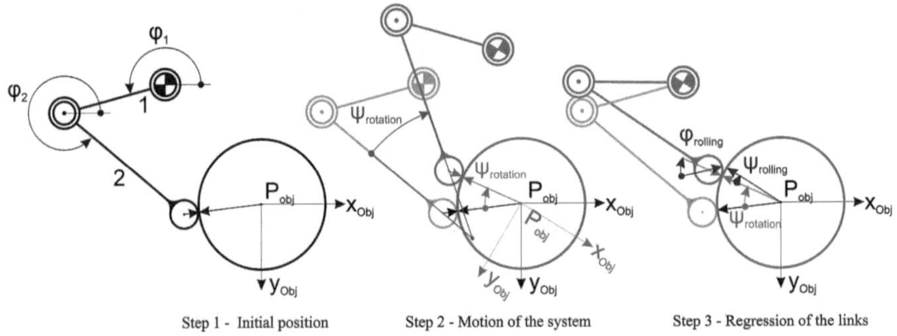

Fig. 2 Motion sequence of the system

Table 1 Variables to describe the motion

Variable	Description
φ_1	Input angle of link 1
φ_2	Rotation angle of link 2
$\varphi_{rolling}$	Rolling angle of the contact element
ψ	Total rotation angle of the object around its z-axis
$\psi_{rolling}$	Rolling angle of the object
$\psi_{rotation}$	Rotation angle of the object
v_1	Vector describing the end-effector centre point from the frame point
c_1	Vector from the object origins to the curvature centre of the contact element

The contact point is fixed and the revolute joints are locked. Subsequently the new base joint is moved into the old one. Unlocking the revolute joints leads to a rolling in the contact point with the angle $\varphi_{\mathrm{rolling}}$ the link 2 and with the angle ψ_{rolling} for the object.

For describing the motion several new variable angels are introduced (Table 1). φ_{i} generally describes the configuration of the manipulator (Fig. 1). With both angles φ_1 and φ_2 known the vector v_1 can be calculated. The second way to calculate v_1 is by the knowing the relative position of the object and its orientation, described by the angle ψ. Then the contact point with the end-effector can be used to calculate v_1.

The rolling condition is found by using the constraint of a rolling contact point between the end-effector and the object shown in Eq. (2).

$$r \cdot \varphi_{rolling} = R \cdot \psi_{rolling} \tag{2}$$

A rotation of the object is described by the angle ψ. During a rotation the contact point moves as described in Fig. 2. The rotation can be divided into two parts. The first one, $\psi_{rotation}$ describes the rotation of the vector c_1 pointing to the end-effector. The second one, $\psi_{rolling}$ results from the rolling movement between both surfaces.

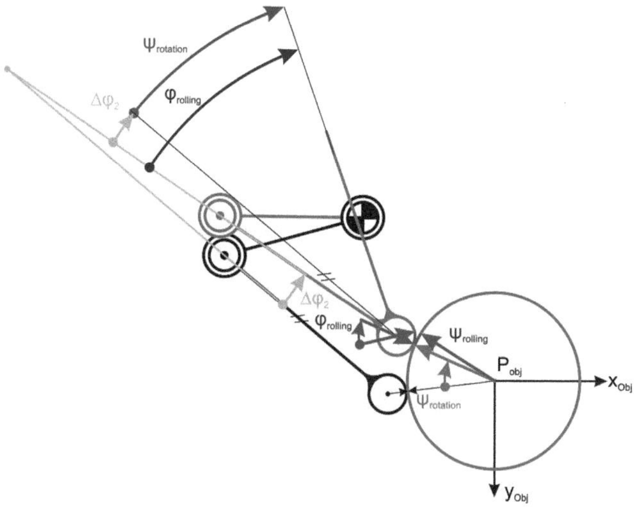

Fig. 3 Geometrically dependence of the different angels for a rotation of the system

$$\psi = \psi_{rotation} + \psi_{rolling} \tag{3}$$

Figure 3 shows the superposed steps shown in Fig. 2 to finally describe the new orientation of c_1 for a rotation the single steps 2 and 3. It is evident that the new orientation link two is relaying on $\psi_{rotation}$ and $\varphi_{rolling}$ and described trough the following Equations.

The calculation of the new position of link 2 bases on the equation $\Delta\varphi_2 = \varphi_2^i - \varphi_2^{i-1}$, where the position i − 1 always is known and starts with the contact position. Thus, the calculation of the input angles becomes trajectory dependent and the inverse kinematics is solved incrementally. The change of the orientation $\Delta\varphi_2$ is equal to the subtraction of the angel described by the motion of the contact point on the end-effector surface $\varphi_{rolling}$ and the change $\psi_{rotation}$ of the orientation of c_1 (Eq. 4).

$$\varphi_{rolling} = \Delta\varphi_2 + \psi_{rotation} \tag{4}$$

Inserting Eqs. (2) and (4) in Eq. (3) leads to Eq. (5) whereas the angel ψ is described with:

$$\psi_{rotation} = \frac{R}{R+r} \cdot \psi - \frac{r}{R+r} \cdot \Delta\varphi_2 \tag{5}$$

With the constraint of the system from Eq. (2) and the identified relation between the movements from Eq. (4), c_1 is described by Eq. (6b).

$$c_{1,rotation} = \begin{pmatrix} (R+r)\cdot\cos(\psi_{Start}+\psi_{rotation}) \\ (R+r)\cdot\sin(\psi_{Start}+\psi_{rotation}) \end{pmatrix}_{rotation} \tag{6a}$$

$$c_{1,rotation} = \begin{pmatrix} (R+r)\cdot\cos\left(\psi_{Start}+\frac{R}{R+r}\cdot\psi-\frac{r}{R+r}\cdot\Delta\varphi_2\right) \\ (R+r)\cdot\sin\left(\psi_{Start}+\frac{R}{R+r}\cdot\psi-\frac{r}{R+r}\cdot\Delta\varphi_2\right) \end{pmatrix}_{rotation} \tag{6b}$$

The angle ψ_{Start} describes the initial grasp point in the objects coordinate system.

3.2 Translation in X- and Y-Direction

For a translation of the object the approach stays the same. Through the motion towards the or away from the manipulator it has to move both links and reposition them with new angels of φ_1 and φ_2. That results in a new orientation of the end-effector and a movement of the contact point. In Fig. 4 the operation in positive x-direction is shown.

The displacement of the centre of the object is described by Δx_{Pobj} and Δy_{Pobj}. Since the object does not rotate for a pure translation $\psi = 0$. Hence, Eq. (5) is reduced to one term, Eq. (7):

$$\psi_{rotation} = -\frac{r}{R+r}\cdot\Delta\varphi_2 \tag{7}$$

The reorientation of c_1 is directly linked with the rotation of link two. As just a translation is calculated, the rotation of c_1 is just relaying on the change of link two and thus on φ_2. For a superposition of both operations c_1 is calculated as Eq. (8b).

$$c_{1,translation} = \begin{pmatrix} (R+r)\cdot\cos(\psi_{Start}+\psi_{rotation}) \\ (R+r)\cdot\sin(\psi_{Start}+\psi_{rotation}) \end{pmatrix}_{translation} \tag{8a}$$

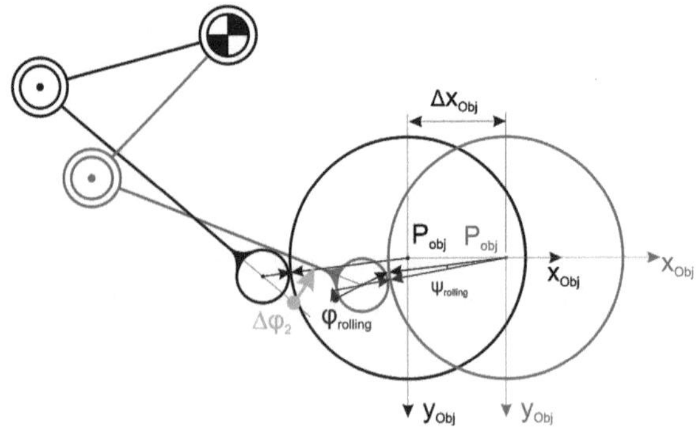

Fig. 4 Sketch of a translation in positive x-direction

$$c_{1,translation} = \begin{pmatrix} (R+r)\cdot\cos\left(\psi_{Start} - \frac{r}{R+r}\cdot\Delta\varphi_2\right) \\ (R+r)\cdot\sin\left(\psi_{Start} - \frac{r}{R+r}\cdot\Delta\varphi_2\right) \end{pmatrix}_{translation} \tag{8b}$$

3.3 Superposed Motion

The general motion of the object is a superposed motion containing translational as well as rotational parts. To calculate the new positions of the input link in terms of $v_{1,nP}$ (new Position), the motion task is known leading to boundary conditions for the displacement ($\Delta x_{Pobj}, \Delta y_{Pobj}$) and rotation of the object (ψ). Starting from the initial position p_{obj}, the following Eq. (9) can be derived from superposing:

$$v_{1,nP} = p_{obj} + \Delta p_{obj} + c_{1,translation} + c_{1,rotation} \tag{9}$$

The vector of the two bar linkage describing the new position $v_{1,nP}$ can be expressed using the link lengths and the input angles, Eq. (10):

$$v_{1,nP} = \begin{pmatrix} l_1\cdot\cos(\varphi_1) + l_2\cdot\cos(\varphi_2) \\ l_1\cdot\sin(\varphi_1) + l_2\cdot\sin(\varphi_2) \end{pmatrix} = p_{obj} + \Delta p_{obj} + c_{1,translation} + c_{1,rotation} \tag{10}$$

This equation depends on known parameters like the link lengths, the initial and final position and orientation of the object, the grasping point. The solution of this equation results in input angles φ_i of the two bar linkage. Since the object motion is a 3-DOF motion, the procedure must be repeated for all two bar linkages resulting in a solution vector φ_i. Using the complete equation system, the inverse kinematics can be solved trajectory dependent and in incremental steps.

4 Design of the Planar Manipulator

The final design of the manipulator can be taken from Fig. 5. The rolling contact elements are docked on to the cylinder. The constrained motion is not considered in this paper. The three single Two-Bar linkages are mounted on a stiff frame. Via belt drive the input moment is transmitted to the input links.

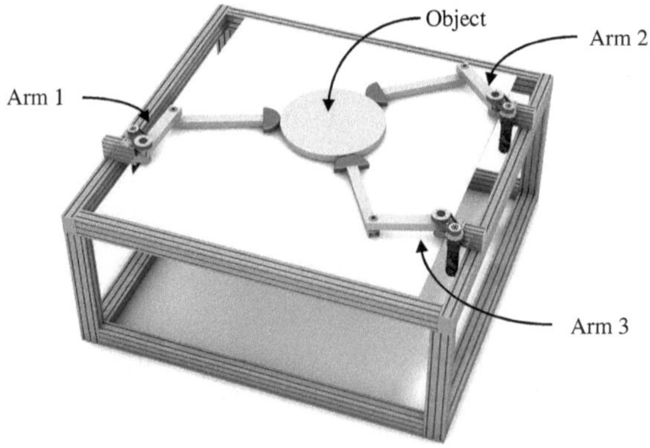

Fig. 5 Final design of the manipulator

5 Conclusions

The paper introduces a manipulator with three arms handling an object using rolling contact elements. Thus, the orientation workspace can be decreased since every translation or rotation is superposed through a rolling of the contact elements. The approved concept can be transported into spatial robotics to support handling tasks in this field of application.

Acknowledgments As part of this work was carried out at the Cluster of Excellence at RWTH Aachen "Integrative production technology for high-wage countries (EXC 128)"; the authors would like to thank The German Research Foundation for their support.

References

1. Müller, R., Riedel, M., Vette, M., Corves, B., Esser, M., Hüsing, M.: Reconfigurable self-optimising handling system. In: Ratchev, S. (ed.) IPAS 2010. IFIP AICT, vol. 315, pp. 255–262. Springer, Heidelberg (2010). ISBN 978-3-642-11597-4
2. Cole, A.B.A., Hauser, J.E., Sastry, S.S.: Kinematics and control of multifingered hands with rolling contact. In: IEEE Transactions on Automatic Control, vol. 34. No. 4 (1989)
3. Ponce, J., Faverjon, B.: On computing three-finger force-closure grasps of polygonal objects. In: IEEE Transactions on Robotics and Automation, vol. 11. No. 6 (1995)
4. Han, L., Guan, Y., Li, Z., Shi, Q., Trinkle, J.: Dextrous manipulation with rolling contacts. In: Proceedings of IEEE International Conference on Robotics and Automation, vol. 2, pp. 992–997 (1997)
5. Hasegawa, T., Mukrakami, K.: Finger motion planning of a multi-fingered manipulation with rolling contact. In: Proceedings of the IEEE International Symposium on Assembly and Task Planning, pp. 196–201 (2001)

6. Yoshida, M., Arimoto, S., Bae, J.-H., Luo, Z.-W.: Stable grasp of a 2D rigid object through rolling with soft fingers. In: IEEE International Conference on Robotics and Biomimetics, pp. 870–876 (2007)
7. Jeong, H., Cheong, J.: In-hand rolling motion planning using independent contact region (ICR) with guaranteed grasp quality margin. In: IEEE International Conference on Robotics and Automation (ICRA). Karlsruhe, Germany, 6–10 May 2013

The Bennett Linkage as a Hinge Application

Uwe Hanke, Jana Ehlig, Cornelia Fischer, Karl-Heinz Modler and Niels Modler

Abstract Parametric commercial CAD environments can be used as interactive geometry software (IGS). Therefore they are the best tool for the implementation of graphical synthesis methods. This paper gives an introduction of using graphical synthesis methods solving two pose spatial guidance tasks with the BENNETT linkage. The synthesis is presented for a hinge application.

Keywords Two pose synthesis · Spatial guidance · Bennett linkage

1 Introduction

The aim of this paper is to show a geometrical way for the mechanism synthesis of spatial guidance tasks with a simple mechanism structure using robust joints. Robust joints are rotatory joints (R-joints). They are easy to implement, have no slip stick or locking problems. The simplest spatial mechanism build up only by R-joints is the BENNETT linkage [1]. Therefore the focus of this paper lies on geometrical synthesis of a BENNETT mechanism for a spatial guidance task. The CAD environment is ideal for the implementation of geometrical synthesis of motion problems. Especially the use of CAD gives a great benefit in finding good compromises during the modelling phase. This will lower the amount of development loops.

U. Hanke (✉) · C. Fischer · K.-H. Modler
Dresden University of Technology, Institute of Solid Mechanics, Dresden, Germany
e-mail: uwe.hanke@tu-dresden.de

J. Ehlig · N. Modler
Dresden University of Technology, Institute of Lightweight Engineering and Polymer Technology, Dresden, Germany

B. Corves et al. (eds.), *Mechanisms, Transmissions and Applications*,
Mechanisms and Machine Science 31, DOI 10.1007/978-3-319-17067-1_8

2 Theoretical Background

Mechanism analysis and synthesis in two dimensional space is a well researched field of study with geometric approaches that goes back to BURMESTER in the late 19th century [2]. Besides the technological development since then a visual representation is still of great importance for the understanding of mechanism motion and its suitability for solving given motion tasks. Planar graphical synthesis algorithms allow a targeted modification of a set of free parameters combined with a visual feedback concerning the mechanism position and the required space for the motion. Within this section this synthesis approach will be adapted to spatial mechanisms.

A 4R with parallel or intersecting axes is movable by one degree of Freedom. Using a general screw axes configuration the 4R gets immovable. For the special dimensions given in Fig. 1b the spatial 4R-chain gets mobile. This linkage is known as Bennett-linkage [1]. The amount of links and the four R-joints makes the Bennett linkage robust and therefore attractive for industrial applications. The linkage is widely discussed in research publications [3–7]. Synthesis approaches for two pose and three pose synthesis is presented in [8–10]. Three poses fully determine the dimension of the Bennett linkage. Two poses offer the choice of 5 parameters to determine the linkage and is therefore valid for developing a pure graphical approach with the direct use of the CAD environment.

The synthesis of the Bennett linkage is based on two important properties (Fig. 2, [6, 11]):

1. the four axes of the Bennett linkage form a truncated pyramid with the symmetry axis n,
2. the screw axis describing the motion from a reference position into a second position intersect the symmetry axis n perpendicularly.

Therefore the Bennett mechanism synthesis for the 2-pose task can be done into two steps:

- Synthesis of a 2R-chain (4 parameters) and
- synthesis of the second 2R-chain by using the common normal n (1 parameter).

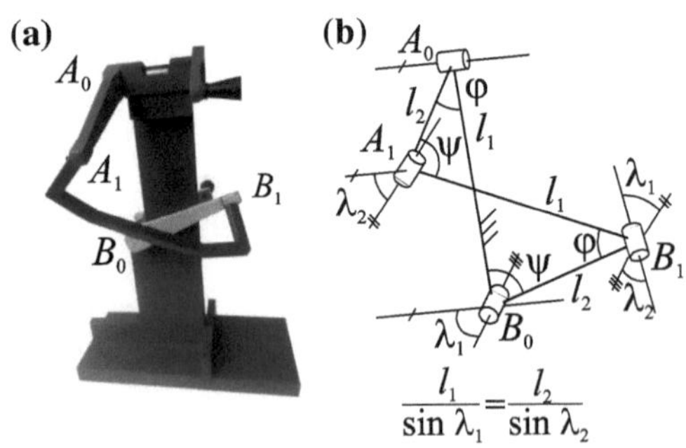

Fig. 1 Bennett mechanism: **a** characteristic dimensions **b** linkage model

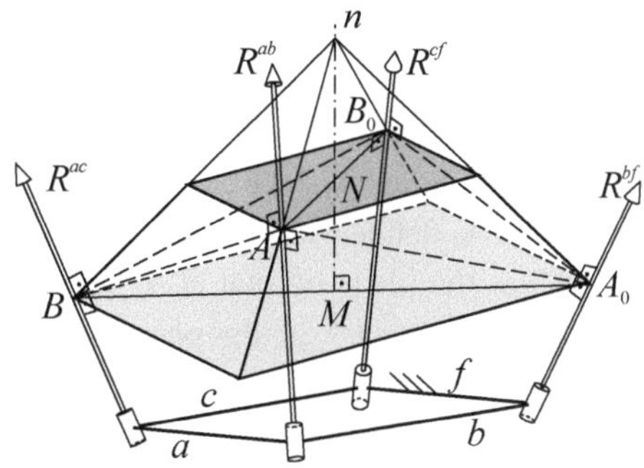

Fig. 2 BENNETT linkage and the according truncated pyramid

2.1 Synthesis of the 2R-Chain

For a given two pose task with the screw axis $S_{12}^{af}(s_{12}^{af}, \vartheta_{12}^{af})$ (s_{12}^{af} translation along S_{12}^{af}; ϑ_{12}^{af} rotation angle in reference to S_{12}^{af}: indices see Appendix A.1; construction of the screw axis see Appendix A.2) the 2R-chain is fully defined by four parameters. For the case using a graphical synthesis approach it is useful to start the synthesis by choosing a point for one of the revolute joint axes (3 parameters). A fourth parameter is the joint location of one link line of the screw triangle. The step wise synthesis of the 2R-chain is according to Fig. 3 as follows:

- define a point A (3 parameters) of the revolute joint R_{12}^{ab},
- the orthonormal projection of A onto S_{12}^{af} gives the point K and therefore line l^a perpendicular to S_{12}^{af},

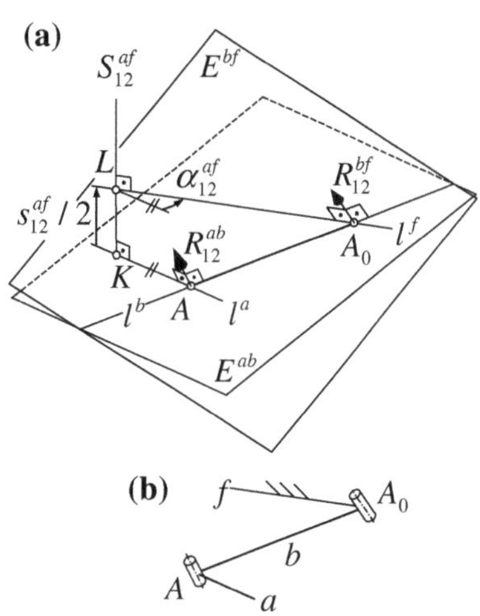

Fig. 3 Synthesis of the 2R-chain: **a** construction **b** kinematic scheme of the 2R-chain

- the line l^f is the result of shifting l^a by half of the translational distance s_{12}^{af} of the screw motion along the screw axis S_{12}^{af}, followed by a rotation with the rotation angle $\alpha_{12}^{af} = 1/2\vartheta_{12}^{af}$ around S_{12}^{af},
- define a point A_0 on l^f,
- A and A_0 define l^b. The lines l^a and l^b define the plane E^{ab} with the revolute joint R_{12}^{ab} as the plane normal at A. The lines l^b and l^f define the plane E^{bf} with the revolute joint R_{12}^{bf} as the plane normal at A_0.

2.2 Synthesis of the Bennett Linkage

The Bennett mechanism is build up by two pairs of equal links. The orientation of these pairs is rotation symmetric in reference to the symmetry axis n. Each location of the counterpart is the result of a half turn in reference to this axis. This line is also the symmetry axis of the Plücker conoid. Therefore all screw axes describing the motion from this reference position to a second position of the Bennett mechanism intersect n perpendicularly. This can be used for deriving a graphical synthesis approach.

Analysing the tetrahedron A_0ABB_0 and the symmetry line n, we see that n intersects the plane AA_0B_0 with the plane normal R_{12}^{bf} in N (see Fig. 2). The point N is the mid point of AB_0 with AN perpendicular to n. As well the plane AA_0B with the plane normal R_{12}^{ab} has a piercing point M as midpoint of A_0B. The line through A_0M is also perpendicular to n. Hence the synthesis of the Bennett-mechanism by using a given 2R-chain is the construction of n, with n intersecting the screw axis S_{12}^{af} perpendicularly and intersecting the plane E^{ab} at point N that way, that AN is as well perpendicular to n.

The graphical synthesis using CAD focuses on the projection onto a plane E through L perpendicular to the screw axis S_{12}^{af} (Fig. 4a). The construction of n is as follows (Fig. 4b):

- With the given projection of the screw triangle $S_{12}^{af\prime} R_{12}^{bf\prime} R_{12}^{ab\prime}$ and its characteristic points $A'A_0'K'L'$ the projected symmetry line n' can be directly drawn as a line intersecting $L' = K'$ with an angle δ. The points $N'M'$ are the intersecting points of the Thales circles k^f given by $L'A_0'$ and k^a given by $K'A'$.
- The line $A'N'$ intersects $L'A_0'$ in D'.
- The location of n along S_{12}^{af} can be found by using the projection onto a plane through the points KLA_0 with points $A''K''A_0''$ and $D'' = D$.
- The point N' on $A'D'$ directly gives the location of N'' on $A''D''$ and therefore with O'' the location of n along S_{12}^{af}.

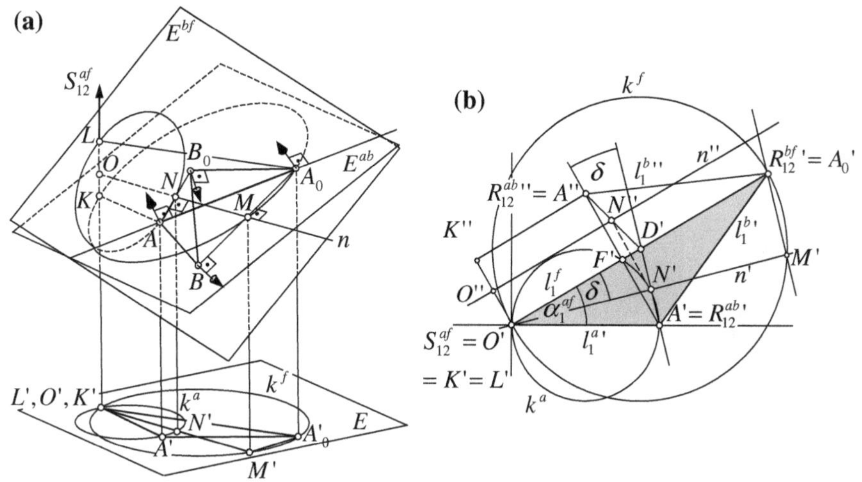

Fig. 4 Construction of the common normal n **a** isometric **b** projection

The revolute joint axis R_{12}^{cf} with the point B_0 is the result of the half turn of R_{12}^{ab}, A around n (Fig. 4a). The revolute joint axis R_{12}^{ac} with B is the result of the half turn of R_{12}^{bf}, A_0 around n. The link a can be represented by the section AB. As well b by A_0A, c by B_0B and f by A_0B_0.

3 Application

3.1 Dimensional Synthesis

A very interesting field of spatial guidance tasks are hinge applications. The dimension synthesis of the Bennett-Mechanism as hinge application shall be presented at the door guidance for a vehicle door (see Fig. 5a). This vehicle is used as demonstrator to show the main achievements in function integration, processing and manufacturing of light weight material, collected in the collaborative research center 639, founded by the German Research Foundation.

The guidance task is defined by two positions of the door wing shown in Fig. 5b, with the screw axis S_{12}^{af} (construction of the screw axis using CAD is given in Appendix A.2). The screw parameters are the rotation angle $\vartheta_{12}^{af} = 91.7°$ and the sliding distance $s_{12}^{af} = 121.6\,\text{mm}$. The dimensional synthesis is divided into the synthesis of the 2R-chain followed by the construction of the common normal n to build up the Bennett-mechanism.

The boundary conditions to this problem are related to the positions of the rotation axes. The axes R_{12}^{bf}, R_{12}^{cf} shall be mounted close to the vehicle frame. The moving axes

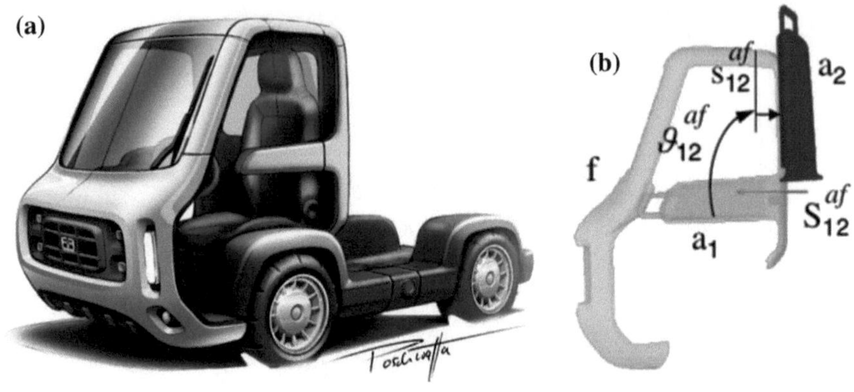

Fig. 5 FiF demonstrator vehicle: **a** design model **b** 2 pose guidance task

R^{ab}_{12}, R^{ac}_{12} should be mounted as close a possible to the door. A useful compromise can only be found by the intelligent use of the 4 parameters defining the 2R-chain. In this case we start defining R^{bf}_{12} by A_0 close to the screw axis S^{af}_{12} and the vehicle frame f. The fourth parameter is defined by the location of R^{ab}_{12} by A_1 on the link line l^a. With this resultant screw triangle $R^{ab}_{12}R^{bf}_{12}S^{af}_{12}$ the Bennett mechanism can be constructed by defining the direction of the common normal n. The choice of δ has an influence of the location of the revolute axes R^{cf}_{12}, R^{ac}_{12} and the character of the door guidance (shown through door flock given in Fig. 6). A satisfactory solution will be the result of an iterative process varying the parameters of the 2R-chain and the location of the common normal n. Here we can see the advantages of the 2 pose task compared to the 3 pose task. All the 5 free parameters can directly used to locate the linkage axes and find the best compromise.

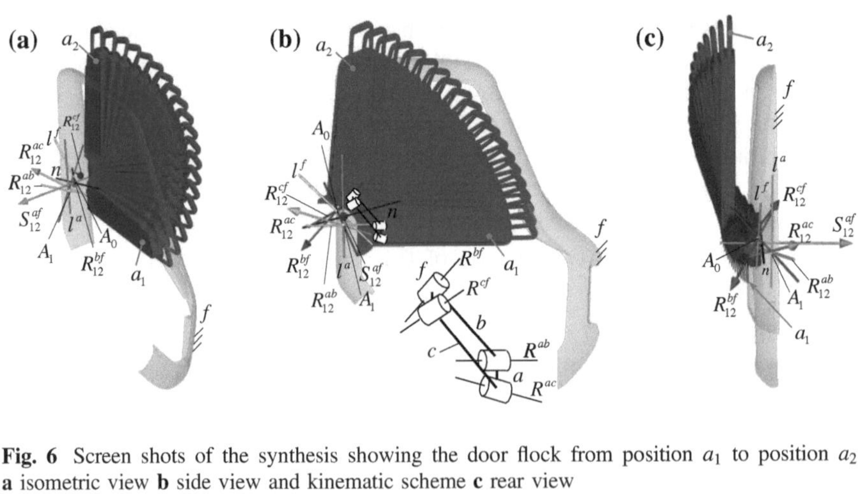

Fig. 6 Screen shots of the synthesis showing the door flock from position a_1 to position a_2: **a** isometric view **b** side view and kinematic scheme **c** rear view

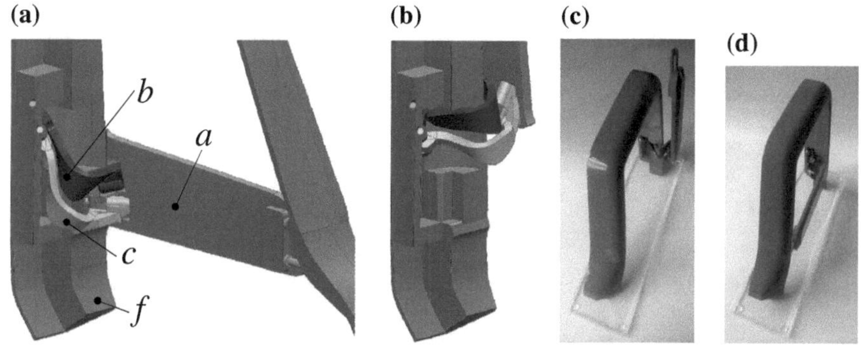

Fig. 7 Hinge: CAD model in **a** open and **b** closed position; rapid prototyping model in **c** open and **d** closed position

3.2 Design

The intermediate links $b = R^{bf}_{12}R^{ab}_{12}$ and $c = R^{cf}_{12}R^{ac}_{12}$ connect frame link f and door a. The design of both links should avoid collisions with the functional surfaces of frame f and door a as well to their neighbor. Hence the design is an iterative process as well. In this special case it is useful to start the design process designing the revolute joints in the frame f and door a. Hence the cylindrical bodies of b and c are given (Fig. 7a). In this special case it is useful to use the open position for the second design stage for designing the connection between the joint cylinders of b and c. Avoiding a collision to the frame surface the material has to be wrapped around the frame for the joint connection (Fig. 7b). This base design has to be validated by moving the linkage between open and close position. Any collision has to be handled by removing material or a redesign of the base design. For the model validation the model is printed in a scaled form using rapid prototyping (Fig. 7c, d).

4 Conclusion

CAD environments can be directly used for the synthesis of simple spatial guidance tasks by using graphical synthesis methods. The benefit is the direct coupling of kinematic parameters to the linkage design. The synthesis shown at the Bennett linkage gives an example solving spatial motion tasks graphically by a simple and robust linkage structure. The aim is to encourage design engineers to recover graphical methods for their CAD implementation. The result in this investigation is that not the synthesis is the main challenge. The real difficulties arises in the part design. For a more efficient design process further investigations has to follow.

Acknowledgments The authors would like to express their gratitude towards the Deutsche Forschungsgemeinschaft (DFG), which supports this research within the scope of the subproject D2 of the Collaborative Research Centre SFB 639 "Textile-Reinforced Composite Components in Function-Integrating Multi-Material Design for Complex Lightweight Applications".

Appendix

A.1 *Indexing Rules*

Using graphical synthesis methods it is useful to use a special nomenclature. Lines defined by small letters and screw axes by big letters. Motions are defined by link positions and the reference link from where this movement is measured. In case of definite positions the link pairs and position pairs are marked with indices:

$$P_{12}^{af} \quad \begin{array}{l} \rightarrow \text{link index: } a \text{ moves in reference to } f \\ \rightarrow \text{pose index: position 1 moves to position 2} \end{array}$$

Exemplarily a screw motion from position a_1 to a_2 in reference to f is defined by the screw axis S_{12}^{af} with ϑ_{12}^{af} for the rotation angle around S_{12}^{af} and s_{12}^{af} for the translation along S_{12}^{af}. This nomenclature leads to the rules given in Fig. 8.

A.2 *Screw Axis*

Using two specific screws p_1, q_1 for the initial and p_2, q_2 for the second position, the construction of the screw axis can be done by using the screw perpendicular bisectors $p_2 - p_1$ and $q_2 - q_1$ [7]. The screw axis s_{12} for the displacement is the common normal of the screw perpendicular bisectors

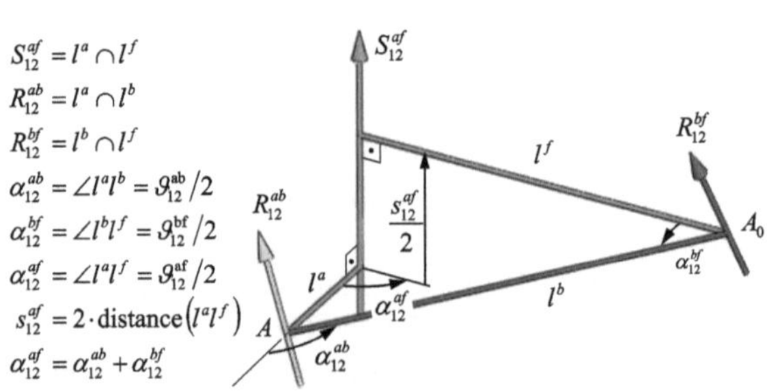

Fig. 8 Rules and nomenclature for the graphical synthesis demonstrated at a two pose 2R-chain example

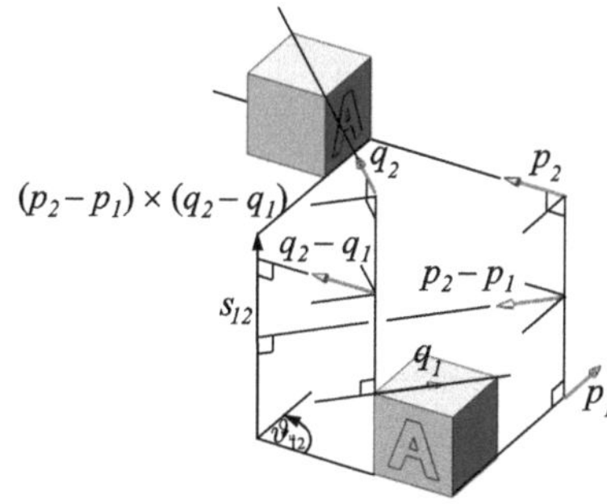

Fig. 9 Construction of the screw axis

$$s_{12}=\frac{(p_2-p_1)\times(q_2-q_1)}{|(p_2-p_1)\times(q_2-q_1)|},\tag{1}$$

and therefore the rotation angle ϑ_{12} around s_{12} and translation s along s_{12} (see Fig. 9).

References

1. Bennett, G.T.: A new mechanism. Engineering **76**, 777–778 (1903)
2. Luck, K., Modler, K.-H.: Getriebetechnik: Analyse-Synthese- Optimierung. Akademie-Verlag, Berlin (1990)
3. Baker, J.E.: The axodes of the Bennett linkage. Mech. Mach. Theory **36**, 105–116 (2001)
4. Baker, J.E.: On the motion geometry of the Bennett linkage. In: Proceedings of the 8th International Conference on Engineering Computer Graphics and Descriptive Geometry, Austin, Texas, USA, vol. 2, pp. 433–437 (1998)
5. Bottema, O., Roth, B.: Theoretical Kinematics. North-Holland Publishing Company, Amsterdam (1979)
6. Huang, C.: The Screw Triangle Now and Then. Bernhard Roth Symposium, Stanford (2003). http://synthetica.eng.uci.edu/BernieRothCD/Papers/Huang.pdf
7. McCarthy, J.M.: Geometric Design of Linkages. Springer, New York (2000)
8. Alba, P., McCarthy, J.M.: Bennetts linkage and the cylindroid. Mech. Mach. Theory **37**, 1245–1260 (2002)
9. Alba, P., McCarthy, J.M.: Dimensional synthesis of Bennett linkages. In: ASME Design Engineering Technical Conferences, Baltimore, Maryland, USA, Sept 10–13 2000
10. Brunnenthaler, K., Schröcker, H.-P., Husty, M.: An new method for the synthesis of Bennett mechanisms. In: International Workshop on Computational Kinematics, Cassino, May 4–6 2005
11. Yu, H.C.: The Bennett linkage, its associated tetrahedron and the hyperboloid of its axes. Mech. Mach. Theory **16**, 105–114 (1981)

Distance from Conic to Point, Plane or Line

Paul Zsombor-Murray

Abstract To avoid close proximity between building envelope and a nearby power line, geometric methods to compute normal distances from spatial point, plane and line to a parabola, approximating the catenary, are developed using projection onto ideal planes. Then line geometry is applied for the first time to reveal a unified approach.

Keywords Stationary distance · Conics · Transformation · Line geometry

1 Introduction

"Flat" catenaries can be satisfactorily approximated by parabolæ specified by two given supporting points P, Q and only the height t_2 of the lowest point or vertex T of the curve. First the problem will be reduced to a planar model by taking the point or line (expressed in point view) on the origin of a Cartesian frame while the parabola or its subsequent projection on the plane taken normal to the line initially in standard form is translated, after possible rotation, to assume its required position relative to the origin. Planar point or line—The line may represent an edge or line view of an intruding plane to parabola distances are computed using products of homogeneous planar point or line coordinate vectors and matrices representing conic coefficients, rotation, translation and orthogonal projection. Line geometry is introduced. All solution cases are treated as a product of a 5×6 homogeneous matrix of constraint equation coefficients and a Plücker coordinate vector (Fig. 1).

P. Zsombor-Murray (✉)
Centre for Intelligent Machines and Mechanical Engineering, McGill University, Montreal, Canada
e-mail: paul@cim.mcgill.ca

B. Corves et al. (eds.), *Mechanisms, Transmissions and Applications*,
Mechanisms and Machine Science 31, DOI 10.1007/978-3-319-17067-1_9

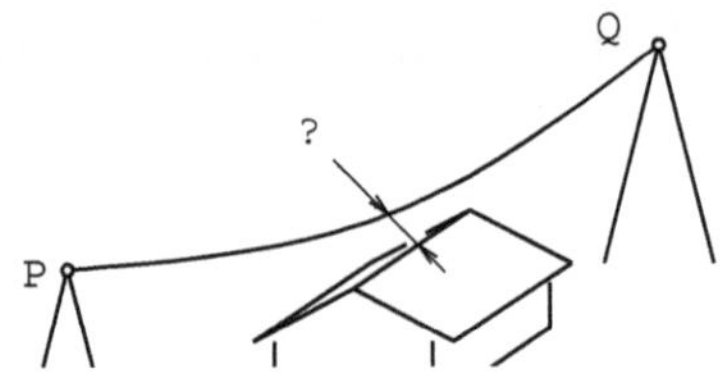

Fig. 1 Minimum distance from ridge pole to power line

1.1 Origin to Translated Parabola

The missing coefficient a and coordinate t_1 are computed given $P(p_1, p_2)$, $Q(q_1, q_2)$ and t_2 of $T(t_1, t_2)$. The solution with the closest point on the left appears in Fig. 2. Constraint equations, $(\mathbf{p}^\top, \mathbf{q}^\top)\mathbf{T}\mathbf{M}_{sp}\mathbf{T}^\top(\mathbf{p}, \mathbf{q}) = 0$, Eq. 1, are set up as follows.

$$\begin{bmatrix} 1 & p, q_1 & p, q_2 \end{bmatrix} \begin{bmatrix} 1 & -t_1 & -t_2 \\ 0 & 1 & 0 \\ 0 & 0 & 1 \end{bmatrix} \begin{bmatrix} 0 & 0 & -1/2 \\ 0 & a & 0 \\ -1/2 & 0 & 0 \end{bmatrix} \begin{bmatrix} 1 & 0 & 0 \\ -t_1 & 1 & 0 \\ -t_2 & 0 & 1 \end{bmatrix} \begin{bmatrix} 1 \\ p, q_1 \\ p, q_2 \end{bmatrix}$$
$$= a(p_1 - t_1)^2 + t_2 - p_2 = 0, \quad a(q_1 - t_1)^2 + t_2 - q_2 = 0 \tag{1}$$

Eliminating a produces, Eq. 2, a quadratic in t_1, its values being given for the numerical example shown in Fig. 2.

$$(p_2 - q_2)t_1^2 + 2[(q_2 - t_2)p_1 + (t_2 - p_2)q_1]t_1 + (t_2 - q_2)p_1^2 + (p_2 - t_2)q_1^2 = 0 \tag{2}$$

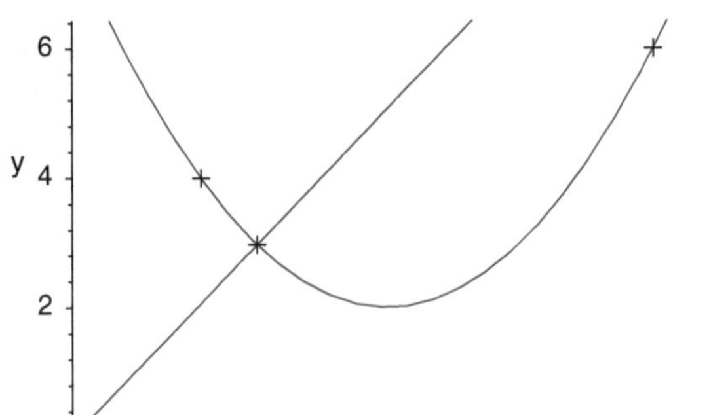

Fig. 2 A numerical example, with $P(2,4)$, $Q(9,6)$, $T(-5 + 7\sqrt{2}, 2)$, Closest Point at (2.8722, 2.9777) with $a = 2/\left[49(3 - 2\sqrt{2})\right]$

The parabola size coefficient a is obtained with either one of Eq. 1, linear in a, having chosen the positive root of t_1. The other root places P, Q both in the right hand branch of the parabola. Equation 3 is the equation of the displaced parabola, like either of Eq. 1 but in terms of an arbitrary point (x, y).

$$t_2 - y + a(x - t_1)^2 = 0 \tag{3}$$

Forming the squared distance from origin to parabola, $x^2 + y^2$, and taking the derivative set to zero of this with respect to x, having eliminated y with Eq. 3, yields a cubic, Eq. 4, in x-coordinates of stationary points.

$$2a^2x^3 - 6a^2t_1x^2 + \left[2a\left(at_1^2 + t_2\right) + 1\right]x - 2at_1\left(at_1^2 + t_2\right) = 0 \tag{4}$$

1.2 Distance Between Parabola and Line

The same numerical parameters will be used together with the line given by Eq. 5.

$$W + Xx + Yy = 0 \tag{5}$$

One finds a tangent line on the parabola that is parallel to the given one and measures the length of the normal line segment on the point of tangency that spans to the given line as shown in Fig. 3. The tangent line and the normal line are given by Eq. 6.

$$U + Xx + Yy = 0 \quad \text{and} \quad V - Yx + Xy = 0 \tag{6}$$

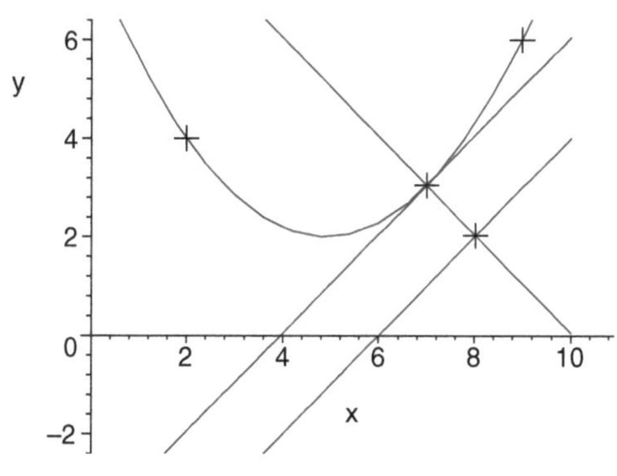

Fig. 3 Minimum distance from sloping roof plane, line $60 - 10x + 10y = 0$ to power line, Eq. 4. The length of the segment, given the original parabola values, is 1.4993

Coincident point on the parabola and tangent line is expressed by Eq. 7.

$$\begin{bmatrix} t_2+at_1^2 & -at_1 & -1/2 \\ -at_1 & a & 0 \\ -1/2 & 0 & 0 \end{bmatrix} \rightarrow \begin{bmatrix} 0 & 0 & 2a \\ 0 & -1 & 2at_1 \\ 2a & 2at_1 & 4at_2 \end{bmatrix} \begin{bmatrix} U \\ X \\ Y \end{bmatrix} = \begin{bmatrix} 2aY \\ 2at_1Y - X \\ 2aU + 2at_1X + 4at_2Y \end{bmatrix} \tag{7}$$

Substituting the dehomogenized point vector above, the last term in Eq. 7, into the parabola equation, Eq. 4, provides a linear equation in U. This symbolic value of U completely defines the tangent line, Eq. 8.

$$X^2 - 4aY(t_1X + t_2Y) + 4aY(Xx + Yy) = 0 \tag{8}$$

Similarly by substituting the point coordinates into the normal line equation the value of V is determined to yield the equation of the normal line, Eq. 9.

$$2at_1(Y^2 - X^2) - 2aX(U + 2t_2Y) - XY + 2aY(Xy - Yx) = 0 \tag{9}$$

Intersecting the normal line and the edge view of the roof plane yields

$$[V \quad -Y \quad X]^{\top} \times [W \quad X \quad Y]^{\top} = Q\{q_0 : q_1 : q_2\}$$

the other end point of the line segment between given and tangent line.

1.3 Orthogonal Projection

Because the plane of the catenary is not generally normal to the ridge line an image of the parabola, projected orthogonally on to a plane normal to the ridge line or building wall or roof, is usually required. Operations proceed as follows (Fig. 4).

- The standard form parabola coefficient matrix $\mathbf{M}_{sp}$, the rotation operator $\mathbf{R}_\theta$, the translation operator $\mathbf{T}$ and the orthogonal projection operator $\mathbf{M}_o$ are constructed.

$$\mathbf{M}_{sp} = \begin{bmatrix} 0 & 0 & 1/2 \\ 0 & -a & 0 \\ 1/2 & 0 & 0 \end{bmatrix}, \quad \mathbf{R}_\theta = \begin{bmatrix} 1 & 0 & 0 \\ 0 & \cos\theta & -\sin\theta \\ 0 & \sin\theta & \cos\theta \end{bmatrix}$$

$$\mathbf{T} = \begin{bmatrix} 1 & -t_1 & -t_2 \\ 0 & 1 & 0 \\ 0 & 0 & 1 \end{bmatrix}, \quad \mathbf{M}_o = \begin{bmatrix} \cos\phi & 0 & 0 \\ 0 & 1 & 0 \\ 0 & 0 & \cos\phi \end{bmatrix}$$

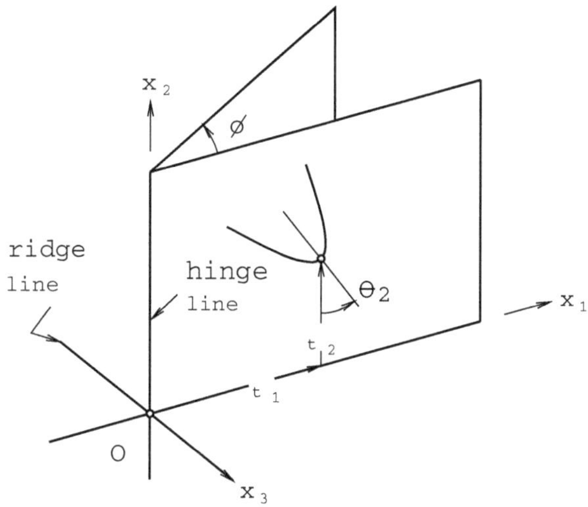

Fig. 4 Roof ridge and hinge lines

- Forming the product $\mathbf{M}_o\mathbf{T}\mathbf{R}_\theta\mathbf{M}_{sp}\mathbf{R}_\theta^\top\mathbf{T}^\top\mathbf{M}_o^\top$ produces the coefficient matrix $\mathbf{M}_{op}$ of the rotated, translated parabola rotated by ϕ out of the fixed plane and projected orthogonally onto it. Let $\mathbf{M}_{op}$ and the orthogonally projected parabola image be represented as

$$\mathbf{M}_{op} = \begin{bmatrix} a_{00} & a_{01} & a_{02} \\ a_{01} & a_{11} & a_{12} \\ a_{02} & a_{12} & a_{22} \end{bmatrix}$$
$$\rightarrow a_{00} + 2a_{01}x_1 + 2a_{02}x_2 + a_{11}x_1^2 + 2a_{12}x_1x_2 + a_{22}x_2^2 = 0$$

where

$$\begin{aligned} a_{00} &= \cos^2\phi\left[t_1\sin\theta - t_2\cos\theta - a(t_1\cos\theta + t_2\sin\theta)^2\right] \\ a_{01} &= \cos\phi[2a\cos\theta(t_1\cos\theta + t_2\sin\theta) - \sin\theta]/2 \\ a_{02} &= \cos^2\phi[2a\cos\theta(t_1\sin\theta + t_2\sin\theta) + \cos\theta + 2at_2]/2 \\ a_{11} &= -\cos^2\theta, \quad a_{12} = -(a\cos\phi\cos\sin\theta)/2, \quad a_{22} = -a\cos^2\phi\sin^2\theta \end{aligned}$$

- Figure 5 shows the standard form parabola, its rotated and translated image and its projection from its natural frame orthogonally onto the fixed frame normal to the ridge line. The narrower image with its vertex farther to the left is the required projection.

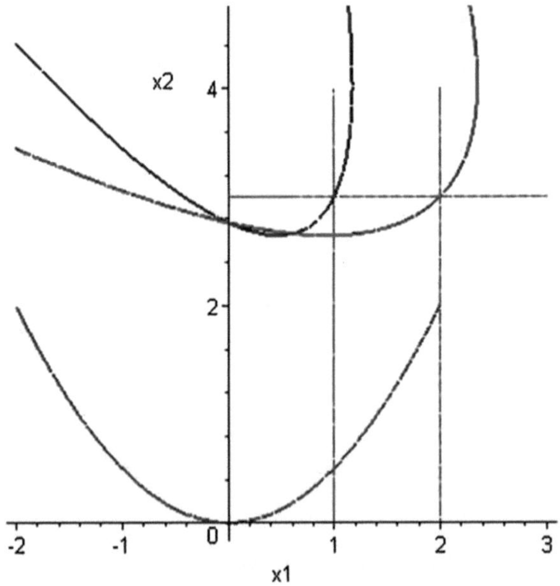

Fig. 5 Image of three parabolæ on plane $x_3 = 0,\ a = 1/2,\ t_1 = 2,\ t_2 = 3,\ \theta = \pi/4,\ \phi = \pi/3$

2 Line Congruence Pairs

Confining analysis to a planar model has its drawbacks. E.g. cases where the ridge line is parallel to the plane of the parabola cannot be accommodated. Imagine an approach that seeks to find all lines common to two line *congruences*. Without loss in generality one contains all lines normal to tangents on a standard form parabola while the other contains those normal to an arbitrary axial line that represents the ridge. Alternately one may choose the second congruence to be all lines normal to a given plane so as to represent a building wall or roof surface or the tip of a lightening rod represented by a point. All these are shown together in Fig. 6. Consider the parabola, an arbitrary point $P(p_1, p_2)$ upon it bearing a tangent line and its normal on P. Taking an auxiliary view normal to the tangent shows the normal pencil of lines $\mathscr{G}'_1$ on P where $\mathscr{G}_1$ is the line congruence on *all* P. The line one seeks has radial Plücker coordinates $\mathscr{G}_r\{g_{01} : g_{02} : g_{03} : g_{23} : g_{31} : g_{12}\}$. $\mathscr{G}_2$ is shown thrice. $Q \in \mathscr{G}_2$ depicts the congruence on given point $Q\{1 : q_1 : q_2 : q_3\}$. $\mathscr{G}_2 \perp e$ shows parallel lines normal to all given planes $e\{E_0 : E_1 : E_2 : E_3\}$ in a parallel pencil. This situation is equivalent to $Q\{0 : q_1 : q_2 : q_3\}$ an absolute point where $q_1 = E_1, q_2 = E_2, q_3 = E_3$. This example will not be detailed below because it is just a special case of a congruence on a given point. Finally $\mathscr{G}_2 \perp, \cap \mathscr{H}$ where line $\mathscr{H}_a\{H_{01} : H_{02} : H_{03} : H_{23} : H_{31} : H_{12}\}$ is given by its axial coordinates.

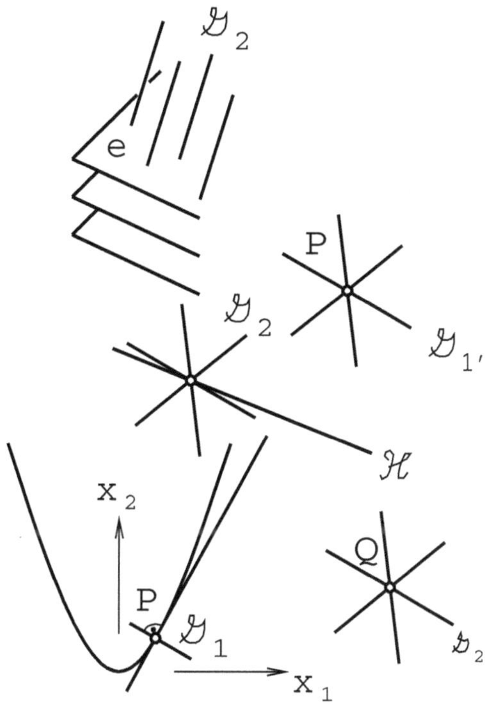

Fig. 6 Four line congruences

2.1 Line Geometry

What follows makes use of elementary line geometry. Recall points $P(p_1, p_2)$ on the standard form parabola on the plane $x_3 = 0$.

$$p_2 - ap_1^2 = 0 \tag{10}$$

Tangent lines on this curve have slope $dp_2/dp_1 = 2ap_1$. Spatial direction vectors of lines $\mathscr{G}$ in the congruence normal to tangents on the parabola can be expressed as

$$[g_{01} \quad g_{02} \quad g_{03}]^\top \cdot [1 \quad 2ap_1 \quad 0]^\top = 0 \tag{11}$$

The parabola is taken in standard form while Q, e, $\mathscr{H}$ are given as expressed in this frame. The case where $dp_2/dp_1 \to \infty$ can be safely ignored because it occurs indefinitely far up the branches of the parabola. Therefore all radial lines $\mathscr{G}\{g_{01} : g_{02} : g_{03} : g_{23} : g_{31} : g_{12}\}$ must satisfy the normality condition Eq. 12,

$$g_{01} + 2ap_1 g_{02} = 0. \tag{12}$$

In addition the condition $P \in \mathscr{G}$, recalling that spatially $P\{p_0 : p_1 : p_2 : p_3\}$, $p_0 = 1, \quad p_3 = 0$, provides the two middle equations chosen from the doubly singular set, expressed in *axial* coordinates of $\mathscr{G}$ such that $G_{ii} = 0$, is given by Eq. 13. In synopsis, $P \in \mathscr{G}$. If $\mathscr{G} \cap P = e$ then $\sum_{j=0}^{3} G_{ij}p_j = E_i \neq 0$, $G_{ji} = -G_{ij}$, $G_{ii} = 0$. If $\mathscr{G} \cap e = P$ then $\sum_{j=0}^{3} g_{ij}P_j = p_i \neq 0$.

$$\begin{aligned} G_{00}p_0 + G_{01}p_1 + G_{02}p_2 + G_{03}p_3 = 0, \quad -G_{01}p_0 + G_{11}p_1 + G_{12}p_2 - G_{31}p_3 = 0 \\ -G_{02}p_0 - G_{12}p_1 + G_{22}p_2 + G_{23}p_3 = 0, \quad -G_{03}p_0 + G_{31}p_1 - G_{23}p_2 + G_{33}p_3 = 0 \end{aligned} \tag{13}$$

which lead to Eq. 14

$$-g_{23} + g_{03}p_2 = 0, \quad -g_{31} - g_{03}p_1 = 0 \tag{14}$$

because of the term by term proportional equivalence, Eq. 15.

$$\{g_{01} : g_{02} : g_{03} : g_{23} : g_{31} : g_{12}\} \propto \{G_{23} : G_{31} : G_{12} : G_{01} : G_{02} : G_{03}\} \tag{15}$$

The orthogonality between the first and second vector element triads of line coordinates, Eq. 16, called the Plücker condition or quadric, is used to get p_1, hence p_2, to yield coordinates of point P where $\mathscr{G}$ intersects the parabola.

$$g_{01}g_{23} + g_{02}g_{31} + g_{03}g_{12} = 0 \tag{16}$$

2.2 Parabola to Line, Point and Plane

Lines $\mathscr{G}$ that are normal to and intersect the ridge line $\mathscr{H}$ constitute the second congruence. These provide the following two necessary constraint equations Eq. 17.

$$\begin{aligned} aligned H_{23}g_{01} + H_{31}g_{02} + H_{12}g_{03} &= 0, H_{01}g_{01} + H_{02}g_{02} + H_{03}g_{03} + H_{23}g_{23} \\ &+ H_{31}g_{31} + H_{12}g_{12} \\ &= 0 aligned \end{aligned} \tag{17}$$

All this can be arranged in Eq. 18, a system of five homogeneous linear equations in g_{ij}, in detached coefficients form. Recall that the first three rows represent Eqs. 12 and 14 while the last two are, respectively, normality of directions $\mathscr{H} \perp \mathscr{G}$ and intersection $\exists \mathscr{H} \cap \mathscr{G}$, the two equations Eq. 17.

$$\begin{bmatrix} 1 & 2ap_1 & 0 & 0 & 0 & 0 \\ 0 & 0 & ap_1^2 & -1 & 0 & 0 \\ 0 & 0 & -p_1 & 0 & -1 & 0 \\ H_{23} & H_{31} & H_{12} & 0 & 0 & 0 \\ H_{01} & H_{02} & H_{03} & H_{23} & H_{31} & H_{12} \end{bmatrix} \begin{bmatrix} g_{01} \\ g_{02} \\ g_{03} \\ g_{23} \\ g_{31} \\ g_{12} \end{bmatrix} = \begin{bmatrix} 0 \\ 0 \\ 0 \\ 0 \\ 0 \\ 0 \end{bmatrix} \tag{18}$$

Solving homogeneously yields all six $g_{ij} = g_{ij}(p_1)$. Inserting these into the Plücker condition results in a cubic, Eq. 19, in p_1, as was Eq. 4, and a trivial solution (factor).

$$H_{12}(2aH_{23}p_1 - H_{31})\left(c_3p_1^3 + c_2p_1^2 + c_1p_1 + c_0\right) = 0 \tag{19}$$

Using the first and second equations from the set Eq. 15, with Q replacing P, to form the last two rows in the 5×6 matrix (or alternately with e replacing P) one obtains with Eq. 20 a linear, rather than a cubic, equation in p_1, Eq. 21.

$$\begin{bmatrix} 0 & 0 & 0 & q_1 & q_2 & q_3 \\ 0 & -q_3 & q_2 & 0 & 0 & 0 \end{bmatrix} \begin{bmatrix} g_{31} \\ g_{12} \end{bmatrix} = \begin{bmatrix} 0 \\ 0 \end{bmatrix} \tag{20}$$

Solving homogeneously yields all six $g_{ij} = g_{ij}(p_1)$. Inserting these into the Plücker condition produces the linear equation Eq. 21 in p_1 and a trivial factor.

$$-aq_3^2(q_1 + 2aq_2p_1)p_1^2 = 0 \tag{21}$$

3 Conclusion

This paper grew from difficulties experienced by a colleague in Innsbruck who ran afoul of municipal authorities while building her house. Although it is unconventional to put references here, rather than at the beginning, one may see relevance to wider application by realizing that [3] was written to help in dynamic simulation of a falling chain. Klien's [1] little book covers many of the projective geometry transformations used but I could not find material describing conics (and quadrics) in terms of their (symmetric) coefficient matrices. This was acquired by osmosis from exposure to my many Austrian geometer friends. Pottmann and Wallner's [2] line geometry text treats that subject thoroughly while [4] is helpful in grasping the basics. Though I was asked to consider removing the approach and analysis connected with the first three problems I decided to retain it in the interest of illustrating the geometric thinking the leads to selection of ideal frames in problem formulation. This was carried through to conceive Fig. 6 which explains the novel application of line geometry in the unification of these types of problem.

Acknowledgments The reviewers, who clearly worked hard to help me improve, deserve my gratitude. I hope they are not disappointed. This work is supported by Natural Sciences and Engineering Canada "Discovery" grant 228162.

References

1. Klein, F.: Elementary Mathematics from an Advanced Standpoint, Geometry. Dover, Delaware (2004)
2. Pottmann, H., Wallner, J.: Computational Line Geometry. Springer, Heidelberg (2001)
3. Zsombor-Murray, P.J., Hayes, M.J.D., Husty, M.L.: Extreme distance to a spatial circle. Trans. Canandian Soc. Mech. Eng. **28**, 221–235 (2004)
4. Zsombor-Murray, P.J.: Line Geometry Primer, Course Notes, Geometry in Mechanics MECH 576 (2006) http://www.cim.mcgill.ca/~paul/LG.pdf

Motion Synthesis of a Planar Watt II Type Six-Bar Mechanism with Two End-Effectors

Gökhan Kiper

Abstract The study deals with motion generation with closed-loop mechanisms with several end-effectors. As a case study a single degree-of-freedom planar Watt II type six-bar mechanism with two end-effectors is worked on. Dyad formulation with complex numbers is made use of for the mathematical model. It is found that the motion synthesis is possible for at most three poses of the two end-effectors. The formulations are illustrated with numerical examples.

Keywords Watt II type six-bar mechanism · Motion generation · Dyad formulation

1 Introduction

The formulations for the dimensional synthesis of mechanisms with single end-effector are well-known [1–3]. These methods may as-well be used for mechanisms with several end-effectors. By this way, mechanisms with less actuators may be utilized in applications which require multiple simultaneous operations. Some examples for such a system are a gripper with several fingers which have non-symmetric motions, a pick-and-place manipulator for relocation of several different objects and a surgery robot with several end-effectors for different operations. Recently Simo-Serra et al. devised methodologies for designing mechanisms with multiple end-effectors [4], specifically applied to multi-fingered robotic hands with several serial chains and fingers meeting in a common palm [5]. Shen et al. [6, 7] worked on the design of Watt I type planar six-bar mechanisms to be used as a mechanical finger, where prescribed motions of the middle and distal sections of the

G. Kiper (✉)
Izmir Institute of Technology, Izmir, Turkey
e-mail: gokhankiper@iyte.edu.tr

B. Corves et al. (eds.), *Mechanisms, Transmissions and Applications*,
Mechanisms and Machine Science 31, DOI 10.1007/978-3-319-17067-1_10

finger. Also, Wobrecht et al. [8] designed a planar 8-link single degrees-of-freedom (dof) exoskeleton finger mechanism for which path and motion generation problems are solved for two of the links. In [8], the authors use numerical optimization techniques for the synthesis.

In [6, 7], the motion of one of the links is described with respect to the other link with prescribed motion. That is, the relative motion of the dependent link is used for the synthesis. However, in an application with multiple end-effectors, the end-effector motions with respect to the base may be independently described. This study is an initial attempt for working out dimensional synthesis of such mechanisms with several end-effectors. As the first step, analytical motion synthesis formulation is developed for a single dof Watt II type planar six-bar mechanism with two end-effectors for two and three poses of the end-effectors. The dyad formulation with complex numbers is presented in Sect. 2. Two and three pose synthesis is the subject of Sects. 3 and 4, respectively. Examples are provided for both two and three pose synthesis. Further studies are discussed in Sect. 5.

2 Dyad Formulation

The Watt II type planar six-bar mechanism shown in Fig. 1 has end-effectors P and Q attached to the coupler links of the two-four bar loops. The task is to relocate the end-effectors from poses P_1, Q_1 to P_j, Q_j for j = 2, 3, … etc. For the mathematical model of this problem we make use dyad formulation following the notation in [2]. Accordingly, one of the two four bar loops consists of the dyads ($\mathbf{W}_1$, $\mathbf{Z}_1$) and ($\mathbf{W}_3$, $\mathbf{Z}_3$), and the other loop consists of the dyads ($\mathbf{W}_2$, $\mathbf{Z}_2$) and ($\mathbf{W}_4$, $\mathbf{Z}_4$). We represent

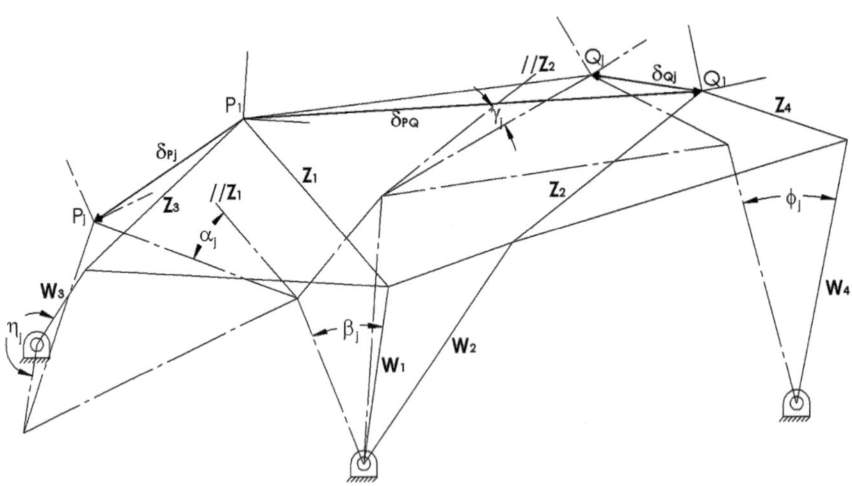

Fig. 1 Planar six-bar mechanism

Table 1 Number of scalar unknowns and free selections for two and three pose synthesis

	Equations (1–3)			Equations (4–5)		
	Unknowns	Equations	Free selections	Unknowns	Equations	Free selections
2 poses	9	6	3	10	4	6
3 poses	10	10	–	12	8	4

all vectors with complex numbers, i.e. $\mathbf{W}_1 = W_{1x} + W_{1y}\hat{\imath}$, etc. We can write the following vector equations from Fig. 1:

$$\mathbf{W}_2 + \mathbf{Z}_2 - \mathbf{W}_1 - \mathbf{Z}_1 = \delta_{PQ} \tag{1}$$

$$\left(e^{\hat{\imath}\beta_j} - 1\right)\mathbf{W}_1 + \left(e^{\hat{\imath}\alpha_j} - 1\right)\mathbf{Z}_1 = \delta_{Pj} \tag{2}$$

$$\left(e^{\hat{\imath}\beta_j} - 1\right)\mathbf{W}_2 + \left(e^{\hat{\imath}\gamma_j} - 1\right)\mathbf{Z}_2 = \delta_{Qj} \tag{3}$$

$$\left(e^{\hat{\imath}\eta_j} - 1\right)\mathbf{W}_3 + \left(e^{\hat{\imath}\alpha_j} - 1\right)\mathbf{Z}_3 = \delta_{Pj} \tag{4}$$

$$\left(e^{\hat{\imath}\phi_j} - 1\right)\mathbf{W}_4 + \left(e^{\hat{\imath}\gamma_j} - 1\right)\mathbf{Z}_4 = \delta_{Qj} \tag{5}$$

where $\delta_{PQ} = \overrightarrow{P_1Q_1}$, $\delta_{Pj} = \overrightarrow{P_1P_j}$, $\delta_{Qj} = \overrightarrow{Q_1Q_j}$; β_j, η_j, ϕ_j are crank link rotations and α_j, γ_j are coupler link rotations as shown in Fig. 1. The angles are directed such that counterclockwise is positive. Actually Eqs. (2), (4) and (3), (5) represent two different solutions for the dyads of a four bar. The difference of the solution from the classical four-bar design comes from Eq. (1).

δ_{PQ}, δ_{Pj}, δ_{Qj}, α_j and γ_j are given in the motion generation problem. The unknowns are $\mathbf{W}_1$, $\mathbf{Z}_1$, $\mathbf{W}_2$, $\mathbf{Z}_2$, $\mathbf{W}_3$, $\mathbf{Z}_3$, $\mathbf{W}_4$, $\mathbf{Z}_4$, β_j, η_j and ϕ_j. The number of scalar unknowns and free selections for two and three pose synthesis are listed in Table 1. Exact motion synthesis for more than three poses of the end-effectors is not possible. In Sects. 3 and 4 we formulate the solution for two and three pose problems.

3 Two Pose Synthesis

For the two pose synthesis problem, three scalar parameters out of $\mathbf{W}_1$, $\mathbf{Z}_1$, $\mathbf{W}_2$, $\mathbf{Z}_2$, and β_2 can be selected freely. It is wise to select β_2 as one of the specified parameter values, because it appears in trigonometric functions, which causes nonlinearity. Two more scalars from $\mathbf{W}_1$, $\mathbf{Z}_1$, $\mathbf{W}_2$ and $\mathbf{Z}_2$ should be selected. In general, there is no superiority of one over other possible selections. However, from numerical examples we found out that two parameters among the nine cannot be selected arbitrarily. For example selecting W_{1x} and W_{1y} does not work. Depending on the application, the designer needs to decide on which parameters to be assumed. For illustration of the formulation let's select W_{1x} and Z_{1x}. From real and imaginary parts of Eq. (2)

$$(\cos\beta_j - 1)W_{1x} - \sin\beta_j W_{1y} + (\cos\alpha_j - 1)Z_{1x} - \sin\alpha_j Z_{1y} = \delta_{Pjx} \tag{6}$$

$$\sin\beta_j W_{1x} + (\cos\beta_j - 1)W_{1y} + \sin\alpha_j Z_{1x} + (\cos\alpha_j - 1)Z_{1y} = \delta_{Pjy} \tag{7}$$

W_{1y} and Z_{1y} can be solved linearly from Eqs. (6–7) to obtain:

$$W_{1y} = \frac{\begin{bmatrix} (\cos\alpha_j - 1)\delta_{Pjx} + \sin\alpha_j \delta_{Pjy} \\ -(\cos(\alpha_j - \beta_j) - \cos\alpha_j - \cos\beta_j + 1)W_{1x} - 2(2 - \cos\alpha_j)Z_{1x} \end{bmatrix}}{\sin(\alpha_j - \beta_j) - \sin\alpha_j + \sin\beta_j} \tag{8}$$

$$Z_{1y} = -\frac{\begin{bmatrix} (\cos\beta_j - 1)\delta_{Pjx} + \sin\beta_j \delta_{Pjy} \\ -(2 - \cos\beta_j)W_{1x} - (\cos(\alpha_j - \beta_j) - \cos\alpha_j - \cos\beta_j + 1)Z_{1x} \end{bmatrix}}{\sin(\alpha_j - \beta_j) - \sin\alpha_j + \sin\beta_j} \tag{9}$$

Once $\mathbf{W}_1$ and $\mathbf{Z}_1$ are set, $\mathbf{W}_2$ and $\mathbf{Z}_2$ can be linearly solved from Eqs. (1) and (3) to obtain:

$$\mathbf{W}_2 = \frac{\left(e^{\hat{i}\gamma_2} - 1\right)(\mathbf{W}_1 - \mathbf{Z}_1 + \delta_{PQ}) - \delta_{Q_2}}{e^{\hat{i}\gamma_2} - e^{\hat{i}\beta_2}} \tag{10}$$

$$\mathbf{Z}_2 = \frac{\delta_{Q_2} - \left(e^{\hat{i}\beta_2} - 1\right)(\mathbf{W}_1 - \mathbf{Z}_1 + \delta_{PQ})}{e^{\hat{i}\gamma_2} - e^{\hat{i}\beta_2}} \tag{11}$$

For the other two dyads, it is necessary to select three parameters per dyad. Let's select η_2 and $\mathbf{W}_3$ for one dyad and ϕ_2 and $\mathbf{W}_4$ for the other dyad. Then $\mathbf{Z}_3$ and $\mathbf{Z}_4$ can be easily solved from Eqs. (4–5) as:

$$\mathbf{Z}_3 = \frac{\delta_{Pj} - \left(e^{\hat{i}\eta_j} - 1\right)\mathbf{W}_3}{e^{\hat{i}\alpha_j} - 1} \tag{12}$$

$$\mathbf{Z}_4 = \frac{\delta_{Qj} - \left(e^{\hat{i}\phi_j} - 1\right)\mathbf{W}_4}{e^{\hat{i}\gamma_j} - 1} \tag{13}$$

Since the formulation is analytical, the unique result is obtained almost instantly once the assumed parameter values are specified. By changing the assumed parameter values, the designer can converge to viable solutions and obtain a satisfactory mechanism.

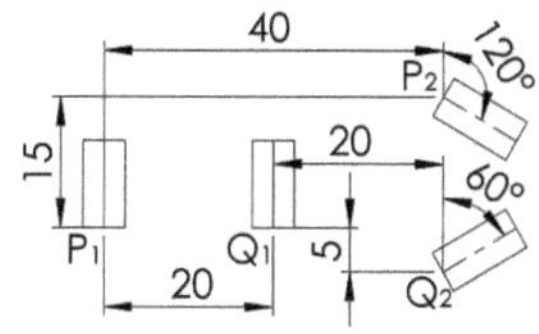

Fig. 2 Two pose pick-and-place application for two objects

As an example consider two poses of the two rectangular objects shown in Fig. 2. From the figure, $\delta_{PQ} = 20$, $\delta_{Pj} = 40 + 15\hat{\imath}$, $\delta_{Qj} = 20 - 5\hat{\imath}$; $\alpha_2 = -120°$, $\gamma_j = -60°$. The computations are done using Maple® software. After several trials, considering link length ratios and link collision avoidance, a proper selection for the assumed parameters is done as $\beta_2 = -90°$, $W_{1x} = -35$, $Z_{1x} = 6$, $\eta_2 = -60°$, $\mathbf{W}_3 = -30 + 5\hat{\imath}$, $\phi_2 = -100°$ and $\mathbf{W}_4 = -5 + 10\hat{\imath}$. Using Eqs. (8–13), the remaining parameters are computed as $W_{1y} = 12.902$, $Z_{1y} = 1.268$, $\mathbf{W}_2 = -3.170 + 12.268\hat{\imath}$, $\mathbf{Z}_2 = -5.830 + 1.902\hat{\imath}$, $\mathbf{Z}_3 = -7.887 + 10.207\hat{\imath}$ and $\mathbf{Z}_4 = -3.711 + 2.804\hat{\imath}$. Hence resulting link lengths are $|\mathbf{W}_1| = 37.302$, $|\mathbf{Z}_1| = 6.133$, $|\mathbf{W}_2| = 12.671$, $|\mathbf{Z}_2| = 6.133$, $|\mathbf{W}_3| = 30.414$, $|\mathbf{Z}_3| = 12.899$, $|\mathbf{W}_4| = 11.180$, $|\mathbf{Z}_4| = 4.651$, $|\mathbf{Z}_1 + \mathbf{Z}_3| = 16.515$, $|\mathbf{Z}_2 + \mathbf{Z}_4| = 16.515$ and fixed joint coordinates with respect to P_1: $(29, -14.170)$, $(37.887, -15.207)$, $(28.711, -12.804)$ in order of dyads 1, 3 and 4. The two poses of the resulting mechanism is illustrated in Fig. 3.

Fig. 3 Two poses of the designed mechanism

4 Three Pose Synthesis

For j = 2, 3, Eqs. (1–3) constitute a nonlinear set of equations. However, Eqs. (2–3) are linear in $\mathbf{W}_1$, $\mathbf{Z}_1$, $\mathbf{W}_2$ and $\mathbf{Z}_2$. Solving $\mathbf{W}_1$ and $\mathbf{Z}_1$, from Eq. (2):

$$\mathbf{W}_1 = \frac{\left(e^{\hat{i}\alpha_3} - 1\right)\delta_{P_3} - \left(e^{\hat{i}\alpha_2} - 1\right)\delta_{P_2}}{\left(e^{\hat{i}\beta_2} - 1\right)\left(e^{\hat{i}\alpha_3} - 1\right) - \left(e^{\hat{i}\alpha_2} - 1\right)\left(e^{\hat{i}\beta_3} - 1\right)} \tag{14}$$

$$\mathbf{Z}_1 = \frac{\left(e^{\hat{i}\beta_2} - 1\right)\delta_{P_3} - \left(e^{\hat{i}\beta_3} - 1\right)\delta_{P_2}}{\left(e^{\hat{i}\beta_2} - 1\right)\left(e^{\hat{i}\alpha_3} - 1\right) - \left(e^{\hat{i}\alpha_2} - 1\right)\left(e^{\hat{i}\beta_3} - 1\right)} \tag{15}$$

$$\mathbf{W}_2 = \frac{\left(e^{\hat{i}\gamma_3} - 1\right)\delta_{Q_3} - \left(e^{\hat{i}\gamma_2} - 1\right)\delta_{Q_2}}{\left(e^{\hat{i}\beta_2} - 1\right)\left(e^{\hat{i}\gamma_3} - 1\right) - \left(e^{\hat{i}\gamma_2} - 1\right)\left(e^{\hat{i}\beta_3} - 1\right)} \tag{16}$$

$$\mathbf{Z}_2 = \frac{\left(e^{\hat{i}\beta_2} - 1\right)\delta_{Q_3} - \left(e^{\hat{i}\beta_3} - 1\right)\delta_{Q_2}}{\left(e^{\hat{i}\beta_2} - 1\right)\left(e^{\hat{i}\gamma_3} - 1\right) - \left(e^{\hat{i}\gamma_2} - 1\right)\left(e^{\hat{i}\beta_3} - 1\right)} \tag{17}$$

Substituting Eqs. (14–17) in Eq. (1) results in a complex equation in terms of the unknowns β_2 and β_3, only:

$$\mathbf{A}e^{\hat{i}2\beta_2} + \mathbf{B}e^{\hat{i}2\beta_3} + \mathbf{C}e^{\hat{i}\beta_2}e^{\hat{i}\beta_3} + \mathbf{D}e^{\hat{i}\beta_2} + \mathbf{E}e^{\hat{i}\beta_3} + \mathbf{F} = \mathbf{0} \tag{18}$$

where

$$\begin{aligned}
\mathbf{A} &= \left(e^{\hat{i}\gamma_3} - 1\right)\delta_{P_3} - \left(e^{\hat{i}\alpha_3} - 1\right)\delta_{Q_3} + \left(e^{\hat{i}(\alpha_3+\gamma_3)} - e^{\hat{i}\alpha_3} - e^{\hat{i}\gamma_3} + 1\right)\delta_{PQ} \\
\mathbf{B} &= \left(e^{\hat{i}\gamma_2} - 1\right)\delta_{P_2} - \left(e^{\hat{i}\alpha_2} - 1\right)\delta_{Q_2} + \left(e^{\hat{i}(\alpha_2+\gamma_2)} - e^{\hat{i}\alpha_2} - e^{\hat{i}\gamma_2} + 1\right)\delta_{PQ} \\
\mathbf{C} &= -\left(e^{\hat{i}\gamma_3} - 1\right)\delta_{P_2} - \left(e^{\hat{i}\gamma_2} - 1\right)\delta_{P_3} + \left(e^{\hat{i}\alpha_3} - 1\right)\delta_{Q_2} - \left(e^{\hat{i}\alpha_2} - 1\right)\delta_{Q_3} \\
&\quad - \left(e^{\hat{i}(\alpha_2+\gamma_3)} + e^{\hat{i}(\alpha_3+\gamma_2)} - e^{\hat{i}\alpha_2} - e^{\hat{i}\alpha_3} - e^{\hat{i}\gamma_2} - e^{\hat{i}\gamma_3} - 2\right)\delta_{PQ} \\
\mathbf{D} &= \left(e^{\hat{i}\gamma_3} - 1\right)e^{\hat{i}\alpha_3}\delta_{P_2} - \left(e^{\hat{i}(\alpha_2+\gamma_3)} - e^{\hat{i}\alpha_2} - e^{\hat{i}\gamma_2} + e^{\hat{i}\gamma_3}\right)\delta_{P_3} \\
&\quad - \left(e^{\hat{i}\alpha_3} - 1\right)e^{\hat{i}\gamma_3}\delta_{Q_2} - \left(e^{\hat{i}(\alpha_3+\gamma_2)} + e^{\hat{i}\alpha_2} - e^{\hat{i}\alpha_3} - e^{\hat{i}\gamma_2}\right)\delta_{Q_3} \\
&\quad + \left(e^{\hat{i}(\alpha_2+\gamma_3)} + e^{\hat{i}(\alpha_3+\gamma_2)} - 2e^{\hat{i}(\alpha_3+\gamma_3)} - e^{\hat{i}\alpha_2} + e^{\hat{i}\alpha_3} - e^{\hat{i}\gamma_2} + e^{\hat{i}\gamma_3}\right)\delta_{PQ}
\end{aligned}$$

$$\begin{aligned}\mathbf{E} = &-\left(e^{\hat{i}(\alpha_3+\gamma_2)} - e^{\hat{i}\alpha_3} + e^{\hat{i}\gamma_2} - e^{\hat{i}\gamma_3}\right)\delta_{P_2} + \left(e^{\hat{i}\gamma_2} - 1\right)e^{\hat{i}\alpha_2}\delta_{P_3} \\ &+ \left(e^{\hat{i}(\alpha_2+\gamma_3)} + e^{\hat{i}\alpha_2} - e^{\hat{i}\alpha_3} - e^{\hat{i}\gamma_2}\right)\delta_{Q_2} - \left(e^{\hat{i}\alpha_2} - 1\right)e^{\hat{i}\gamma_2}\delta_{Q_3} \\ &- \left(2e^{\hat{i}(\alpha_2+\gamma_2)} - e^{\hat{i}(\alpha_2+\gamma_3)} - e^{\hat{i}(\alpha_3+\gamma_2)} - e^{\hat{i}\alpha_2} + e^{\hat{i}\alpha_3} - e^{\hat{i}\gamma_2} + e^{\hat{i}\gamma_3}\right)\delta_{PQ} \\ \mathbf{F} = &\left(e^{\hat{i}\gamma_2} - e^{\hat{i}\gamma_3}\right)\left(e^{\hat{i}\alpha_3}\delta_{P_2} - e^{\hat{i}\alpha_2}\delta_{P_3}\right) - \left(e^{\hat{i}\alpha_2} - e^{\hat{i}\alpha_3}\right)\left(e^{\hat{i}\gamma_3}\delta_{Q_2} - e^{\hat{i}\gamma_2}\delta_{Q_3}\right) \\ &+ \left(e^{\hat{i}(\alpha_2+\gamma_2)} - e^{\hat{i}(\alpha_2+\gamma_3)} - e^{\hat{i}(\alpha_3+\gamma_2)} + e^{\hat{i}(\alpha_3+\gamma_3)}\right)\delta_{PQ}\end{aligned}$$

Multiplying Eq. (18) by $e^{-\hat{i}\beta_2}e^{-\hat{i}\beta_3}$ results in a simpler form:

$$\mathbf{A}e^{\hat{i}\beta_2}e^{-\hat{i}\beta_3} + \mathbf{B}e^{-\hat{i}\beta_2}e^{\hat{i}\beta_3} + \mathbf{F}e^{-\hat{i}\beta_2}e^{-\hat{i}\beta_3} + \mathbf{E}e^{-\hat{i}\beta_2} + \mathbf{D}e^{-\hat{i}\beta_3} + \mathbf{C} = \mathbf{0} \tag{19}$$

Writing real and imaginary parts of Eq. (19):

$$K_x c_2 c_3 + L_x s_2 s_3 + M_y s_2 c_3 + N_y c_2 s_3 + E_x c_2 + E_y s_2 + D_x c_3 + D_y s_3 + C_x = 0 \tag{20}$$

$$K_y c_2 c_3 + L_y s_2 s_3 - M_x s_2 c_3 + N_x c_2 s_3 + E_y c_2 - E_x s_2 + D_y c_3 - D_x s_3 + C_y = 0 \tag{21}$$

where c_2, s_2, c_3, s_3 stand for $\cos\beta_2$, $\sin\beta_2$, $\cos\beta_3$, $\sin\beta_3$, respectively, and $\mathbf{K} = K_x + K_y\hat{i} = \mathbf{A} + \mathbf{B} + \mathbf{F}$, $\mathbf{L} = L_x + L_y\hat{i} = \mathbf{A} + \mathbf{B} - \mathbf{F}$, $\mathbf{M} = M_x + M_y\hat{i} = -\mathbf{A} + \mathbf{B} + \mathbf{F}$ and $\mathbf{N} = N_x + N_y\hat{i} = \mathbf{A} - \mathbf{B} + \mathbf{F}$. Either of β_2 or β_3 can be eliminated from Eqs. (20–21). Let's eliminate β_3: First linearly solve for c_3 and s_3 from Eqs. (20–21). Then using $c_3^2 + s_3^2 = 1$, an equation in terms of only β_2 is obtained. This equation is of fourth order in terms of c_2 and s_2, which indicates that there are at most eight real solutions for β_2. After determining β_2, the corresponding $\beta_3 = \text{atan2}(s_3, c_3)$ are found and $\mathbf{W}_1$, $\mathbf{Z}_1$, $\mathbf{W}_2$ and $\mathbf{Z}_2$ are determined from Eqs. (14–17).

Table 1 indicates that for the three pose synthesis the 10 parameters are to be solved from Eqs. (1–3) without any free selected parameters. This implies that for given three poses of the two end-effectors, there are finitely many solutions. It is well-known from the five pose synthesis of a four-bar mechanism that in case of finitely many solutions, finding the solution is computationally problematic and usually it is hard to find a practically applicable solution. Also all poses may not be attained in the same assembly mode of the mechanism. These problems were also encountered in our computational studies as well.

For the solution for the remaining dyads we can choose two parameters freely per dyad. When we choose η_2, η_3, ϕ_2 and ϕ_3 the rest of the unknowns can be easily solved linearly. We may use Eqs. (14–17) by substituting η instead of β in Eqs. (14–15) for $\mathbf{W}_3$ and $\mathbf{Z}_3$ and ϕ instead of β in Eqs. (16–17) for $\mathbf{W}_4$ and $\mathbf{Z}_4$.

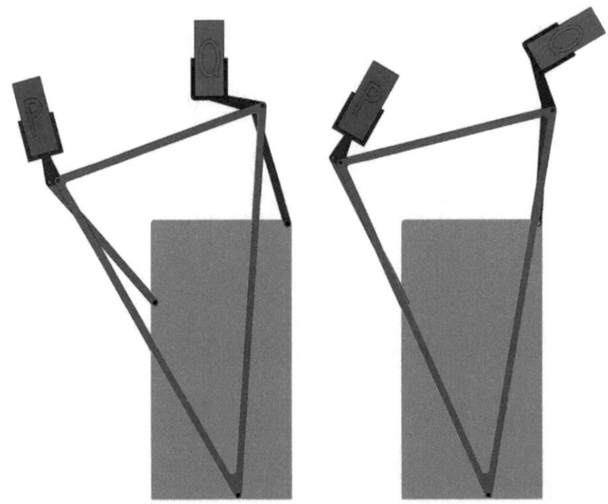

Fig. 4 Three poses of the designed mechanism

As an example consider the case where $\delta_{P_2} = 15 + 5\hat{i}$, $\delta_{P_3} = 25 + 10\hat{i}$, $\delta_{Q_2} = 20 + 10\hat{i}$, $\delta_{Q_3} = 30 + 5\hat{i}$, $\alpha_2 = -45°$, $\alpha_3 = -30°$, $\gamma_2 = -60°$ and $\gamma_3 = -45°$. Using Maple®, one of the solutions for β_2 is solved numerically resulting $\beta_2 = -5.908°$, $\beta_3 = -13.576°$, $\mathbf{W}_1 = -49.057 + 83.162\hat{i}$, $\mathbf{Z}_1 = -49.057 + 83.162\hat{i}$, $\mathbf{W}_2 = 6.848 + 102.869\hat{i}$ and $\mathbf{Z}_2 = -14.468 + 2.556\hat{i}$. For the other two dyads, assuming $\eta_2 = -15°$, $\eta_3 = -20°$, $\phi_2 = -20°$ and $\phi_3 = -45°$, the link vectors are solved as $\mathbf{W}_3 = -28.833 + 31.755\hat{i}$, $\mathbf{Z}_3 = -1.233 + 7.690\hat{i}$, $\mathbf{W}_4 = -8.850 + 30.196\hat{i}$ and $\mathbf{Z}_4 = -12.185 + 3.517\hat{i}$. The three poses of the resulting mechanism is illustrated in Fig. 4.

5 Conclusions

This study is a first attempt for kinematic synthesis of closed-loop mechanisms with several end-effectors, motions of which are independently described with respect to the base. Specifically, the motion synthesis problem for a Watt II mechanism with two end-effectors is issued. With dyad formulation it was shown that nine parameters may be selected freely for two pose synthesis and four parameters are free for three pose synthesis. Examples showed that three pose synthesis is prone to computational problems and the solutions are hardly practical. However, the two pose synthesis proved itself quite successful.

The methods discussed in this study can easily be adapted to other two loop planar mechanisms which comprise prismatic joints as well. Also similar formulation can be devised for spherical mechanisms. The dyad formulation may also be used for path generation problem for mechanisms with two end-effectors.

More than two end-effectors in a mechanism is quite rare, nevertheless it seems that at least two pose synthesis is possible for a mechanism with three end effectors. The main course of progress in our study will be focused on synthesis methods for multi-dof mechanisms with several end-effectors.

References

1. Suh, C.H., Radcliffe, C.W.: Kinematics and Mechanisms Design. Wiley, New York (1978)
2. Erdman, A., Sandor, G.N.: Advanced Mechanism Design: Analysis and Synthesis, vol. 2. Prentice Hall, New Jersey (1984)
3. McCarthy, J.M., Soh, G.S.: Geometric Design of Linkage, 2nd edn. Springer, Berlin (2010)
4. Simo-Serra, E., Perez-Gracia, A.: Kinematic synthesis using tree topologies. Mech. Mach. Theory **72**, 94–113 (2014)
5. Simo-Serra, E., Perez-Gracia, A., Moon, H., Robson, N.: Kinematic synthesis of multi-fingered robotic hands for finite and infinitesimal tasks. In: Lenarcic, J., Khatib, O. (eds.) Advances in Robot Kinematics. Springer, Berlin (2014)
6. Shen, Q., Lee, W.-T., Russell, K., Sodhi, R.S.: On motion generation of Watt I mechanisms for mechanical finger design. Trans. CSME /de la SCGM **32**(3–4), 411–421 (2008)
7. Shen, Q., Russell, K., Lee, W.-T., Sodhi, R.S.: On mechanical finger design for expanded prescribed grasping poses. J. Adv. Mech. Des. Syst. **2**(5), 903–914 (2008)
8. Wolbrecht, E.T., Reinkensmeyer, D.J., Perez-Gracia, A.: Single degree-of-freedom exoskeleton mechanism design for finger rehabilitation. In Proceedings of ICORR 2011: IEEE International Conference on Rehabilitation Robotics, Zurich (2001)

Part II
VDI-Guideline

Motion Conversion with the Crank-Slider Mechanism Regarding Transfer Quality (Part 1)

A.J. Klein Breteler

Abstract The paper discusses the dimension synthesis procedure of the crank-slider mechanism, matching a given input angle and a desired output stroke, for the best possible transfer quality (highest minimum value of the transmission angle), according to the German guideline VDI-2126 (1989). A modified approach is proposed in which the transmission angle just needs to be acceptable. This leads to a much simpler synthesis procedure that covers the most relevant design criteria, like the minimum transmission angle, space occupation and occurrence of dead points. The link dimensions can be obtained from a diagram or can be calculated using simple formulas.

Keywords Dimension synthesis · Transmission angle · Space occupation · Dead point · VDI-2126

1 Introduction

Conversion of oscillating rotation into translational motion can be done easily with a rack and a pinion, providing a linear kinematic transfer function. Occasionally however a designer prefers a different mechanism, for instance to avoid the backlash that is typical for a pair of gears. The planar crank-slider mechanism is the alternative with the simplest kinematic structure [1, 2]. Applying the crank angle as the input, the dimension synthesis of this mechanism, for the best possible transmission angle, is the topic of [1]. Because a part of the current theory is complicated, the aim of this paper is to discuss the theory of the synthesis procedure and to propose new ideas for improvement of the guideline.

The mechanism is depicted in Fig. 1, drawn in the three positions that play a role in the procedure. The synthesis problem can be described as follows.

A.J. Klein Breteler (✉)
University of Technology Delft, Delft, The Netherlands
e-mail: a.j.kleinbreteler@tudelft.nl

B. Corves et al. (eds.), *Mechanisms, Transmissions and Applications*,
Mechanisms and Machine Science 31, DOI 10.1007/978-3-319-17067-1_11

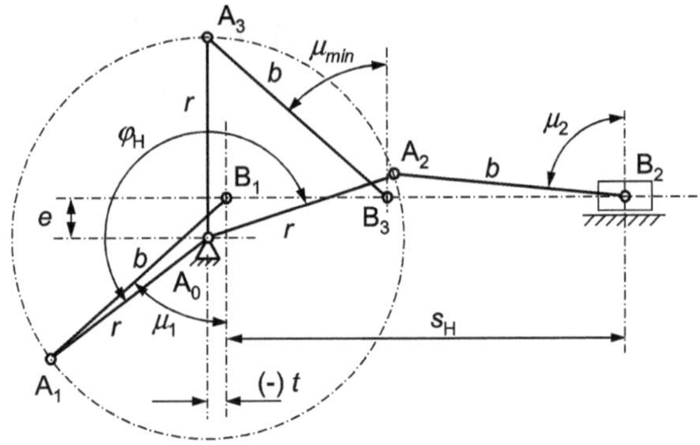

Fig. 1 Crank-slider mechanism in the three design positions

Given are the angular input stroke φ_H and a desired output stroke s_H, for which the four kinematic dimensions (bar lengths r and b, and the co-ordinates of the fixed pivot e and t) must be determined. Further condition is that the transmission angle μ is "good" during the whole motion range. In the VDI-guideline this is expressed by applying the conditions $\mu_1 = \mu_2 = \mu_3$ ($=\mu_{min}$), extra to the design objective equation that input angle φ_H corresponds to output stroke s_H.

In both end positions the transmission angle μ must be equal to the minimum value, while this minimum value should be as high as possible. The position 3, where the crank is perpendicular to the slider path, holds here the μ_{min}-value.

The ideal situation is that $\mu_{min} = 90°$, but in general it is impossible to achieve this as the synthesis result. Which value of μ_{min} is acceptable depends also on the use of the mechanism and on dynamic forces that are however unknown yet in the early design stage of the mechanism. That is why the acceptable value of μ_{min} is usually chosen from experience with previous design cases. In practice frequently a minimum value of 60° or 45° will be adopted. Anyhow it is useful to know which μ_{min}-value can maximally be achieved.

A side condition is that the transfer function $s(\varphi)$ must not show backward motion: the output displacement must be monotonic during the whole motion of the input angle at interval φ_H. A dead point at one or both boundaries of the interval will be included in the theory. Dependent on the application, such a dead point can be in favour (e.g. as output lock) or needs to be avoided (e.g. in a controlled drive with output feedback). The precise behaviour of the transfer function on the interval will not be subject of discussion. In case that an approximated linear behaviour is wanted, some design freedom can be exploited to meet this requirement.

In case that $\varphi_H < 180°$, the problem can be solved easily using the so-called symmetric solution. This case has adequately been described in [1]. The crank positions 1 and 2 are then symmetric to the middle position 3 and the crank end-points A_1 and A_2 have the same distance, but opposite, to the guiding line as A_3. Theoretically coupler length $b = \infty$ would provide the best transfer quality ($\mu_{min} = 90°$), but this solution is not practical. The smallest possible value of b includes a dead point in the end-position. By choosing a proper finite value of b a compromise between transfer quality and space occupation must be accepted.

In case that $\varphi_H > 180°$ the symmetric solution is no longer possible due to monotony failure, but a non-symmetric solution demanding $\mu_1 = \mu_3$ ($=\mu_{min}$), while $\mu_2 > \mu_{min}$, can still be obtained. It is the intention of this paper to describe the design problem of this case and to present the solution options for a designer. This will be done by means of a graphical approach (Chap. 2), after which the required parameter calculations will be specified (Chap. 3). To support the user a diagram will be proposed that overviews the design options and that also provides the parameter values of design cases with and without dead points (Chap. 4).

2 The Non-symmetric Solution, Graphical Approach

This chapter deals with a non-symmetric solution and includes the case of larger input angles ($180° < \varphi_H < 270°$). We split up the input angle φ_H into two angles α and β, see also Fig. 2:

$$\varphi_H = \alpha + \beta, \text{ with } \alpha > \beta \tag{1}$$

We assume that $\varphi_H /2 < \alpha < 180°$ and that $0° < \beta < 90°$. Instead of the transmission angle μ we consider the distance d of point A to the guiding line g in the three positions involved. The maximum value of d should be as low as possible: $d_1 = d_3$ ($=d_{max}$), while $d_2 < d_{max}$.

For any choice of α the configuration of Fig. 2 can be drawn. Start for instance by drawing the crank in the intermediate position 3 (the crank length $A_0A_3 = r$ is arbitrary and is considered as the drawing unit) and apply the angles α and β according Eq. (1) to find the points A_1 and A_2. Then the guiding line g can be determined: perpendicular to A_0A_3 and at equal distance to A_1 and A_3. Now the dimension $e = A_0Y$ is known and the distance $d_1 = d_3 = r - e$. Alternatively the value of e/r can be chosen from which point A_1 and thus α can be constructed.

Fig. 2 Non-symmetric solution: choose α and β

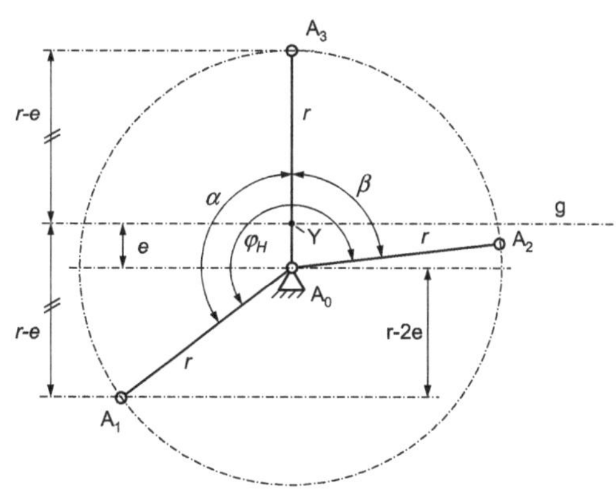

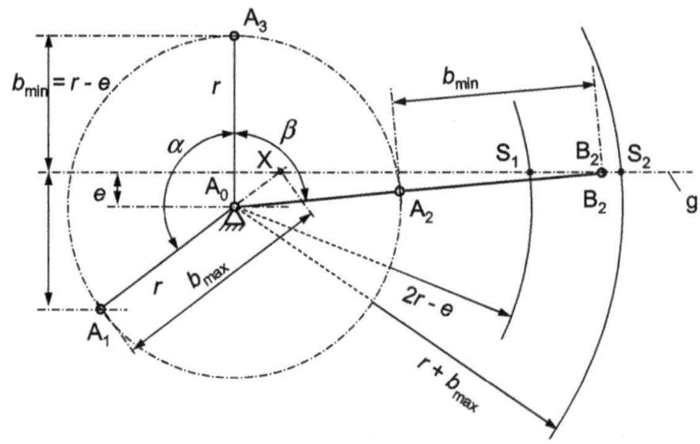

Fig. 3 Range of coupler length *b*

Next step is to determine the range of length *b*, see Fig. 3. The largest possible value $b_{max} = A_1X$, otherwise the crank will surpass the dead point at position 1. The smallest possible value is at least $b_{min} = r - e$, for which $\mu_{min} = 0$.

Applying $r + b_{min} = 2r - e$ and $r + b_{max}$ as radii of concentric circles around A_0, we find the intersection points with the guiding line *g* named S_1 and S_2. When, for the chosen α, the angle β is such that the elongation of A_0A_2 intersects *g* between S_1 and S_2, the minimum value of *b* must be increased to A_2B_2 as drawn (solution with dead point at end position 2). Smaller values of coupler length *b* would cause backward motion of the slider. In case that angle β is smaller (the elongation of A_0A_2 intersects *g* before S_1) a valid solution may still exist, yet without a dead point in position 2. With a larger value of angle β, such that A_0A_2 intersects *g* beyond S_2, no valid range of *b* remains.

It is obvious that the initial choice of α has great effect on the range to choose coupler length *b*. In case of an invalid result the procedure must be repeated with a better (larger) value of α. In general it can be stated that a larger *b*-value will provide a higher transfer quality.

The third step concerns the drawing of the mechanism, with the dimensions of *e* and *b* obtained during the previous two steps, in the three positions as depicted in Fig. 1. This determines also the start position of the slider (point B_1, dimension *t*) and the end-position (point B_2). The output stroke s_H/r is thus known, that means it can be measured on the drawing relative to crank length *r*. Scaling the drawing, such that s_H corresponds to the demanded value (applying the ratio of s_H and s_H/r), provides crank length *r* and thus the other dimensions.

3 The Non-symmetric Solution, Parameter Calculation

In the previous chapter it appeared that a simple synthesis procedure can be followed when all dimensions are initially taken relative to crank length *r*. This length can be considered as the scaling value to achieve the desired output stroke s_H. We introduce thus the parameters e/r, b/r and t/r for the dimensional synthesis.

At least the equation $d_1 = d_3$ must be satisfied, yielding the following relation between angle α and parameter e/r:

$$\cos\alpha = 2 \cdot (e/r) - 1 \tag{2}$$

Applying Eq. (2) two parameters still can be chosen freely. Each required dead point reduces the number of free parameters by one. Several situations on dead points can be distinguished and they will be described below. Focus will be laid on the determination of the two parameters e/r and b/r. The calculation of the third parameter t/r and the relative output stroke s_H/r is straightforward and parallel to the graphical description in the previous chapter.

In case that a dead point both at start and at end is required, the parameter values can be calculated as follows. The dead point at the start position requires that

$$\cos\alpha = \frac{e - r}{b} \tag{3}$$

Combination of the Eqs. (2) and (3) yields a relation between e/r and b/r:

$$b/r = \frac{r - e}{r - 2 \cdot e} = \frac{1 - e/r}{1 - 2 \cdot e/r} \tag{4}$$

The dead point at the end requires that (see Fig. 3, position 2)

$$\cos\beta = \frac{e}{r + b} = \frac{e/r}{1 + b/r} \tag{5}$$

Using Eq. (1) and substituting Eqs. (2), (4) and (5) an equation in the unknown parameter e/r results:

$$\varphi_H = \arccos(2 \cdot e/r - 1) + \arccos\left(\frac{(e/r) \cdot (1 - 2 \cdot e/r)}{2 - 3 \cdot e/r}\right) \tag{6}$$

This equation needs to be solved numerically (root calculation, e.g. regula falsi). Extensive trials have learned that a start value $e/r = 0$ will always do and that only one solution will be found. The result of Eq. (6) can be applied in Eq. (4) to find the value of b/r. It is the solution with the *highest possible* value of both e/r and b/r and also with the *highest transfer quality*, see later.

If only a dead point at the end (position 2) is required, one parameter can be chosen freely. In the previous chapter it became already clear that the value of e/r can be chosen within a certain small range. The highest possible value of e/r has been found by solving Eq. (12). The lowest possible value of e/r is determined by coincidence of the points B_2 and S_1, see Fig. 3:

$$b/r = 1 - e/r \tag{7}$$

Using again Eq. (1) and substituting Eqs. (2), (5) and (7) the following equation in the unknown *lowest possible* value of *e*/*r* can be obtained:

$$\varphi_{\mathrm{H}} = \arccos(2 \cdot e/r - 1) + \arccos\left(\frac{1 - e/r}{2 - e/r}\right) \tag{8}$$

This equation has to be solved numerically. Extensive trials have learned that a start value *e*/*r* = 0 will always do and that only one root will be found.

When a user has chosen a certain value of *e*/*r*, within the range as specified above, the corresponding value of *b*/*r* can be calculated, using Eq. (2), see Fig. 3:

$$b/r = \frac{e/r}{\cos(\varphi_H - \alpha)} - 1 \tag{9}$$

Now the range of both parameters *e*/*r* and *b*/*r* has been determined for a given value of crank angle φ_{H}. Note that the better solutions are close to the highest possible value of *e*/*r*. Approximate maximum and minimum values of *e*/*r*, like available through a diagram (see later) are usually sufficient for practical use. It is not required then to solve Eq. (6) or Eq. (8). To obtain an accurate solution, the application of Eqs. (2) and (9) are the only calculations required!

In case that a dead point only in the start position is required, there is one parameter free. Equation (4) specifies a direct relation between the parameters *e*/*r* and *b*/*r*. One of them can be chosen and the other can be calculated, no matter the value of φ_{H}. Regarding the range of *e*/*r* it can be stated that any value lower than the result of Eq. (6) provides a valid solution. Equation (8) does not play a role here.

In case that no dead points are required, the two parameters *e*/*r* and *b*/*r* can be chosen freely within certain limits. The area of free choice appears in the diagram of Fig. 4, *below* the φ_{M}-line indicating the solution with a dead point at the end.

4 Diagram

Based on the theory as described above, a diagram has been developed, the two parameters *e*/*r* and *b*/*r* along the axes, that provides an overview of possible solutions. The symmetric solutions are included in this diagram as well. Area I contains the symmetric solutions, area II the non-symmetric ones.

The $\mu_{\min}$-lines are straight lines with direction cos $\mu_{\min}$, intersecting at (0,1). This can be derived immediately from Fig. 1, mechanism in position 3:

$$e/r = 1 - (b/r) \cdot \cos\mu_{\min} \tag{10}$$

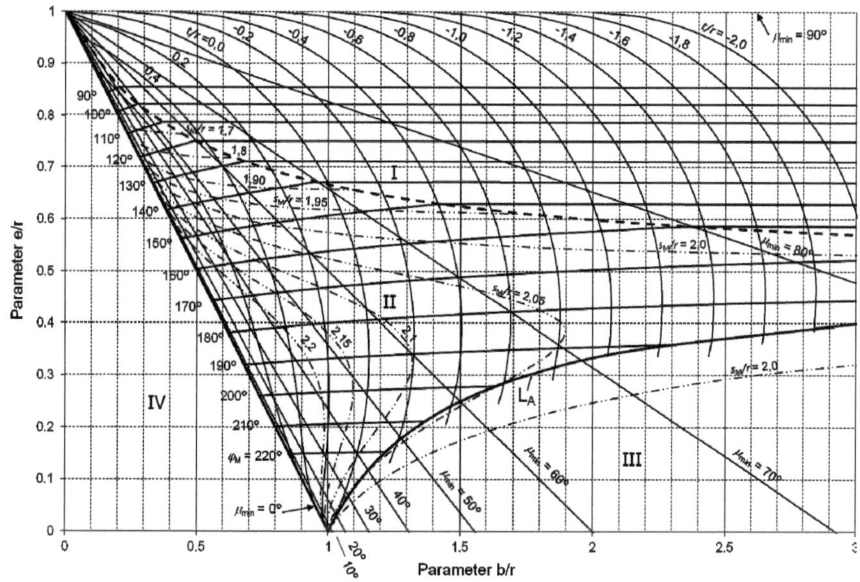

Fig. 4 Diagram

The boundary line L_A: Eq. (4) provides the relation between the parameters *e*/*r* and *b*/*r*. The equation can be rewritten as Eq. (11), in which the value of *b*/*r* is between one (*e*/*r* = 0) and infinity (*e*/*r* = 0.5).

$$e/r = \frac{b/r - 1}{2 \cdot b/r - 1}; \quad 0 \le e/r \le 0.5 \tag{11}$$

The boundary line $L_{\text{I-II}}$ between symmetrical and unsymmetrical solutions (dashed): In accordance with [1], the symmetrical solution, Eq. (12) can be derived. Note that $\mu_{min} = \varphi_H/2$ and that for *b*/*r* in infinity *e*/*r* = 0.5.

$$e/r = \frac{1 + b/r}{1 + 2 \cdot b/r}; \quad 0.5 \le e/r \le 1 \tag{12}$$

The φ_M-lines: For region II these lines have been described in Chap. 3 (solution with a dead point at the end). Note that such solutions also exist for $\varphi_M < 180°$! In region I of symmetric solutions these lines continue horizontally, but they identify here ordinary solutions without dead point.

The *t*/*r*-lines: Clearly parameter *t*/*r* is completely determined by a point in the diagram: the values of *e*/*r* and *b*/*r* can be applied in the start position 1 of the mechanism, using also Eq. (2). A constant value *t*/*r* is thus a line in the diagram.

The s_M/r-lines in region II (dash-dotted): They deal with the special situation that a dead point at the end will occur. For a given point in the diagram (all parameters known!) the s_M/r-value is also known. A constant value of s_M/r is thus a line in the diagram, showing the output stroke relative to the mechanism size.

5 Conclusions

The synthesis procedure to obtain a crank-slider mechanism as described in [1] has been investigated. It deals with monotonic transfer of motion from the crank (input) to the slider (output) such that the transmission angle is as good as possible. The current procedure for larger input angles requires that a user specifies a demanded transmission angle exactly. This leads to a complicated equation system that needs (nested) iterative calculations. The category of solutions has been limited to those mechanisms that have a dead point at the end of the motion interval. The existence of a dead point at the start of the interval has completely been neglected.

The paper shows that a much simpler procedure can be developed when the transfer quality does not need to be specified exactly, but is used to guide the possible design options for the user like choosing or avoiding a dead point at the start or at the end of the motion interval, the transfer quality and the size of the mechanism. To calculate the mechanism dimensions a set of simple equations will do. A diagram has been developed that overviews the design options.

The newly proposed synthesis procedure should be preferred and the current guideline VDI-2126 can be improved with the results of this paper.

References

1. NN: Planar mechanisms, Transfer of a rocker motion into a slider motion with regard to optimum transmission angle. VDI-guideline 2126, May 1989 (VDI-Richtlinie 2126, German)
2. NN: Planar Mechanisms, Transfer of a slider motion into a rocker motion with regard to optimum transmission angle. VDI-guideline 2125, May 2014 (VDI-Richtlinie 2125, German)

Motion Conversion with the Crank-Slider Mechanism Regarding Transfer Quality (Part 2)

A.J. Klein Breteler

Abstract The paper demonstrates the use of a diagram for synthesis of the crank-slider mechanism, matching a given input angle and a desired output stroke, for the best possible transfer quality (transmission angle). The diagram is part of a new proposal of the German guideline VDI-2126 (1989). Examples concerning non-symmetric solutions—applicable to larger input angles—are presented that show the procedure to achieve sufficient transfer quality and to choose or avoid dead points. The transfer functions of the solutions of a typical example are drawn to discuss the various design options.

Keywords Dimension synthesis · Transmission angle · Space occupation · Dead point · VDI-2126

1 Introduction

In [1] the development of the synthesis procedure of the mechanism has been described. The procedure is supported by a diagram, repeated here as Fig. 1, that provides a quick overview of the various design options of both symmetric and non-symmetric solutions. In case of larger input angles (angular stroke $\varphi_H > 180°$) only a non-symmetric solution can be applied. Because that part of the synthesis theory concerns a new proposal for [2], examples will be presented here to discuss this type of synthesis problem.

The diagram has been constructed such that a point (a value for the two parameters e/r and b/r) represents a mechanism solution. For a drawing of the mechanism and the explanation of parameters see [1]. The user can obtain the following information from the diagram:

A.J. Klein Breteler (✉)
University of Technology Delft, Delft, The Netherlands
e-mail: a.j.kleinbreteler@tudelft.nl

B. Corves et al. (eds.), *Mechanisms, Transmissions and Applications*,
Mechanisms and Machine Science 31, DOI 10.1007/978-3-319-17067-1_12

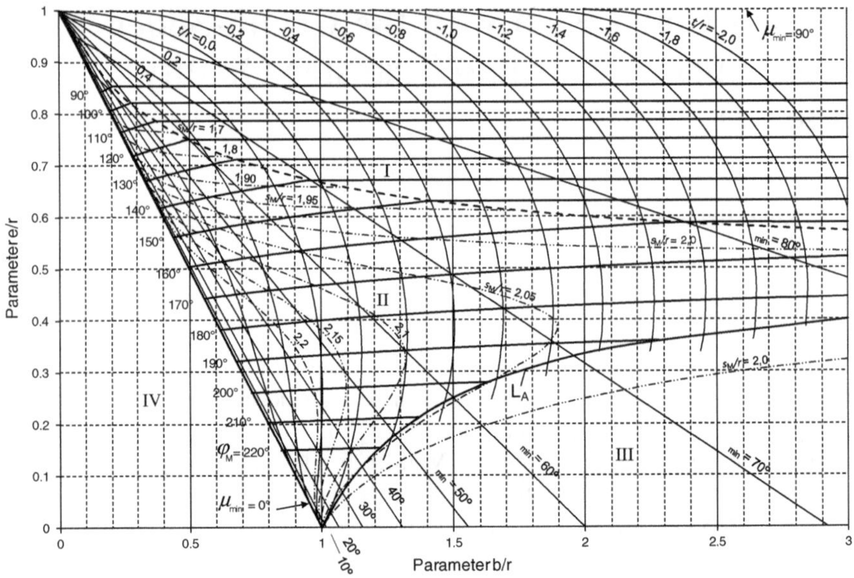

Fig. 1 Diagram

- The smallest value of the transmission angle (lines $\mu_{\min}$),
- The occurrence of a dead point at start of the motion interval (line L_A),
- The occurrence of a dead point at the end of a given interval (line φ_M),
- The third parameter value *t*/*r* assuming optimum transmission angle,
- Relative size of the output stroke (line s_M, only valid for the case of a dead point at end).

Discussion on how to choose a point in the diagram will be done by means of examples (Chap. 3). How to complete the synthesis procedure by determination of the mechanism dimensions will be explained first in Chap. 2.

2 Calculation of the Mechanism Dimensions

The diagram of Fig. 1 concerns the synthesis parameters: the dimensions are taken relative to crank length *r*. When the two parameter *e*/*r* and *b*/*r* are known (chosen or calculated, guided by the diagram) the mechanism dimensions still need to be determined. This can be done with the Eqs. (1–5) below that can be used in all situations. For explanation of the various parameters and angles see the figures in [1]. Instead of calculation with Eq. (4), the approximated value of parameter *t*/*r* can also be obtained from the diagram. The value of s_H/r can be obtained from the diagram only in case of a dead point at the end of the interval ($s_H = s_M$). Equation (5) provides s_H in the general case.

$$\cos \mu_{\min} = \frac{1 - e/r}{b/r} \tag{1}$$

$$\cos \alpha = 2 \cdot e/r - 1 \tag{2}$$

$$\cos \mu_2 = \frac{\cos(\varphi_{\mathrm{H}} - \alpha) - e/r}{b/r} \tag{3}$$

$$t/r = \sin \alpha - b/r \cdot \sin\mu_{\min} \tag{4}$$

$$s_{\mathrm{H}}/r = \sin(\varphi_{\mathrm{H}} - \alpha) - b/r \cdot \sin\mu_2 + t/r \tag{5}$$

The crank length r follows by dividing the given value of s_{H} and the result of Eq. (5), after which all other dimensions (e, b and t) are thus known.

3 Examples

A crank-slider mechanism has to be designed that, for a given input angle $\varphi_{\mathrm{H}} = 200°$, shows a slider displacement $s_{\mathrm{H}} = 1000$ mm. Four options will be considered:

(a) With a dead point both at start and at end. In the diagram this concerns the intersection point of the lines $\varphi_{\mathrm{M}} = 200°$ and $\mathrm{L_A}$. The point is marked with character (a) and a blue dot in Fig. 2, which is an outline of the diagram. Obviously this is the solution with the best possible $\mu_{\min}$—value. To find this point accurately Eq. (6) in [1] must be solved: $e/r = 0.2801$. Applying Eq. (4) in [1] yields $b/r = 1.6366$.

(b) With a dead point at start only. This solution will exist for any point on the line $\mathrm{L_A}$, where $\varphi_{\mathrm{M}} > 200°$, for instance with a chosen value $e/r = 0.2$. That point is marked with character (b) and a green dot in Fig. 2. Applying Eq. (4) in [1] yields $b/r = 1.3333$.

(c) With a dead point at the end. Any point on the line $\varphi_{\mathrm{M}} = 200°$ will have this property, like the one marked with character (c) and a light blue dot in Fig. 2. The simplest way to determine the point precisely is by choosing the e/r value (although this choice is very sensitive). With $e/r = 0.275$ the b/r —value can be obtained with Eq. (9) from [1] and Eq. (2): $b/r = 1.3419$ The diagram shows clearly how much the $\mu_{\min}$—value will be reduced for any choice on that line.

(d) With no dead point at all. Any point below the line $\varphi_{\mathrm{M}} = 200°$ will do. The point marked with character (d) and a red dot shows such a point. Now both parameters can be chosen, like for instance here $e/r = 0.25$ and $b/r = 1.3$.

Table 1 below shows the results of calculating the remaining parameters and intermediate results, Eqs. (1–5), followed by the mechanism dimensions.

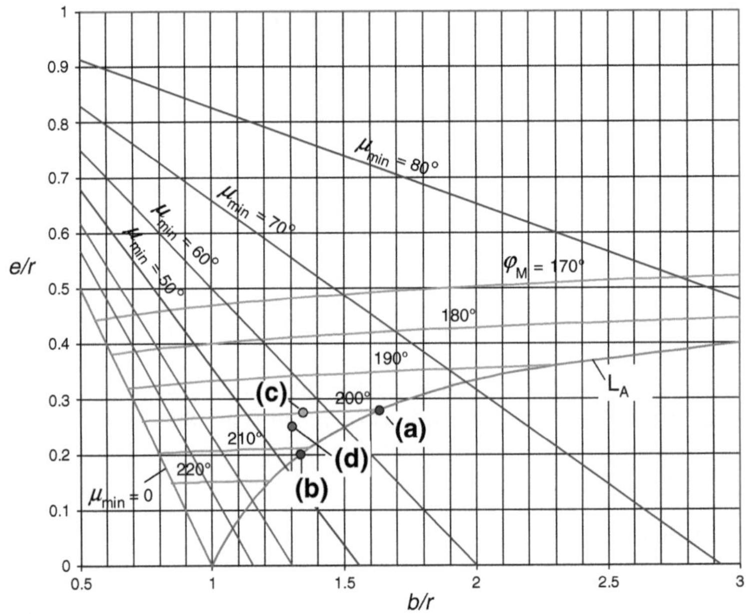

Fig. 2 Characteristic design solutions

Table 1 Calculated results of the examples

Quantity	Eq. No	Case (a)	Case (b)	Case (c)	Case (d)
μ_{min}	(1)	63.9°	53.1°	57.3°	54.8
α	(2)	116.09°	126.87°	116.74°	120°
μ_2	(3)	83.9°	86.1°	83.3°	86.6
t/r	(4)	−0.5717	−0.2667	−0.2361	−0.1958
s_H/r	(5)	2.05	2.0206	2.09	2.0868
r		487.8 mm	494.9 mm	478.6 mm	479.2 mm
b		798.4 mm	659.9 mm	642.2 mm	623.0 mm
e		136.6 mm	99.0 mm	131.6 mm	119.8 mm
t		−278.9 mm	−132.0 mm	−113.0 mm	−93.8 mm

4 Analysis of Results

The transfer functions of the sample mechanisms as calculated in the previous chapter have been drawn using a proven numerical method for kinematic analysis [3]. Figure 3 shows the transfer functions and Fig. 4 the first derivatives (transfer functions of order one). These figures show the effect of the chosen design options.

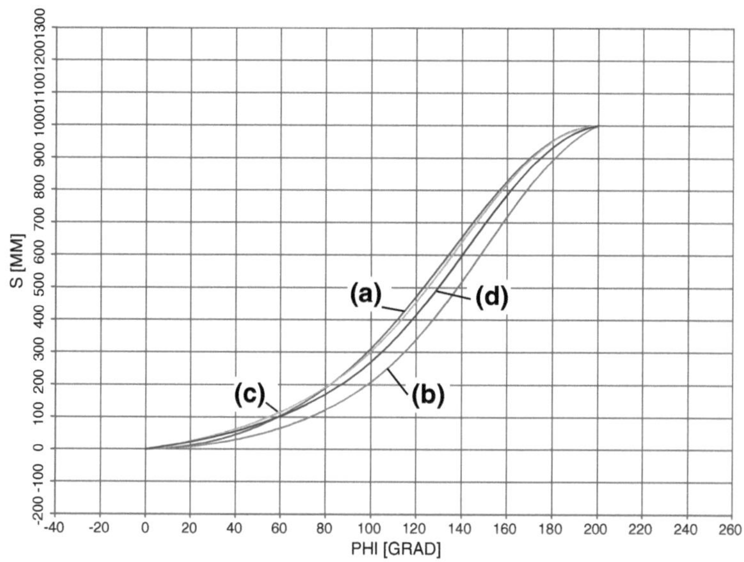

Fig. 3 Transfer functions (order 0)

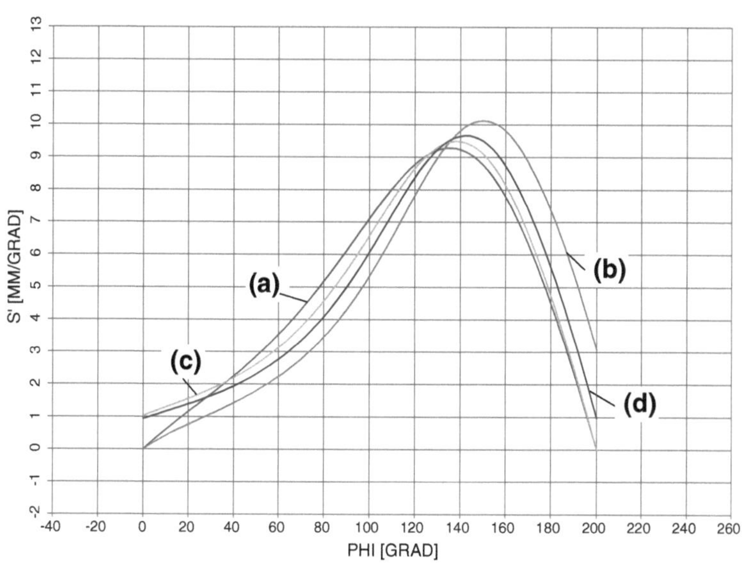

Fig. 4 Transfer functions (1st order)

Especially in Fig. 4 it is clear where the dead points occur: the start value or the end value of the function will then be zero. Mechanisms (a) and (b) have a dead point at start, mechanisms (a) and (c) have a dead point at the end of the motion.

5 Conclusions

To demonstrate the use of a newly developed diagram for synthesis of the crank-slider mechanism several examples have been worked out. They show that the problem of matching larger input angles and desired output strokes can adequately be solved with so-called non-symmetric solutions. The examples show pretty good transfer quality (μ_{min} is 53° or better) and dead points can be included or not. Synthesis parameters can be estimated with the diagram or they can be calculated using mostly simple formulas. The mechanism dimensions follow then by scaling the parameters such that the demanded output stroke will be generated.

The examples confirm that the new synthesis procedure works well and that the current VDI-guideline can be improved indeed.

References

1. Klein Breteler, A.: Motion conversion with the Crank-Slider mechanism regarding transfer quality (Part 1). In: Proceedings of the 3rd Conference on Mechanisms, Transmissions and Applications MeTrApp2015, Aachen, Germany (2015), vol. 8779. Springer, Berlin, MMS series
2. NN: Planar mechanisms, Transfer of a rocker motion into a slider motion with regard to optimum transmission angle. VDI-guideline 2126, May 1989 (in German, VDI-Richtlinie 2126)
3. Klein Breteler, A.: Lecture notes on Mechanisms (course wb3303), chapter 5: Numerical kinematics with finite elements. TU Delft, The Netherlands, 2004.Website: http://wbmttt.tudelft.nl/cadom (>downloads> program Runmec)

Implementation of VDI Guidelines in Parametric 3D CAD Systems and Their Functional Extension to Dynamically Associative Optimization Tools

Christian Ahl and Rainer Lohe

Abstract Design guidelines, such as the VDI guidelines, help the designer in the synthesis of gears and mechanisms for standard recurring tasks. The guidelines provide graphical and computational methods, leading the designer systematically to solve the task. In addition to the actual synthesis process, they often contain diagrams and formulas to support identification of a "good" mechanism. This paper shows how the synthesis method derived from such guidelines can be integrated into modern 3D parametric CAD systems. In addition, design criteria can be used in the search for "good" mechanisms which are specifically tailored to the particular application and thus cannot be covered in general guidelines. In order to take into account such criteria in the individual optimization, the editor should work dynamically, interactively and associatively. Simultaneously with the modification of the mechanism via the graphical input device of his of her CAD system, the editor has to identify and manage the new mechanism dimensions, space requirement and the quality of its criteria. The paper clearly shows also that by using design guidelines in conjunction with parametric CAD systems, synergies are realized. The design engineer is offered a good guide to the solution of the task in his or her usual working environment. The use of features and parameters in CAD systems is explained and solutions are shown to design the mechanism according to requirements. Typical applications are shown using the 3D CAD systems CATIA V5 and Pro/Engineer Wildfire 5.0.

Keywords VDI guidelines · CAD systems · Mechanism synthesis · Graphical methods · Design criteria

C. Ahl (✉) · R. Lohe
University of Siegen, Siegen, Germany
e-mail: christian.ahl@uni-siegen.de

R. Lohe
e-mail: rainer.lohe@uni-siegen.de

B. Corves et al. (eds.), *Mechanisms, Transmissions and Applications*,
Mechanisms and Machine Science 31, DOI 10.1007/978-3-319-17067-1_13

1 Introduction

During conferences in the past, several models to appropriately synthesize mechanisms using modern CAD systems have been presented, covering the design requirements of VDI guidelines [1]. In this paper, a general procedure is presented and complemented by the specific application of two VDI guidelines. Two recently released official VDI guidelines are presented in this paper, the new CAD-based dynamic associative component is also discussed. In the VDI Guideline 2125 [2], new design and calculation instructions for 4-bar linkage to convert rocker motion into slider motion are described. In the VDI Guideline 2728 Part 2 [3], a guide for the synthesis of a mechanism is presented for approximating linear guiding.

2 VDI Guidelines

VDI guidelines are approved standards that represent the state of the art. They serve the community as manuals, working papers and decision-making support [4]. For the selection, design and calculation of 4-bar linkages, numerous guidelines are available to accompany the product development of type and dimension synthesis up to analysis and simulation.

2.1 Approximating Linear Path-Generating Mechanisms of High Quality, VDI Guideline 2728

Using the VDI Guideline 2728, 4-bar linkages, such as for example those shown in Fig. 1 can be realized by utilizing a symmetrical coupler curve in regions with an approximated linear guiding of high quality.

In the area of linear guiding, a coupler curve has a maximum of six points of intersection with a line. The extremes that the coupler curve, k_k, passes in the area of the approximated linear guiding determine the dimensions of a rectangular tolerance band. The largest deviation from the straight line determines the height h of the tolerance range and the distance of the entry points G_R, G_L determines the length L. The dimensions of the deviations can be unified using the Chebyshev approximation and thus be minimized to h_{opt}. In general, there is exactly one point at the coupler link which corresponds to the coupler curve of the Chebyshev approximation. This coupler curve is also referred to as a balanced straight guide [3].

As a basis for an starting linkage, an isosceles crank rocker or the ROBERTS'sche equivalent, the symmetric double rocker can be used.

For the selection of such four bar mechanisms monograms are available in the VDI Guideline 2728 from which the achievable linear guide quality and the required dimensions can be read off or calculated. The guideline allows an overview

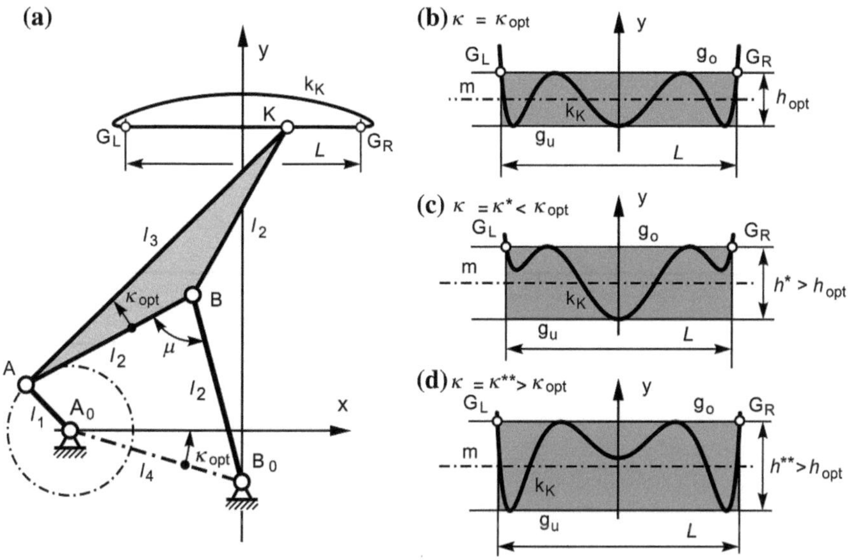

Fig. 1 Curve in the linear guided section of the isosceles crank rocker [3]

and evidence feasibility for one's own task. This is required for the 4-bar mechanism synthesis and is helpful to create a parametric CAD model which allows interactive variation of task and solution.

A four bar mechanism with isosceles bars, ($AB = BB_0 = BK = l_2$,) which is drawn in the Sketcher in 4 different positions, serves as a starting point. These four positions correspond to those positions of the coupler point on one half of the symmetrical coupler curve in which the maximum deviations occur (Fig. 2).

In order to allow variation only within Chebyshev solutions, some restrictions must be imposed on the model:

- In isosceles basic four bar, the length, $l_2 = AB = BB_0 = BK$, occurs three times.
- In the coupler triangle and in the inclination of the frame, the same angle, κ, occurs.
- The vertical positions of the coupler point, K, have to alternate between the upper limit, g_o, and the lower limit, g_u, of the tolerance band.
- In position 1, the coupler point is located straight above B_0 (${}_1K_x = B_{0x}$).
- In the positions 2 and 3, the coupler path has a horizontal tangent. For this purpose, the orbit normal, n_K, of the coupler point is determined via the instantaneous center and constrained vertical.
- In the position 4, the horizontal position of the coupler point is half the length of the straight guide.

The vertical positions of the coupler points 2 and 3 are automatically arranged according to the Chebychev restrictions pointed above. Figure 3 shows a corresponding coupler curve from the linkages shown in Fig. 2. The display of the

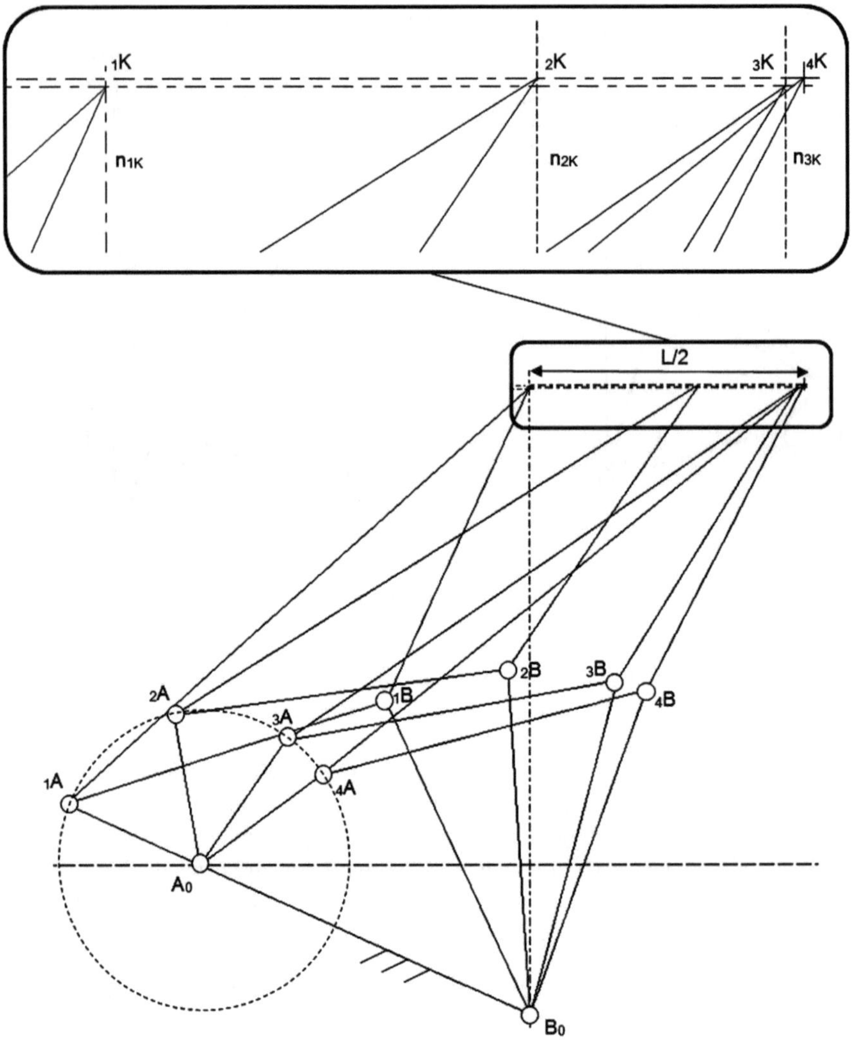

Fig. 2 Isosceles crank rocker in 4 positions as a ProE screenshot

coupler curve in Fig. 3 is required to retain control when varying the linear guide in the CAD system. To clearly define a solution, three parameters can now be varied. For example, moving point ${}_4K$ with the mouse varies two parameters that describe the actual task, i.e. length and width of the tolerance band. Since there is an infinite number of possible four bar mechanisms (∞^1) at any tolerance rectangle, it is possible to choose any bar dimensions and thus the size of the linkage with the third parameter.

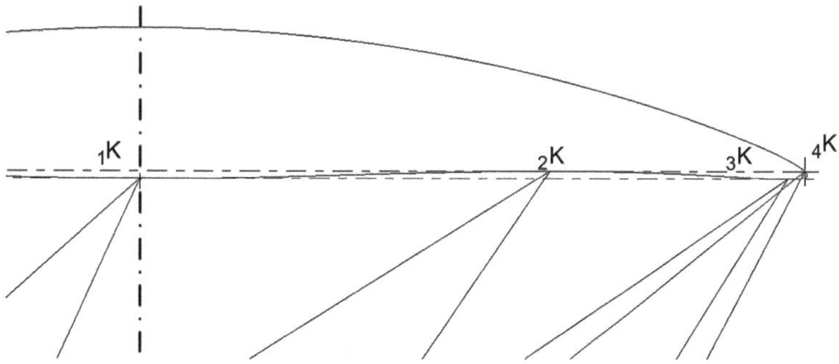

Fig. 3 Coupler curve of the requirements corresponding linkage

The engineering designer is able to realize a mechanism with the help of his or her CAD system which has a predetermined linear guide length and—deviation. As expected, this approach only provides results if the required linear guide qualities and allowable linkage sizes are harmonized. The admissible ranges are clearly laid out in the VDI Guideline 2728 sheet 2 [3].

2.2 Slider Motion into Rocker Motion, VDI Guideline 2125

The VDI Guideline 2125 describes methods for a favourable conversion of slider into rocker motion and provides charts for the selection of a high transmission quality. In the 2014 guideline, the focus is no longer only on the absolute best transmission. The charts now also contain minor deteriorations of the transmission angle in the range of the absolute optimum and thus allow the consideration of additional criteria e.q. the space requirement of the mechanism or the characteristic of the transfer function. The mechanism structure and the equivalent structure of the parametric CAD model are shown in Fig. 4.

This guideline distinguishes between procedures I and II according to the required angular output ψ_H.

If the transmission angle is good enough, other criteria might be improved. For this, Procedure III has been added. It helps to manipulate the transfer function and to monitor the transmission angle in order to find a compromise.

Important equations of the guideline are stored in the knowledge base of CATIA as parameters to be regulated [5]. In **Procedure I**, i.e. for angular output $\psi_H < 76.3°$, the best value achievable for the transmission angle is first determined. For this purpose, the necessary coupler length, b, is either read off from the chart of the guideline, or determined by a parameter study using the Product Engineering Optimizer in CATIA [5]. For the remaining three linkage dimensions, c, e and t, there are simple explicit equations available in the guideline which are easy to use in the CAD system. The linkage lengths, b and c, and the best achievable value of

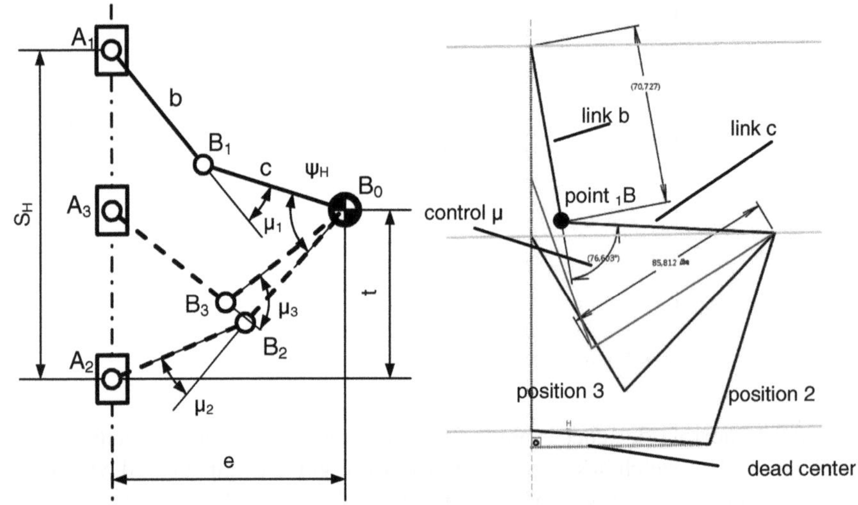

Fig. 4 *Left* Names on the thrust rocker [2]. *Right* sketch of the procedure I

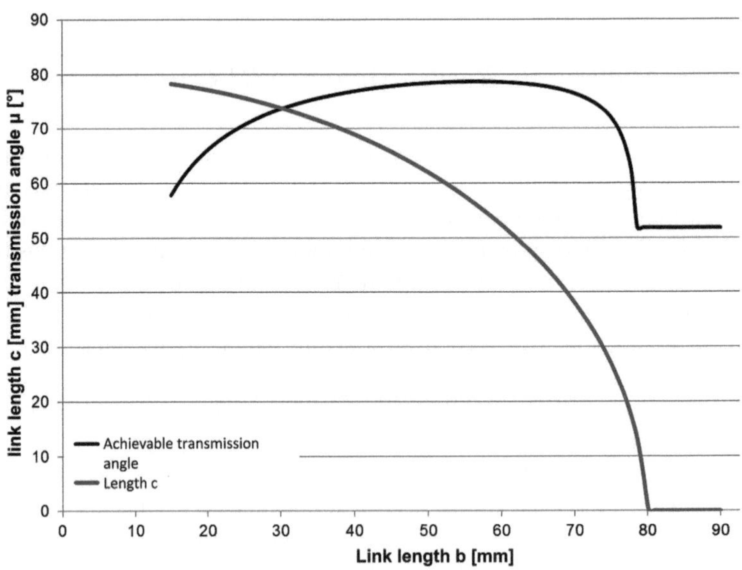

Fig. 5 Transmission angle, μ, and link length, c, for $s_H = 100$ mm and $\psi_H = 75°$

the transmission angle μ_{min} are displayed and illustrated in a diagram (Figs. 4 and 5).

If the angular output of $\psi_H = 75°$ for an input of $s_H = 100$ mm is required, the result of the study is the optimal transmission angle max $\mu_{min} = 76.854°$ with the link length b = c = 52.42 mm.

The design engineer can now adopt this absolute best transmission linkage or, by an initial slight deterioration of the transmission angle, vary the dimensions and take into account his or her own criteria. Therefore he or she must "grab" the coupler length b and modify it dynamically and interactively by dragging point $_1B$. Since the three remaining linkage dimensions, e, t and c, are adjusted automatically (see VDI 2125), the transmission angle only deteriorates recognizably as a result of large manipulations, see Fig. 5. In a range of b = 24 mm to b = 75 mm, the maximum achievable transmission angle max μ_{min} is always above 70°.

The basis for procedure I applies theoretically for angular output up to $\psi_H \leq 90°$. However for an output $\psi_H > 76.3°$ (while b = c) there would be a reversal movement after passing the dead point (cf. [2, 6]).

The requirement of no reverse movement in the range of motion is fulfilled with the help of illustrating the dead point. The transmission is free of return motion if the dead-center as shown in Fig. 6 is on the edge of or outside the range of motion.

In **Procedure II**, i.e. for stroke angle $\psi_H > 76.3°$, the absolute best transmission-friendly linkage has to be determined iteratively. The results are shown in the selection charts of the guideline, in addition to a guide for the numerical calculation. However, because in these linkages the coupler length is always about $b = 0.5 \times s_H$, it gives a very good starting value for an iteration [7, 8]. In general, the iteration is not necessary. Diagram 1 in the VDI Guideline 2125 shows only small improvements for the transmission angle in the case that iteration is applied.

In CATIA, it is advisable to determine the remaining three dimensions in the same manner as given in diagram 2 in VDI Guideline 2125. Varying the base point coordinate, e (eccentricity), the parameter study calculates all linkages that meet the conditions for an optimal transmission angle. From these mechanisms, shown in Fig. 7 the one with the required angular output, ψ_H, is chosen. In our example, an output of $\psi_H = 160°$ at a stroke $s_H = 100$ mm is required. So we have to select the highlighted mechanisms shown in Fig. 7 to find the absolute best transmission.

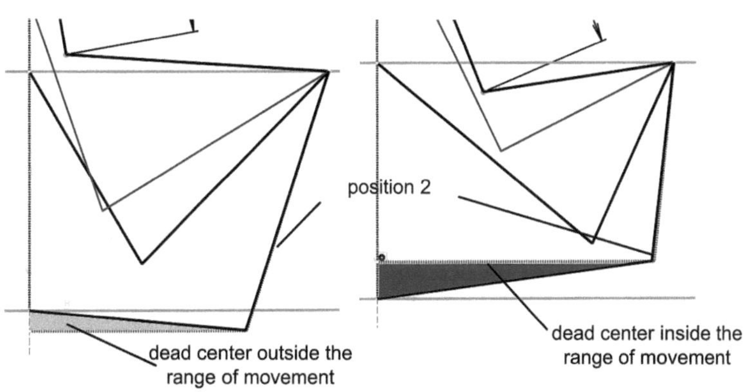

Fig. 6 Parameter variation according to Procedure I and return motion control

e (mm)	c (mm)	t (mm)	psiH (deg	Mue_I (deg)
26,36	33,39	23,58	188,57	29,02
27,12	33,30	24,19	185,42	30,36
27,88	33,23	24,80	182,26	31,71
28,64	33,20	25,41	179,09	33,07
29,39	33,18	26,01	175,92	34,43
30,15	33,20	26,61	172,76	35,79
30,91	33,24	27,21	169,61	37,15
31,67	33,30	27,80	166,47	38,50
32,42	33,39	28,39	163,34	39,85
33,18	33,51	28,98	160,25	41,19
33,94	33,65	29,57	157,18	42,51
34,70	33,81	30,15	154,14	43,83

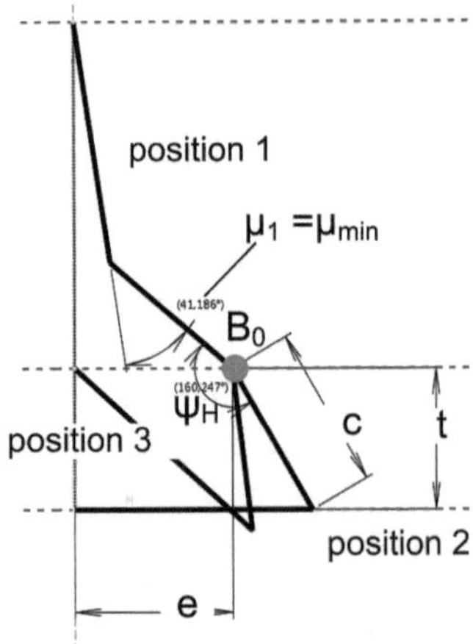

Fig. 7 *Left* VDI 2125 Procedure II—Parameter study for b = 0.5. *Right* Sketch Procedure II

Based on this linkage, Procedure II also gives the possibility to influence positively the angular output ψ_H, the transmission angle μ_{min}, the space requirements or the transfer function (cf. Fig. 7) by dynamically and interactively pulling at point B_0, thus modifying the eccentricity e.

3 Conclusion

By implementing design rules in 3D CAD systems, the designer can crucially simplify the design process for mechanisms.

The dynamic behaviour of CAD systems is ideal for the mechanisms' synthesis. Classical graphical methods gain a new and greater importance because nowadays these graphs can be dynamically, interactively and associatedly varied in CAD systems. Many alternatives are assessed very quickly. Sufficient accuracy is available. Space analysis and collision checks can be simulated.

This dynamic and interactive approach becomes perfect by the associative component, that indicates the search direction in which a better mechanism can be expected to be found. Another positive experience is the required expertise of the user. The CAD models can be applied with standard CAD knowledge and no special programming is required.

References

1. Lohe, R. et al.: Einsatzmöglichkeiten der 3D-CAD Systeme Catia V5 und Pro/Engineer Wildfire in der Getriebetechnik, 9. Kolloquium Getriebetechnik Chemnitz (2011)
2. VDI-Richtlinie 2125: Übertragungsgünstigste Umwandlung einer Schubschwing- in eine Drehschwingbewegung, Gründruck (Entwurf) Beuth Verlag, Berlin (2014)
3. VDI-Richtlinie 2728 Blatt 2: Geradführungsgetriebe mit symmetrischen Koppelkurven, Gründruck (Entwurf), Beuth Verlag, Berlin (2014)
4. VDI-Richtlinie 1000: Richtlinienarbeit Grundsätze und Anleitungen, Beuth Verlag, Berlin (1981)
5. Trzesniowski, M.: CAD mit CATIA V5 Handbuch mit praktischen Konstruktionsbeispielen aus dem Bereich Fahrzeugtechnik 3. Vieweg + Teubner, Wiesbaden, Auflage (2011)
6. Klein Breteler, A.J.: On the conversion of translational into rotational motion with the slider rocker mechanism, regarding transfer quality. In: Proceedings of the Scientific Seminar on Terminology for the Mechanism and Machine Science, Minsk/Gomel, S. 83–91 (2010)
7. Hain, K.: Hydraulische Schubkolbenantriebe für schwierige Bewegungen, In: Oelhydraulik und Pneumatik, 2, Nr. 6, S. 193–199 (1958)
8. Hain, K.: Kräfte und Bewegungen in Krafthebergetrieben, In: Grundlagen der Landtechnik, Nr. 6, S. 45–66 (1955)

Part III
Dynamics of Mechanisms and Machines

Part III

Optimal Motion Cueing Algorithm Selection and Parameter Tuning for Sickness-Free Robocoaster Ride Simulations

Duc An Pham, Sebastian Röttgermann, Francisco Geu Flores and Andrés Kecskeméthy

Abstract Drive simulators using serial robots, such as the KUKA robot "Robocoaster", are becoming attractive for situations in which the workspace of traditional Stewart platforms is not suited to accommodate large target rotations, allowing for a wider range of possibilities. Nevertheless—even when using serial robots—the exact target motion can often not be exactly reproduced. In these cases, motion cueing algorithms (MCA) are used to produce a motion which feels as realistic as possible while remaining in the robot acceleration workspace. This paper analyzes the numerical properties of all currently existing MCA (classical, adaptive, optimal, and model predictive control) and selects the most suitable MCA using objective criteria. It also introduces a new procedure for tuning the optimal MCA such that it behaves as good as and even better than much more involved techniques based on the model predictive control (MPC). The new algorithm, termed ZyRo-K, shows best properties for reproducing the desired linear specific force while reducing the rotational false cues. While the work shown in this paper is restricted to numerical evaluation using state-of-the-art "goodness" metrics, the application and test of the algorithms for human passengers on a Robocoaster is currently being prepared and will be published in the near future.

Keywords Motion cueing algorithm · Robocoaster · Drive simulator · Simulator sickness

D.A. Pham · S. Röttgermann (✉) · F.G. Flores · A. Kecskeméthy
University of Duisburg Essen, Duisburg, Germany
e-mail: sebastian.roettgermann@uni-due.de

D.A. Pham
e-mail: duc.pham@uni-due.de

F.G. Flores
e-mail: francisco.geu@uni-due.de

A. Kecskeméthy
e-mail: andres.kecskemethy@uni-due.de

B. Corves et al. (eds.), *Mechanisms, Transmissions and Applications*,
Mechanisms and Machine Science 31, DOI 10.1007/978-3-319-17067-1_14

1 Introduction

Driving simulators are becoming more popular as affordable systems for both training flight pilots and researching human motion sensation. In recent years, a new generation of driving simulators based on serial robots has been developed in order to reduce the price of the system and, most importantly, to improve the quality of the simulating process.

The aim of MCA in driving simulators is to generate motions which feel as realistic as possible. However, the lack of understanding of human motion perception, the restricted workspace of simulators, and the perception thresholds lead to false cues that induce simulator sickness due to the inconsistency of motion and visual information. Thus, selecting and tuning the best MCA for a particular motion simulation becomes a challenging task.

All current MCA use the tilt coordination technique to simulate the sustained acceleration and translational motion for the high-frequency (HF) part of acceleration. The classical washout filter consists of both linear high-pass filters (HP) and low-pass filters (LP), whose parameters can be adjusted off-line by trial and error. This algorithm is the most widely used in commercial simulators because of its simple properties. However, it exploits the available workspace ineffectively, and produces false cues due to its linear characteristics. In order to exploit the workspace of the Robocoaster, Gio [3] developed a new motion cueing algorithm named "cylindrical classical algorithm", which is the classical washout algorithm using cylindrical end-effector motion for lateral acceleration instead of linear motion, yielding a larger acceleration workspace for serial robots.

A further development of the classical washout filter is the "adaptive washout filter" [7, 8, 10, 11]. In this method, the washout filter parameters are systematically updated in real time to minimize a cost function using steepest descent techniques. As mentioned in [6], although the false cues can be decreased, the stability of the algorithm depends strongly on the selected parameters. An optimal MCA was firstly proposed by Sivan et al. [12] and later improved by Reid and Nahon. Wu and Telban used more involved vestibular system models to develop new algorithms [13] that included higher-order washout filter models and which were fitted to the target motion by optimal control methods. Another optimal algorithm was proposed by Richard Romano and Zywiol [14], and then used with optimized filter parameters by Tändl and Kecskemethy [5]. The algorithm is different from that developed by Sivan [12] in that it only focuses on solving the tracking problem in offline mode by finding the suitable combination of linear acceleration and tilt angle to replicate more realistically the target specific force, leading to less tuning parameters than the other algorithms mentioned above. More recently, model predictive control (MPC) techniques have been used for the MCA strategy, which allows one to handle not only limits of working space, but also the thresholds of motion perceptions.

The aim of the present work is to compare all existing algorithms and to evaluate their performance using a "goodness" metric proposed in the literature [2, 9], which gives a hint on the quality of the motion simulation. The MCA were tested for a virtual ride along an planar s-shaped curve, choosing the lateral acceleration as the target (design) signal, and using the limit parameters of the real ride simulator.

This paper is structured as follows. First the equipment of the driving simulator is described. Then the concepts of the current MCA and the objective criteria are briefly described. Finally, the different methods are compared based on the aforementioned goodness factor, showing which is the best suited for a biofidelic motion cueing of the chosen trajectory.

2 Description of Experimental Setup

Figure 1 shows the physical ride simulator used in the Lab for Mechanics and Robotics (LMR) at the University of Duisburg-Essen. It consists of a *KUKA KR500/1 TÜV* Robocoaster robot with a *Maurer Söhne* roller coaster seat mounted at the robot flange, a *Visette45* head mounted display unit, an *A.R.T.* motion tracking system with two cameras, a visualization PC, a trajectory PC and a control PC. All of these systems are interconnected via ethernet. The robot system is adapted so that it fulfills the safety rules for human passengers provided by the German Technical Control Board *TÜV Süd*.

On the robot controller, the optional *KUKA Robot Sensor Interface* technology package is installed, which allows the needed capability of interacting with an external PC via ethernet. The controller receives the prescribed axis value sets from the trajectory PC at every simulation time step. The simulation program running on

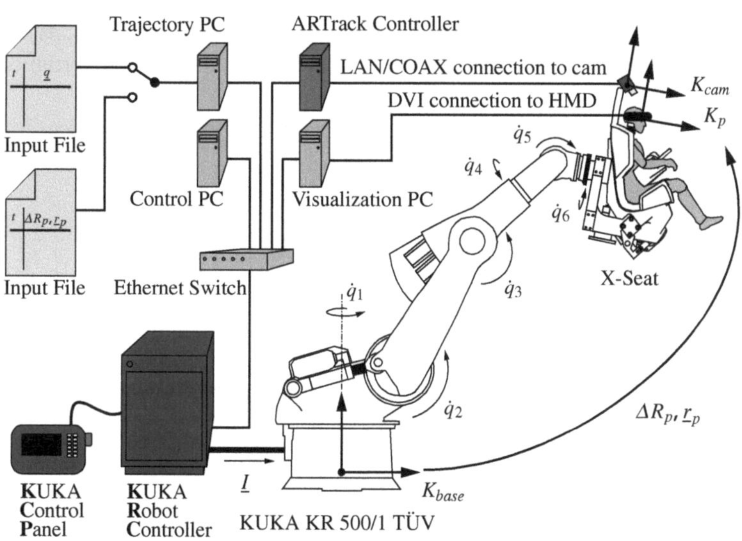

Fig. 1 Architecture of the motion simulator Robocoaster at the LMR of University Duisburg-Essen

that PC is capable of using previously computed input files containing values either in axis or target coordinates. Datasets computed by the different motion cueing algorithms are used as input data on the trajectory PC. The matching virtual reality environment, which is displayed on the head mounted display in stereo 3D with the help of a *more3D* plugin, is computed on the visualization PC. In this virtual reality environment the viewing angle changes depending on the passengers head movement, which is tracked with the *A.R.T.* camera system. Finally, a control PC is included to operate the whole simulation system, which starts all needed devices simultaneously and gives the operator the ability to monitor the simulator while the simulation is in progress. Both the program running on the trajectory PC and the visualization environment and the control environment program are based on the multibody simulation system *MOBILE* [4].

3 Review of Existing MCA

In the literature, a great variety of approaches for MCA exist, which are mainly devoted to linear motion in longitudinal direction. Hereby, "perception" is localized at the vestibular system, and translation is parametrized by the specific force and rotation by the angular velocity.

(a) **Classical algorithm with cylindrical coordinates (CL; Fig. 2a)**. Introduced by Giordano [3], this concept uses the standard classical washout algorithm in cylindrical coordinates $\underline{\xi} = [R, \alpha, z]$ with $R = \sqrt{x^2 + y^2}$ and $\alpha = \text{atan2}(y, x)$. For motion prescription, linear high-pass filters for linear and rotational motion (HPa, HPw), a low-pass filter together with a tilt coordination (TC) which replaces the sustained acceleration by a proper tilted angle, as well as a compensator block for adapting the target motion due to the inertial acceleration of the moving washout frame are used.

(b) **Adaptive algorithms (AD)**. These are developed from the classical MCA by adapting the high-pass filter parameters K (gain), ω_n (natural frequency), and ζ (damping ratio), in real time such that cost function

$$J = \{\frac{1}{2}[w_a(\ddot{\underline{\xi}} - \ddot{\underline{\xi}}_{HP})^2 + w_v \dot{\underline{\xi}}_{HP}^2 + w_s \underline{\xi}_{HP}^2 + w_k(K - K_0)^2 + w_\zeta(\zeta - \zeta_0)^2 + w_\omega(\omega_n - \omega_{n0})^2]\} \quad (1)$$

is minimized [7], where $\ddot{\underline{\xi}}$ represents acceleration in the cylindrical coordinates, the quantities $\ddot{\underline{\xi}}_{HP}, \dot{\underline{\xi}}_{HP}, \underline{\xi}_{HP}$ are high-pass filtered signals of acceleration, velocity and position, respectively, $K_0, \zeta_0, \omega_{n0}$ are the reference anchor values of the filter parameters, and $w_{(\ldots)}$ are weighting parameters which penalize the difference in response between the target motion and the simulator, as well as restraining the translational velocity and displacement in the simulator.

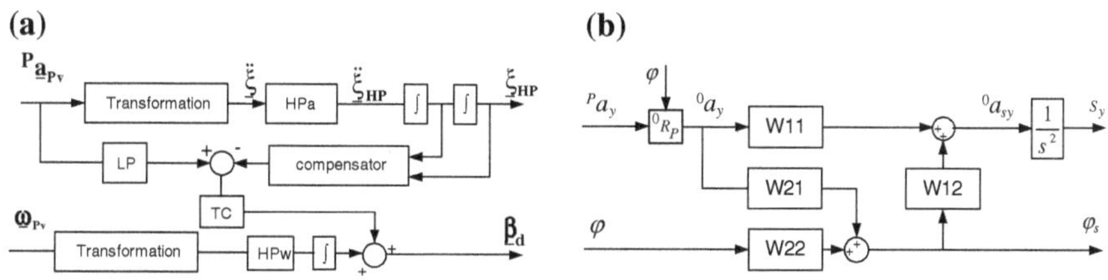

Fig. 2 Control diagrams: **a** classical algorithm in cylindrical coordinates, **b** optimal algorithm

(c) **Optimal Washout Algorithms (OpS, OpRN, Tel-YoMe, Tel; Fig. 2b).** These models, first developed by Sivan [12], determine the higher-order filters $W(s)$ by solving the linear quadratic optimal control problem

$$J \triangleq E\left\{ \int \underline{e}(t)^T Q \underline{e}(t) + \underline{u}^s(t) R \underline{u}^s(t) + \underline{x}_c^s(t)^T R_c \underline{x}_c^s(t) \right\} \quad (2)$$

which includes a linearized mathematical model of the human vestibular system and uses as measure for the cost function the error $\underline{e}(t)$ between the expected vestibular system's output of the passenger and its counterpart coming from the target motion. The optimal MCA was further developed, e.g., by Reid & Nahon [10], Telban [13] by modifying the linearized model of system, e.g. by using different models of the vestibular system, different input signals $\underline{u}^s$ and simulator state variables $\underline{x}_c^s$. The weighing matrices Q, R, R_c were tuning by trial and error to find the suitable values.

(d) **Optimal Control Algorithms (ZyRo-Z, ZyRo-K; Fig. 3a).** This approach, first developed by Zywiol and Romano [14], solves the problem as an optimal tracking problem, whose cost function

$$J = \int \left\{ \left[\underline{y} - \underline{r}\right]^T Q \left[\underline{y} - \underline{r}\right] + \underline{u}^T R \underline{u} \right\} dt \quad (3)$$

is targeted to find the appropriate control input $\underline{u} = [u_1, u_2]$ (u_1 for specific force, u_2 for angular velocity) which produces as accurately as possible the perceivable quantities, such as the specific force at pilot head (y_1), while keeping other signals, such as the offset position of the cabin with respect to the home configuration (y_2), as close to zero as possible. The algorithm, termed ZyRo-Z, finds—for given parameters Q, R, c and γ—the best combination of linear acceleration and tilt angle such that the desired specific forces over time are replicated as close as possible and works in offline mode. In Tändl and Kecskeméthy [5], an approach for optimizing the values of the optimal control parameters Q, R, c and γ with respect to a given trajectory was proposed, which led to better cues than the originally proposed values of Zywiol and Romano [14]. The result of this optimized optimal control trajectory is termed as "ZyRo-K".

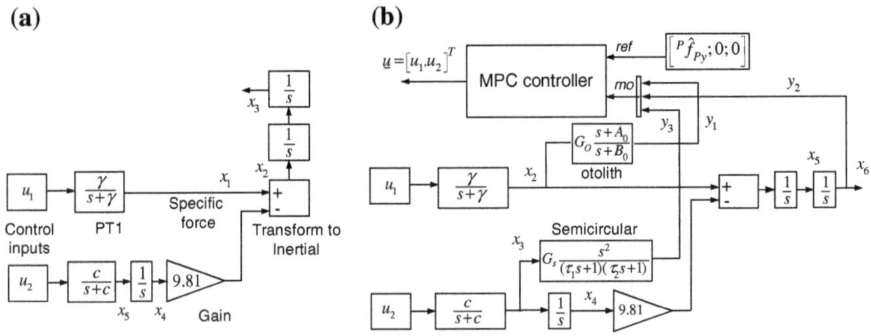

Fig. 3 Control diagrams: **a** ZyRo, **b** MPC-3

(e) **Model predictive control (MPC) algorithms (Fig. 3b)**. These algorithms use multivariable optimization to fit future behavior of the plant output over a future fixed interval (prediction horizon) while respecting the previously defined constraints [1], using a parametrized version of the future input variable $\underline{u}$ time history over the future horizon. We designed three MPC (MPC-1, MPC-2, MPC-3) with different system dynamic modelings, where MPC-1 has a similar model as ZyRo-Z/K, MPC-2 includes an integrated mathematical of the otolith system, and MPC-3 includes a complete vestibular system model based on the concept of [1].

4 Results

The expected sensation of real passengers is difficult to measure, as it will also contain many subjective effects which are difficult to render objectively in a general scheme. However, Fischer [2] proposed a numerical measure for expected subjective sensation which he termed the "good criterion", which is composed of the errors between the target and simulator specific force (acceleration sensation at the vestibular system) as well as between the target and simulator angular velocity, respectively, and their time derivatives. By combining it with a general performance index criterion, such as proposed by Pouliot et al. [9], a refined assessment of motion cue biofidelity can be performed numerically, e.g. to optimize the MCA before its application for a given offset trajectory.

In the performance indicator described in [9], two indicators are used. The first one, λ_1, is intended to yield a single numerical value describing the average error between motion cues generated by the real motion and those produced by the simulator. The second indicator, λ_2, describes the average error in the rate of change of the variable λ_1 [9]. The two indicators are described by the following equations:

$$\lambda_i = 100\left(\frac{\lambda_{if}}{v_{\max}} + \frac{\lambda_{i\omega}}{\omega_{\max}}\right) \quad (i = 1, 2) \tag{4}$$

where index f means specific force and index ω represents angular velocity. A value of $\lambda_i = 0$ means a simulation with perfect expected simulator sensation, while the greater $\lambda_i > 0$ the worse the simulation sensation is expected to be.

The "good criterion" of [2] is obtained by refining the performance indicators according to following expressions

$$\lambda_1 = \lambda_{1f} + \lambda_{1\omega} = \lambda_{1f,sc} + \lambda_{1f,sh} + \lambda_{1\omega,sc} + \lambda_{1\omega,sh} \tag{5}$$

$$\lambda_2 = \lambda_{2f} + \lambda_{2\omega} = \lambda_{2f,sc} + \lambda_{2f,sh} + \lambda_{2\omega,sc} + \lambda_{2\omega,sh} \tag{6}$$

where the index "sc" denotes the errors in scaling (magnitude), while the index "sh" denotes the error in shape as the error average between the corresponding target curve and the optimally scaled simulator curve.

The response of all MCA was tested for a left-right steering maneuver with a constant velocity of 6 m/s [5], thus featuring pure lateral acceleration (which leads to a value of zero for $\lambda_{1\omega,sc}$ and $\lambda_{2\omega,sc}$). All rides were evaluated by the criteria defined above, and the results are shown in Figs. 4 and 5a, b. As can be seen, the optimal control approach ZyRo-K with optimized parameters yields the best results in all criteria and sub-criteria. This is interesting as it is even better than the much more involved model predictive control approaches MPC-1, MPC-2, MPC-3, which take typical computation times of approximately 30 min, as compared to 1.5 min for the ZyRo-K, both on a desktop with quadcore 2.66 GHz. Also, the results show the very poor performance of the classical and adaptive algorithms, which may explain the bad subjective experiences one has when riding this type of simulators.

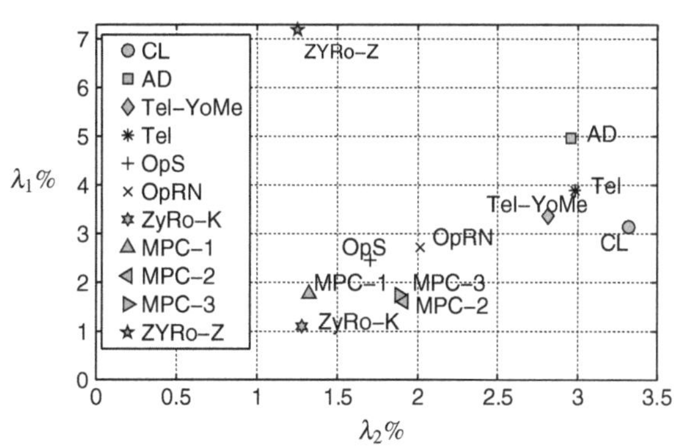

Fig. 4 Comparison of performance index criteria λ_1 and λ_2 for all assessed MCA

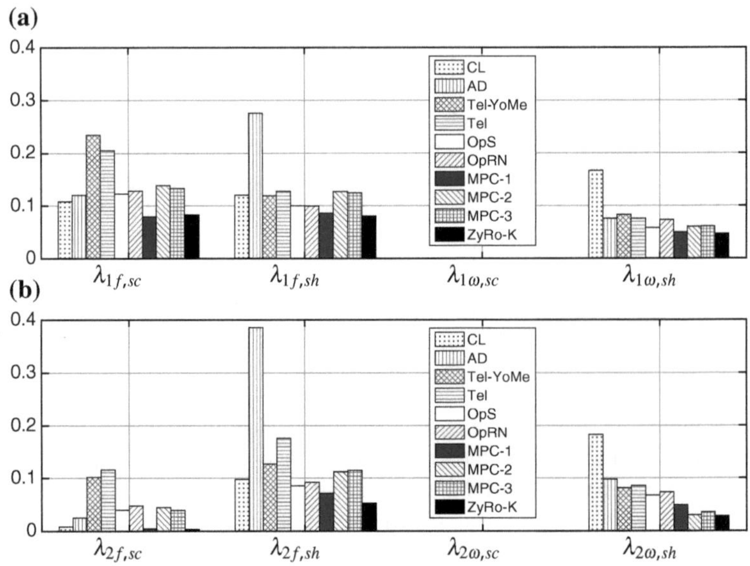

Fig. 5 Comparison of good indicator criterion λ_1, in (**a**) and λ_2, in (**b**)

5 Conclusions

The present paper reviews all (ten) currently existing motion cueing algorithms for one dimensional motion of ride simulators and compares their behaviour using Matlab implementations and virtual rides over a planar s-shaped curve with lateral acceleration only. The performance evaluation is carried out using a "good criterion" in which the errors between the target and simulator specific force and the target and simulator angular velocities are evaluated. The comparison shows that the optimal control approach together with a newly introduced optimization of the optimal control parameters based on the given trajectory yields the best results. Moreover, model predictive control algorithms are shown to be close to the best optimal control approach, but render much longer computational times to obtain the desired motion trajectories. The comparison in this paper is restricted to offline trajectories, as most employed ride motion shaping techniques require the knowledge of the complete future. For online ride simulator control, only the classical washout algorithm with limited adaptive optimization can be used. These show, however, very poor performance in terms of the "good criterion", partly explaining the poor subjective sensation reported to take place in this kind of simulators. While in this paper only the numerical properties of the motion cueing algorithms have been addressed, corresponding MCA evaluation trials with human passengers are currently being prepared, by which the numerical predictions of sensations will be validated. This is planned for future publications.

References

1. Baseggio, M., Beghi, A., Bruschetta, M., Maran, F., Minen, D.: An MPC approach to the design of motion cueing algorithms for driving simulators. In: 2011 14th International IEEE Conference on Intelligent Transportation Systems (ITSC), pp. 692–697. IEEE (2011)
2. Fischer, M.: Motion-cueing-algorithmen für eine realitätsnahe bewegungssimulation. Ph.D. thesis, Deutsches Zentrum für Luft und Raumfahrt in der Helmholz Gemeinschaft, DLR (2009)
3. Giordano, P.R., Masone, C., Tesch, J., Breidt, M., Pollini, L., Bülthoff, H.: A novel framework for closed-loop robotic motion simulation part ii: motion cueing design and experimental validation. In: 2010 IEEE International Conference on Robotics and Automation (ICRA), pp. 3896–3903. IEEE (2010)
4. Kecskeméthy, A.: MOBILE 1.3 Users Guide. Technical report, Lab of Mechanics Und Robotics, University of Duisburg Essen (2002)
5. Kecskeméthy, A., Tändl, M.: Physiologisch optimierte interfaces für die hochdynamische bewegungsgenerierung in physikalischen simulatorsystemen. Technical report, Lab of Mechanics und Robotics, University of Duisburg Essen (2009)
6. Nahon, M.A., Reid, L.D.: Simulator motion-drive algorithms–a designer's perspective. J. Guid. Control Dyn. **13**(2), 356–362 (1990)
7. Nehaoua, L., Mohellebi, H., Amouri, A., Arioui, H., Espié, S., Kheddar, A.: Design and control of a small-clearance driving simulator. Veh. Technol. IEEE Trans. **57**(2), 736–746 (2008)
8. Parrish, R.V., Dieudonne, J.E., Martin Jr, D.J.: Coordinated adaptive washout for motion simulators. J. Aircr. **12**(1), 44–50 (1975)
9. Pouliot, N.A.: Clé, Gosselin, m.M., Nahon, M.A.: Motion simulation capabilities of three-degree-of-freedom flight simulators. J. Aircr. **35**(1), 9–17 (1998)
10. Reid, L., Nahon, M.: Flight Simulation Motion-base Drive Algorithimns: Part 1 Developing and Testing the Equations. Toronto University, Institute for Aerospace Studies (1985)
11. Schweig, S., Kammers, H.: Bewegungssteuerung eines RoboCoaster Kuka Roboters zur Ride Simulation mit Hilfe von Washout Filtern. Project report, University of Duisburg Essen, Duisburg, Germany (2011)
12. Sivan, R., Ish-Shalom, J., Huang, J.K.: An optimal control approach to the design of moving flight simulators. Syst. Man Cybern. IEEE Trans. **12**(6), 818–827 (1982)
13. Telban Robert, J., Cardullo, F.M.: Motion cueing algorithm development: human-centered linear and nonlinear approaches. Tech. rep., NASA/CR-2005-213747 (2005)
14. Zywiol, H.J., Romano, R.: Motion drive algorithms and simulator design to study motion effects on infantry soldiers. In: Driving Simulation Conference, North America 2003 (DSC-NA 2003) (2003)

Structural Body Stiffness Influence on the Vehicle Dynamic Behavior

Antonio Carlos Botosso and Tarcísio A. Hess Coelho

Abstract The influence of vehicle torsional stiffness on the lateral load transfer has been largely studied and are present on many publications. This work aims to go further on this evaluation, not only demonstrating how the lateral load transfer varies when the vehicle is subjected to a lateral acceleration but also evaluating, based on a robust engineering method (Taguchi), in which situation the vehicle response to external factors is significantly influenced by structural stiffness.

Keywords Multibody · Vehicle dynamics · Mechanism stiffness · Body stiffness · Robust engineering

1 Introduction

The analytical approach for engineering problems is fundamental in the product development because of the speed of the analyses and relative low cost of high capacity computers compared to construction of physical prototypes and laboratory tests. The possibility of creating and developing highly optimized products using such equipment, have led companies to invest money and effort in developing Computer Aided Engineering (CAE) areas.

It is well known that vehicle dynamic behavior is influenced by vehicle structural stiffness. Torsional stiffness has been largely study as its influence on vehicle lateral load transfer when cornering [1–3]. This situation is critical for high performance vehicles since suspension tuning must be precisely and torsional stiffness is not measured and accounted easily [4]. In this situation is preferable refined

A.C. Botosso · T.A.H. Coelho (✉)
Polytechnic School of University of Sao Paulo, Sao Paulo, Brazil
e-mail: tarchess@usp.br

A.C. Botosso
e-mail: antonio.botosso@usp.br

B. Corves et al. (eds.), *Mechanisms, Transmissions and Applications*,
Mechanisms and Machine Science 31, DOI 10.1007/978-3-319-17067-1_15

models (multibody models) where vehicle structural stiffness can be easier evaluated.

The analytical modeling of multibody systems that considers a sequence of rigid bodies connected by joints, spring and damper elements has improved to have as the reliability of the physical model as possible and desirable. Although the models of complete vehicles have high complexity, they do not represent a barrier to the current computer processing capacity. This availability and improvement of representativeness have led to the addition of flexible bodies replacing the rigid elements.

The techniques of degrees of freedom reduction applied to the continuous bodies, which allowed flexible elements introduction on multibody systems, dating from the mid-20th century, highlighting the Craig-Bampton method [5] for its publication in the AIAA Journal in July 1968 "Coupling of substructures for Dynamic Analysis". The consideration of the flexible elements in the model may increase its complexity excessively due to the greater number of variables. This might cause troubles on the analysis and the system behavior understanding (it is harder to identify which variable really influences or changes a result).

In order to avoid unnecessary increment of complexity on the model, this paper seeks to understand and discuss comparatively and by robust engineering methodology, from a simplified analytical model and a full car model elaborated in multibody environment, in which situation vehicle dynamic analytical results would be significantly influenced by the consideration of structural stiffness.

2 Lateral Load Transfer Evaluation

Figure 1a, b shows a simplified analytical model (derivate from complete multibody model presented on Sect. 3) sufficient to obtain the equations of motion to evaluate the lateral load transfer as dependent of the vehicle body stiffness.

Where M_f and M_r are the front and rear sprung masses and hs_f and hs_r its height from the ground; m_f and m_r are the front and rear unsprung masses and hu_f and hu_r its height from the ground. Z_f and Z_r are the front and rear heights of the roll center that determines the roll axis. t_f and t_r are the front and rear track of the vehicle. L is the wheel base and K_f, K_r and K_T represents, respectively, front roll stiffness, rear roll stiffness and body torsional stiffness.

The assumptions to the analytical modeling are:

- No inertial forces considered.
- There is no gravity effect on the transversal displacement of the center of gravity due to the body roll.
- Roll center and centers of gravity are considered on XZ plane.
- Normal axis to roll axis are considered vertical (parallel to Z axis).

For lateral dynamic evaluation, Eqs. (1) and (2) are obtained by doing the static equilibrium for the vehicle subjected to a lateral acceleration a_y. Equations (1)

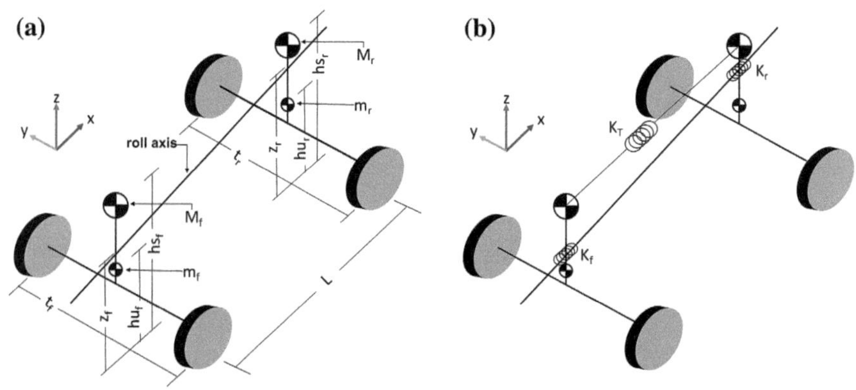

Fig. 1 **a** Masses and dimensions. **b** Springs model to represent stiffness

and (2) describes the front and rear lateral load transfer due to vehicle parameters and lateral acceleration.

$$\Delta N_f = \frac{2}{t_f} \cdot a_y \cdot \left[\left(M_f + m_f\right) \cdot z_f + M_f \cdot d_{sf} \cdot \frac{K_f}{K_f + \frac{K_r \cdot K_T}{K_r + K_T}} + M_r \cdot d_{sr} \cdot \frac{\frac{K_f \cdot K_T}{K_f + K_T}}{K_r + \frac{K_f \cdot K_T}{K_f + K_T}} - m_f \cdot d_{ur} \right] \tag{1}$$

$$\Delta N_r = \frac{2}{t_r} \cdot a_y \cdot \left[\left(M_r + m_r\right) \cdot z_r + M_f \cdot d_{sf} \cdot \frac{\frac{K_r \cdot K_T}{K_r + K_T}}{K_f + \frac{K_r \cdot K_T}{K_r + K_T}} + M_r \cdot d_{sr} \cdot \frac{K_r}{K_r + \frac{K_f \cdot K_T}{K_f + K_T}} - m_r \cdot d_{ur} \right] \tag{2}$$

By filling in the equations with parameters obtained on ADAMS® complete vehicle model (vehicle parameters are not relevant for this evaluation and they will not be described) and assuming a unit acceleration, it is possible to obtain the following graphs to understand how the body stiffness, suspension roll and mass distribution affect the lateral load transfer.

Figures 2 and 3 present how the proportion of front lateral load transfer with respect to the total lateral load transfer varies as a function of the proportion of front suspension roll stiffness for a given value of mass distribution and for 4 different values of proportion between body torsional stiffness and suspension total roll stiffness. This analysis is similar to proposed by [1].

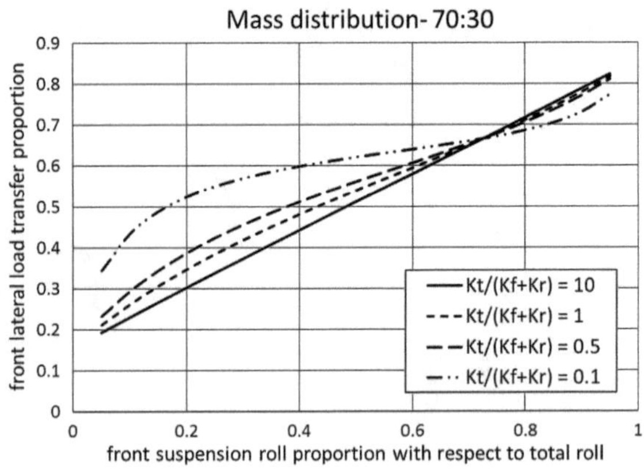

Fig. 2 LLT to 70:30 mass distribution

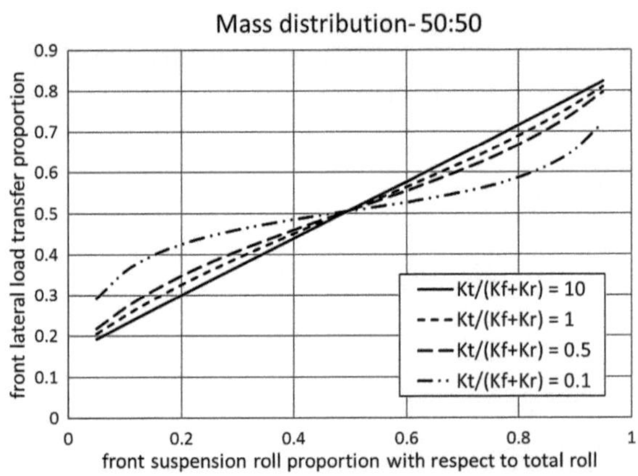

Fig. 3 LLT to 50:50 mass distribution

3 Sensitivity Study of Vehicle Stiffness

Robust Engineering Method, developed by prof. Genichi Taguchi was initially developed to ensure that products or processes were less subject to uncontrollable factors [6]. Similar to Vilela [7], this method can be adapted to the problem analyzed here because the idea of robustness implies the assessment of the influence level of each parameter on the results and how these results are affected by noises (Signal to Noise ratio). In this case, we are interested in identifying how sensible the vehicle is to the external factors by changing the structure stiffness.

The idea is to apply the robust engineering methodology for a stiffness influence evaluation. In that way, two different multibody models of a specific vehicle (in this case a Dodge Neon® with some assumptions due to lack of data specifications) were necessary; one considering only rigid bodies and other with body structure and subframe represented as flexible bodies. Both were built by using Altair Motion

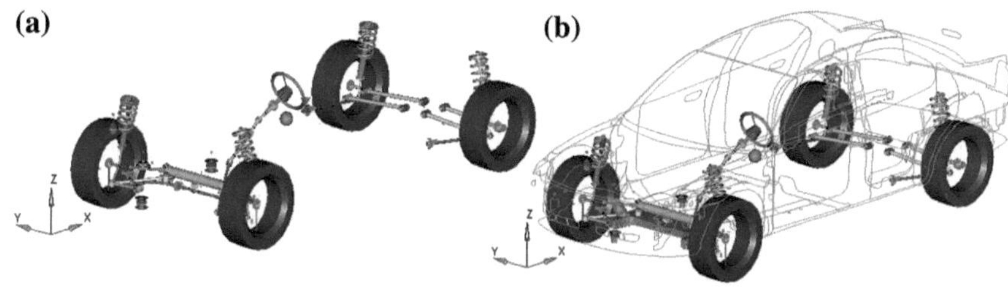

Fig. 4 **a** Rigid multibody vehicle mode. **b** Multibody model with flexible structure

View® as pre-processor and Altair Motion Solve® as solver. Figures below show both models. Figure 4a is the rigid vehicle model and Fig. 4b shows the multibody model with flexibility considerations of body structure and subframe.

3.1 Control Factors

In this problem, control factors will be the external excitations applied to the vehicle in order to evaluate which of them are more sensible to the structural stiffness variation. Table summarizes the variables and its levels (Table 1).

At that point, an arbitrary event is built in order to allow evaluation for static and dynamic condition of the vehicle. At this event, control factors will be applied to the system in the follow way.

While accelerations are linearly applied at time 0 s until the end, wheel travel are applied with a step function starting at 0 s until 0.2 s. Keeping the analyses running until 5 s to reach the system stabilization, will be possible to extract responses from the transient (0–2 s) and static portion (2–5 s) of the event.

Table 1 Control factors considered for robust analyses

	Unit	Level 1	Level 2	Level 3
X acceleration	mm/s2	−4000	0	4000
Y acceleration	mm/s2	−1500	0	1500
Z acceleration	mm/s2	−4000	0	4000
Right front wheel travel	mm	−90	0	90
Left front wheel travel	mm	−90	0	90
Right rear wheel travel	mm	−90	0	90
Left rear wheel travel	mm	−90	0	90

3.2 Noise Factors

Since the study purpose is to determine in which situation the vehicle is more sensible to the body stiffness variation, the noise factor considered is the body and subframe stiffness and the levels are: both rigid and both flexible.

3.3 Experiments

To evaluate seven control factors at three levels for two noises, 4374 number of runs would be necessary as calculated by Eq. (3).

$$3^7 \times 2 = 4374 \tag{3}$$

To avoid unnecessary effort, it will be considered an orthogonal matrix that allows a statistic determination of the control factor influence on the system. In this case, to have necessary combinations, a L18 (18 runs combining all the control factors) will be chosen (further explanations about orthogonal matrix are found on [6]). Table 2 presents the considered matrix filled in with the control factors variables.

Table 2 L18 Matrix with all the runs arrangement

Run	Acc X	Acc Y	Acc Z	RF travel	LF travel	RR travel	LR travel
1	−4000	−1500	−4000	−90	−90	−90	−90
2	0	−1500	0	0	0	0	0
3	4000	−1500	4000	90	90	90	90
4	−4000	0	0	90	90	0	−90
5	0	0	4000	−90	−90	90	0
6	4000	0	−4000	0	0	−90	90
7	−4000	1500	−4000	90	0	90	0
8	0	1500	0	−90	90	−90	90
9	4000	1500	4000	0	−90	0	−90
10	−4000	−1500	4000	−90	0	0	90
11	0	−1500	−4000	0	90	90	−90
12	4000	−1500	0	90	−90	−90	0
13	−4000	0	4000	0	90	−90	0
14	0	0	−4000	90	−90	0	90
15	4000	0	0	−90	0	90	−90
16	−4000	1500	0	0	−90	90	90
17	0	1500	4000	90	0	−90	−90
18	4000	1500	−4000	−90	90	0	0

Where, RF, LF, RR and LR means, respectively, Right Front, Left Front, Right Rear and Left Rear.

For each run (each matrix line arrangement), responses from static and transient portion will be extracted. The responses are:

- From static portion:
  Vehicle roll (rotation around X axe), pitch (rotation around Y axe) and yaw (rotation around Z axe); and displacements in the X, Y and Z directions. All the responses measured at the vehicle center of gravity.
- From transient portion:
  The impulse at the front and rear top mounts measured by the sum of the vertical forces applied to the top mount from 0 to 1.2 s of the event.

In order to evaluate the noises, all the 10 responses for each one of the 18 runs will be performed twice, one for rigid vehicle model and other for the flexible one.

3.4 Noise Factor Consideration

The synthesis proposed by Taguchi says that it is possible to evaluate the noises effects statistically by the mean and its variance. This analysis is done by calculating a factor called Signal to Noise Ratio (SNR).

To adapt methodology to the actual system, we will look for to the highest value of SNR, where SNR is the variance between rigid and flexible bodies responses for the same run. In order to have a unique combined metric for static responses and another unique metric for transient response, all the values obtained on the runs will be normalized to a range of 0–1. In this way, it is possible to combine them by doing a simple sum for static metrics and for transient ones.

As we are evaluating the SNR between rigid and flexible bodies and we know that the variances has a low order compared to the nominal values, we will consider a function which gives a weight to the variances in order to evidence higher variances in detriment of lower values.

Therefore, the SNR function considered is:

$$SNR = -10/\log s^2 \tag{4}$$

where the S^2 is the variance between the metrics for both noises in a same run.

The effect of each level for all the control factors on the dynamic behavior will be calculated by the average of the responses were the level appears on the run arrangement. The control factor with highest influence on the variance between rigid and flexible bodies models will be that one with the highest differences among the averages calculated for their levels.

3.5 Sensitivity Response

By doing the approach presented before and representing the means graphically, we obtain the graphs presented on the Figs. 5 and 6.

First graph (Fig. 5) shows that the vehicle static behavior evaluation is more affected by the structural stiffness when the vehicle is subjected to external accelerations, specially the vertical acceleration. It means that in a static condition with

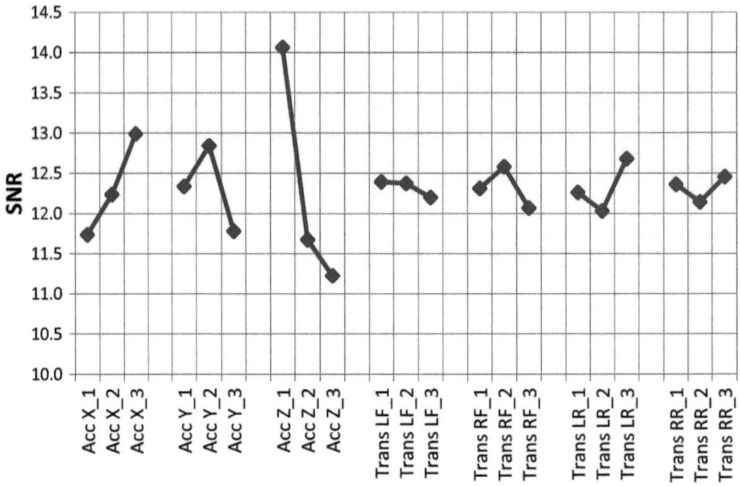

Fig. 5 Sensibility of SNR for static portion (displacement and rotation)

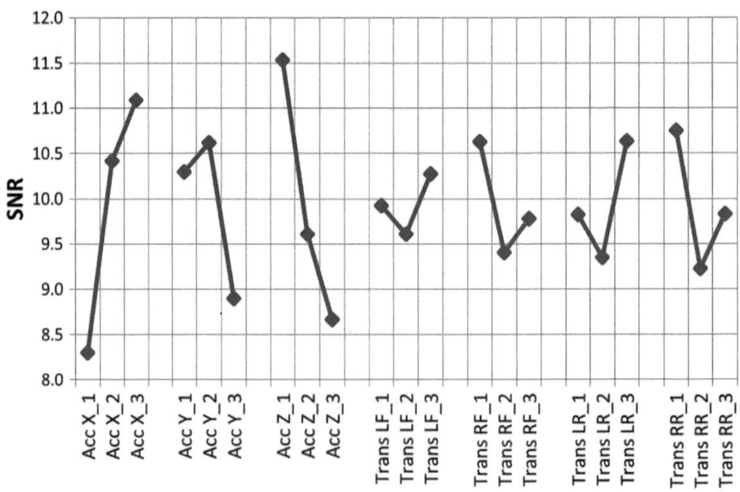

Fig. 6 Sensibility of SNR for dynamic portion (impulse on *top* mounts)

applied vertical acceleration, the measured CG position will vary between different stiffness of structure. Similarly, for the transient condition as we see at the Fig. 6, the impulse responses are more sensible to the structural stiffness variation when the vehicle is subjected to external accelerations.

4 Conclusions

Based on the lateral load transfer analyses, two important information were obtained by graphs on Figs. 2 and 3. First is that, for a given proportion of front suspension roll, lateral load transfer vary as a function of the body stiffness. In the other hand, as much close to the mass distribution proportion the value of proportion of front suspension roll stiffness is, less influence the body stiffness has on the lateral load transfer. Then, the effect of the lateral load transfer variation, evidenced by a low structural stiffness, can be minimized by keeping the suspension roll stiffness ratio approximately equal to mass distribution ratio. Furthermore, this condition is generally observed on the vehicles because springs definition, and consequently the suspension roll, are generally proportional to the mass distribution. It means that an adequate suspension tuning should drives a reasonable ratio equilibrium between suspension roll stiffness and mass distribution.

Seeing that the mechanism stiffness, in this case the body torsional stiffness, influences lateral load transfer and that the amount of influence depends on the mechanism configuration, a more generic evaluation, considering a full model vehicle, is necessary to determine how the vehicle dynamic behavior (not only lateral load transfer) can be influenced by mechanism stiffness.

Based on that, the sensitivity study showed that the dynamic behavior is more influenced by structural stiffness when it is subjected to accelerations, mainly the vertical. It means that, for maneuver that the vertical acceleration is high or comparatively higher than other external factors, like pothole or ride events, the responses can vary depending on the body stiffness. For handling analyses, with low accelerations and displacements, the considerations of structural stiffness are not relevant and would increase model complexity unnecessarily. Although results are specific for evaluated car, the methodology can be extended for other vehicles.

References

1. Deakin, A., Ramirez, P.-R., Hanley, R.: The effect of chassis stiffness on race car handling balance (SAE 2000-01-3554). In: SAE Congress, Dearborn (2000)
2. Arant, M.-O.: Assessing the Effect of Chassis Torsional Stiffness on the Accuracy of Heavy Vehicle Understeer and Rollover Modeling. School of Clemson University [S.l.] (2010)
3. Vargas, V.-A.-D.: Efeitos da Flexibilidade Estrutural em Simulações de Dinâmica Lateral de Veículo de Transporte de Carga. Universidade Federal do Rio Grande do Sul, Porto Alegre (2011)

4. Milliken, W.-F., Milliken, D.-L.: Race Car Vehicle Dynamics. SAE, Warrendale (1995)
5. Craig, R.-R., Bampton, M.-C.-C.: Coupling of substructures for dynamic analyses. AIAA J. **6**, 1313–1319 (1968)
6. Myers, R.-H., Montgomery, D.-C., Anderson-Cook, C.-M.: Response Surface Methodology. Wiley, Hoboken (2009)
7. Vilela, D.: Aplicação de Métodos Numéricos de Otimização ao Problema Conjunto da Dirigibilidade e Conforto Veicular, p. 358. Universidade de São Paulo, São Paulo (2010)

Case Study Regarding a New Knee Orthosis for Children with Locomotion Disabilities

Cristian Copilusi, Alexandra Margine and Nicolae Dumitru

Abstract This research addresses to children locomotion system for designing special orthosis and prosthesis mechanical systems. The research aim is to obtain the motion laws developed by a child locomotion system and with these, the knee joint connection forces will be obtained through an analytical method for walking activity. These parameters are useful for an orthotic system design of a 7 years old child. The research is based on an experimental analysis developed with ultra high-speed video equipment on 20 children and a dynamic analysis by using Newton-Euler method completed with Lagrange multipliers. The obtained dynamic parameters are used as input data for a new knee orthosis design. Virtual simulations are performed by using MSC Adams which validates a new knee orthosis prototype.

Keywords Dynamics · Locomotion system · Virtual simulations · Knee orthosis

1 Introduction

Nowadays different rehabilitation systems were designed but these addresses to adults and elderly persons. It is well known that these systems are limited in children case, due to the fact that a child is in a continuous growth. Basically, specific companies and research centers needs to develop parameterized models for any child's age and anthropomorphic data. It can be mentioned similar research work in this field by [1–4]. Similar dynamic analyses with remarkable result can be found in [5–8]. Due to this fact modular orthotics devices and parameterized prosthetics can be developed in order to improve the locomotion system and to satisfy the ability to move in case of

C. Copilusi (✉) · A. Margine · N. Dumitru
University of Craiova, Craiova, Romania
e-mail: cristache03@yahoo.co.uk

A. Margine
e-mail: fam_margine@yahoo.com

N. Dumitru
e-mail: nicolae_dtru@yahoo.com

B. Corves et al. (eds.), *Mechanisms, Transmissions and Applications*,
Mechanisms and Machine Science 31, DOI 10.1007/978-3-319-17067-1_16

children with age between 4 and 7 years. As starting point on this research a state-of-the-art was performed and regards, the dynamic analyses performed by other researchers and also similar rehabilitation solutions especially designed for children have been studied. In order to develop and perform a dynamic analysis of a human locomotion system, it is necessary to use some parameters obtained on experimental way. For this on the second section of this research, an experimental analysis on children locomotion system was performed. With the obtained data, on third section a human locomotion system dynamic model was developed which can be parameterized accordingly to a child between 4 and 7 years size and weight. On this section, a dynamic analysis was performed by using Newton Euler method completed with Lagrange multipliers for obtaining the knee connection forces. With the obtained knee connection forces, virtual simulations of a new knee orthosis concept were performed with the aid of MSC Adams software. These are described in the fourth section of this research. Important data were obtained, and this are synthesized on the final conclusions of this research through the fourth section frame.

2 Locomotion System Experimental Analysis

For the experimental analysis, a high-speed video analysis equipment was used. This equipment called CONTEMPLAS is owned by the University of Craiova-Faculty of Mechanics [9]. The human locomotion system experimental analysis was performed on a number of 20 human healthy subjects with the age between 4 and 7 years. A database was obtained and this consists on angular amplitudes from hip, knee and ankle joints during walking. Thus, on this research frame only essential data were extracted and these are represented by human knee joint angular amplitude in case of a 7 years old child. The experimental analysis principle consists on attaching of a three reflective markers centered on the interest joints and tracking them with the equipment software aid. The attached markers are shown in Fig. 1, and the acquired knee joint angle variation during walking is presented on Fig. 2.

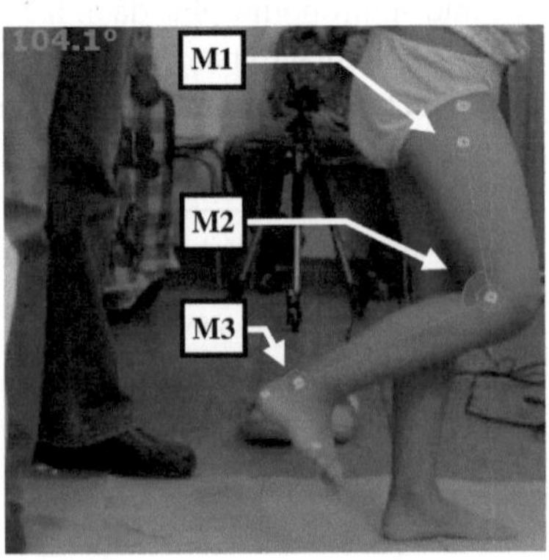

Fig. 1 Markers identification during walking tests in Craiova

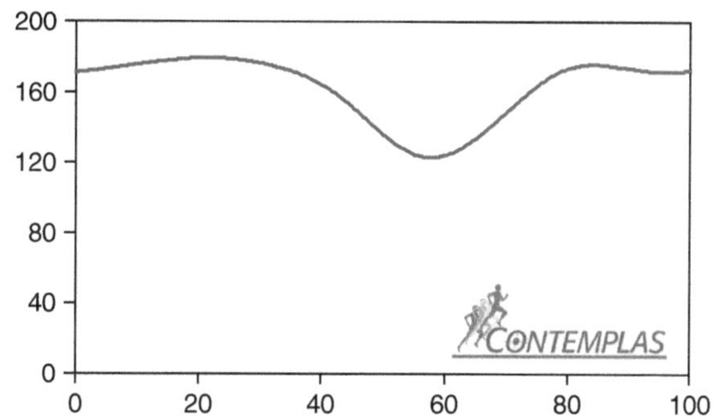

Fig. 2 Acquired knee joint angle (degrees) versus Gait (%) during a test

The analyzed child performs a complete gait cycle during walking in a period of 0.6 s. This period is important to know and to retain for dynamic computations and also virtual simulations.

3 Locomotion System Dynamic Analysis

By having on sight similar dynamic models of the human locomotion system, an inverse dynamic analysis based on Newton-Euler method completed with Lagrange multipliers was performed [10–12]. This was possible by elaborating a dynamic model which is shown on Fig. 3. The input data for the inverse dynamic analysis were considered known and were represented by the geometric elements ($\mathbf{L_{OT}}$, $\mathbf{L_1}$, $\mathbf{L_2}$,…, $\mathbf{L_{16}}$) and generalized coordinates variation laws from kinematic joints: $\mathbf{q_1}$, $\mathbf{q_2}$, $\mathbf{q_3}$,…, $\mathbf{q_{16}}$ obtained with the CONTEMPLAS equipment's aid. A computational algorithm was elaborated with MAPLE software's aid. Through this, it will be followed to obtain the connection forces components, which will appear in the walking activity, for a gait cycle, at the each joint level from the mathematical model structure. This is equivalent to human locomotion system. The constraint equations are:

$$\varphi(q, t) = 0 \tag{1}$$

$\mathbf{q_r}$-Generalized coordinates vector considered when the elements are rigid ones; t-time. The motion equation has the following form:

$$\begin{bmatrix} \boldsymbol{M} & \boldsymbol{J}_q^T \\ \boldsymbol{J}_q & 0 \end{bmatrix} \begin{bmatrix} \ddot{q} \\ \lambda \end{bmatrix} = \begin{bmatrix} \boldsymbol{Q}_a \\ \boldsymbol{a} \end{bmatrix} \tag{2}$$

Where: M, represents the mass matrix with: $M = diag\ (m_i,\ J_i)$; $i = \overline{1, 16}$. We obtain the Lagrange's multipliers are:

$$\lambda = [\mathbf{J_q}]^{-1}[\mathbf{Q_a} - \mathbf{M} \cdot \ddot{\mathbf{q}}] \tag{3}$$

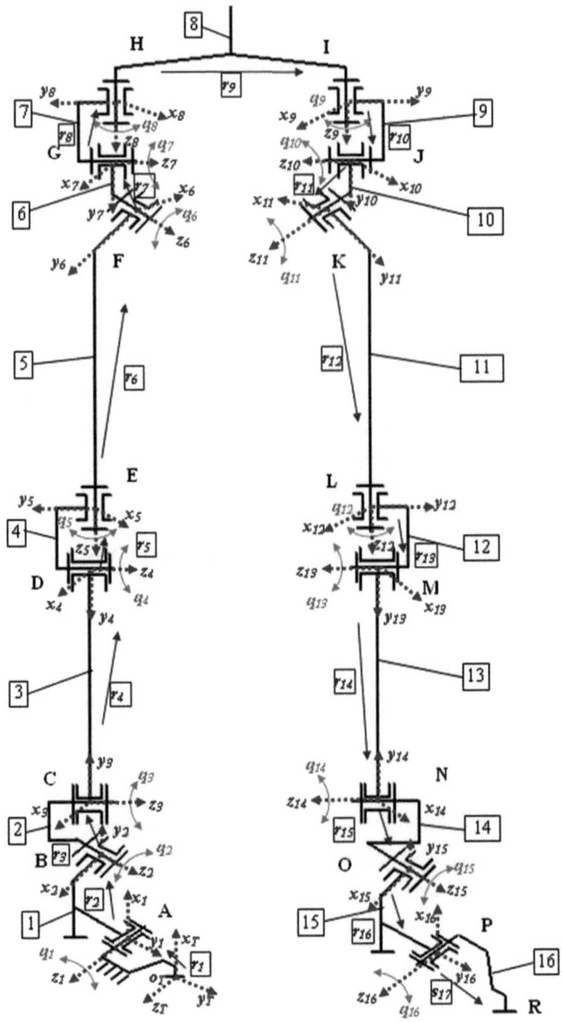

Fig. 3 Human locomotion system dynamic model

The motion laws are known: $\mathbf{q(t)}$, $\dot{\mathbf{q}}(t)$ and $\ddot{\mathbf{q}}(t)$, from the experimentally analysis accomplished in dynamic mode. From Eq. (3) relations we determine λ Lagrange's multipliers, with the aid of a programming algorithm accomplished in MAPLE software. With these it will be process an inverse dynamic analysis from we obtain kinematic joint connection forces. These forces were determined by taking into account the Lagrange multipliers:

$$\mathbf{F}_{\mathbf{i}}^{\prime\prime\mathbf{r}(\mathbf{i},\mathbf{j})} = \left[\mathbf{R}_{\mathbf{i},\mathbf{i}''}\right]^{\mathbf{T}} \cdot \left[\mathbf{A}_{\mathbf{oi}}\right]^{\mathbf{T}} \cdot \left[\boldsymbol{\lambda}\right]^{\mathbf{r}(\mathbf{i},\mathbf{j})} \tag{4}$$

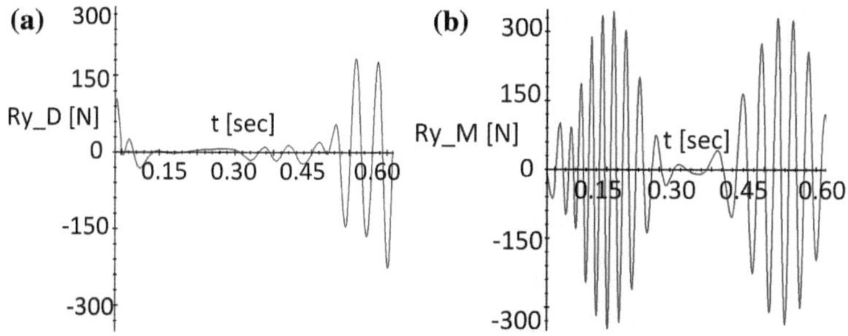

Fig. 4 Computed connection forces for knee joint **a** left lower limb, **b** right lower limb

Based on the elaborated algorithm the connection forces for each joint were obtained.

On this research only the knee joint connection forces were retained and these are shown in Fig. 4.

The force components for knee joint from Fig. 4 have high values and high oscillations due to the foot contact with ground when the analyzed subject performs a single gait during walking. The values presented on the first diagram from Fig. 4 were transformed on a polynomial function which will serve as input data for virtual simulations. Through this will actuate the orthosis and represents the actuator command and control function.

4 CAD Design and Dynamic Simulation

A mechanical design of a new knee orthosis especially designed for a 7 years old child is presented in Fig. 5.

The structure was designed to permit adjustments of mechanism links in accordance with different human body constitution. A dynamic analysis has been worked out by using a proper model for operation tests in ADAMS environment [13] by using GSTIFF solver with an error of 1.0E−003 %. Contact, stiffness, dumping coefficients, and friction force have been defined accordingly as listed in Table 1 with links made of aluminum alloy. This new knee orthosis is composed from 3 modular links which are actuated through a single actuator by pulling two flexible wires. These flexible wires are guided through pulleys in order to perform a real knee flexion motion. The modular elements are in a direct circular contact each other. Basically the orthosis will simulate a roto-translational motion as a natural knee. Also there are two brackets used for fixing this orthosis onto child leg. During simulations, the femur bracket was considered fixed in space and on the opposite component, respectively on tibia bracket, a constant force was applied. The value of this force represents the equivalent weight of the child shank and foot segments, respectively 35 N.

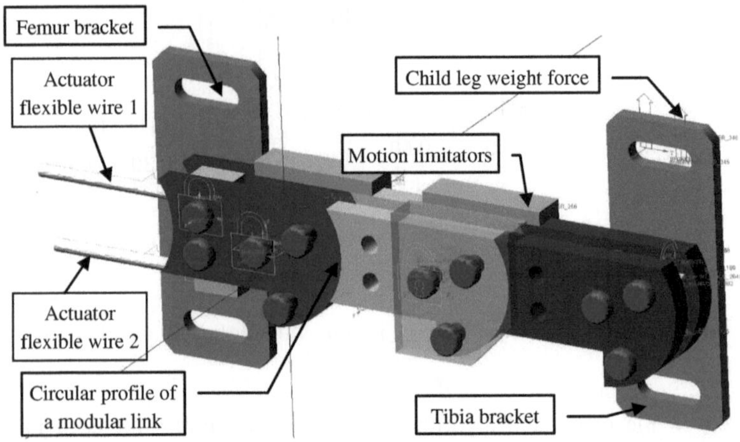

Fig. 5 A mechanical design of the proposed new knee orthosis

Table 1 Input parameters for dynamic simulation

Parameter	Type/value	Units	Parameter	Type/value	Units
Elasticity modulus	7.17054E + 004	N/mm^2	Contact μs	0.7	–
Density	2.74E−006	kg/mm^3	Contact μd	0.5	–
Penetration depth	0.1	mm	Force exponent	1.8	–
Friction force	Coulomb	N	Damping	40	N·s/mm

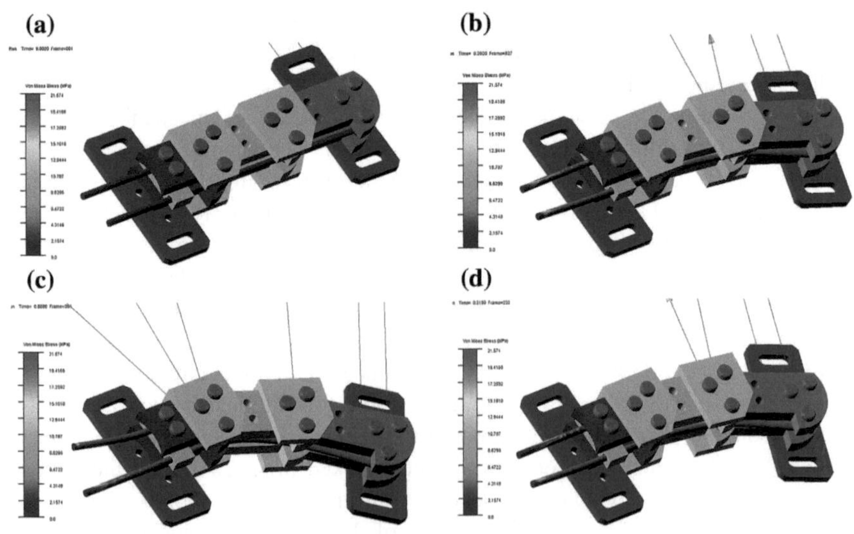

Fig. 6 A simulation sequence with ADAMS software for the new knee orthosis

Computed simulation sequences are represented in Fig. 6 and the main kinematic result is shown in Fig. 7. From Fig. 7 it can be observed that appropriate values are obtained as compared with those from experimental tests. The prototype validation has been made by obtaining the virtual model angular amplitude, respectively the tibia bracket. This is shown on Fig. 7 and it results an angular amplitude of 47°. In addition, on Fig. 8 the variation of the flexible wires Von Misses stress are shown.

The angular amplitude demonstrates the feasibility for manufacturing the new knee orthosis elements. These were manufactured after extracting the CAD drawings and choosing aluminium alloys as base material in order to be lightweight. The flexible wires are usual ones made from steel, provided from market and can hold a weight of 50 kg. The obtained prototype of the proposed knee orthosis is presented on Fig. 9. This was controlled through a servomotor unit GWS S03N 2BB type, with an ARDUINO Duemilanove board.

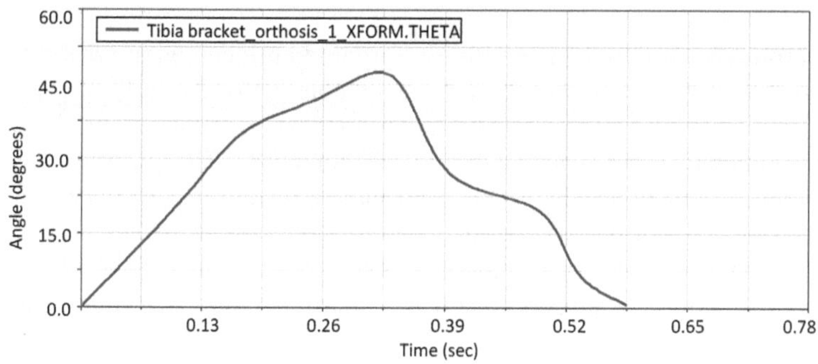

Fig. 7 Tibia bracket angle variation during simulation versus time

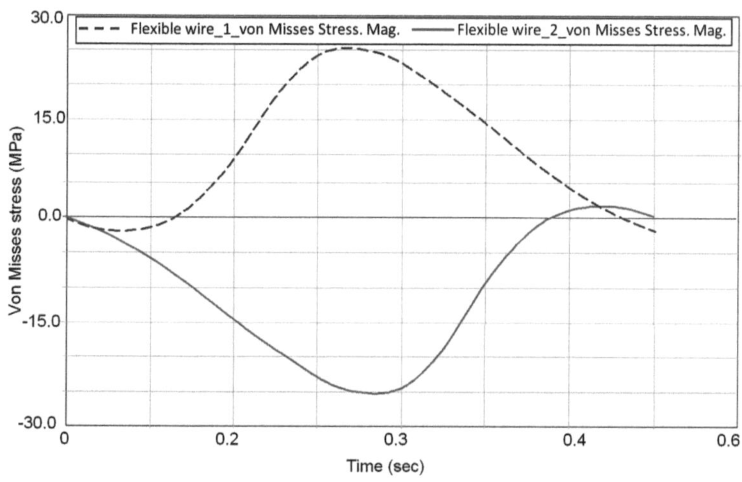

Fig. 8 Computed von Misses stress for flexible wires during simulation versus time

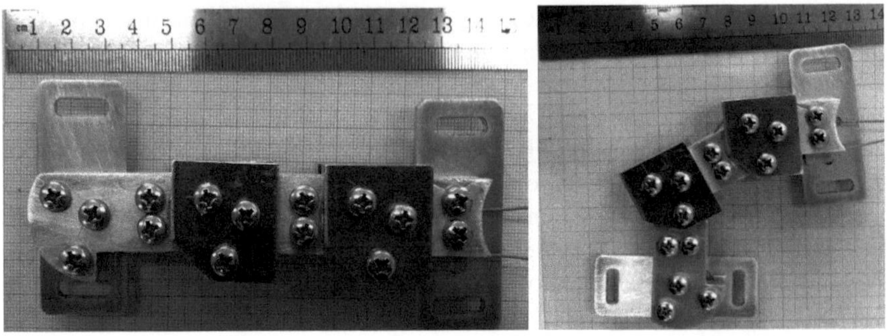

Fig. 9 Prototype of the proposed new knee orthosis

5 Conclusions

A new prototype of a new knee orthosis for children locomotion rehabilitation is designed. Simulation results have outlined suitable performance for suitable user-oriented operation in walking rehabilitation application, although its design may require additional components in future developments. The proposed design is fairly simple wearable and light with adjustable structure and its operation is run with only one actuator that can be regulated to fully help or partially assist a child walking, even with monitoring purposes. The actuator was controlled through a command and control acquisition board with a programming algorithm that can be modified in order to adapt any knee motion law developed by a healthy child with similar anthropomorphic data as the one placed on walking therapeutical recovery programs.

References

1. Copilusi, C., Ceccarelli, M., Carbone, G., Margine, A.: Mechanism of a leg exoskeleton for walking rehabilitation purposes. New advances in mechanisms, transmissions and applications. mechanisms and machine. Science **17**, 107–114 (2014)
2. Copilusi, C., Dumitru, N., Marin, M., Rusu, L.: Children orthotics and prostheses devices designed from cinematic and dynamic considerations. Eng. Lett J. **20**(4), 301–316 (2012)
3. Copilusi, C., Ceccarelli, M., Carbone G.: Design and Numerical Characterization of a New Leg Exoskeleton for Motion Assistance. Robotica Cambridge Press. Robotica. Available on CJO2014. doi:10.1017/S0263574714002069 (2014)
4. Copilusi, C., et al.: Design and simulation of a leg exoskeleton linkage for a human rehabilitation system. In: The 11th IFToMM International Symposium on Science of Mechanisms and Machines, Springer International Publishing, pp. 117–125 (2014)
5. Dejnabadi, H., Jolles, B.M., et al.: Estimation and visualization of sagittal kinematics of lower limbs orientation using body-fixed sensors. IEEE Trans. Biomed. Eng. **53**(7), 1385–1393 (2006)
6. Kiss, R.M., Kocsis, L., Knoll, Z.: Joint kinematics and spatial temporal parameters of gait measured by an ultrasound-based system. J. Med. Eng. Phys. **26**, 611–620 (2004)

7. Sohl, G.A., Bobrow, J.E.: A recursive multibody dynamics and sensitivity algorithm for branched kinematic chains. ASME J. Dyn. Syst. Meas. Control **123**(3), 391–399 (2001)
8. Wang, C.-Y.E., Bobrow, J.E.: Dynamic motion planning for the design of robotic gait rehabilitation. J. Biomech. Eng. **127**(4), 672–679 (2005)
9. CONTEMPLAS Motion Equipment. User manual. http://www.contemplas.com
10. Anderson, F.C., Pandy, M.G.: Dynamic Optimization of Human Walking. J. Biomech. Eng. **123**(5), 381–390 (2001)
11. Copilusi, C.: Researches regarding some mechanical systems applicable in medicine. Ph.D thesis, Faculty of Mechanics, Craiova (2009)
12. Dumitru, N., Nanu, G., Vintilă, D.: Mechanisms and mechanical transmissions. Modern and classical design techniques. Didactic printing house. ISBN 978-973-31-2332-3, Bucharest (2008)
13. Giesbers J.: Contact mechanics in MSC Adams. A technical evaluation of the contact models in multibody dynamics software MSC Adams. Bachelor thesis, University of Twente (2012)

Dynamic Modelling of Lower-Mobility Parallel Manipulators Using the Boltzmann-Hamel Equations

Oscar Altuzarra, Philipp Marcel Eggers, Francisco J. Campa, Constantino Roldan-Paraponiaris and Charles Pinto

Abstract Explicit Dynamic equations are needed for simulation and control in the field of Mechatronics. Several classical methods are available to get Dynamic equations. Regarding those methods that aim exclusively at the equations relating applied torques and motion generated, i.e. avoiding any calculus of joint wrenches, Analytical mechanics offers several approaches. For serial mechanisms, Lagrange equations are very convenient and systematic. However, finding such mathematical expressions can be cumbersome when facing closed-loop mechanisms even with the help of Lagrange multipliers. Moreover, this is quite complex if spatial rotations are considered, and hence, generalized coordinates are very much coupled in the expressions of Lagrange functions to be differentiated. Boltzmann-Hamel equations come to help in this regard. In this paper, authors show the finding of the dynamic equations of the 3*P̲RS* lower-mobility parallel manipulator using the Boltzmann-Hamel equations and exploring the effect of coupled freedoms in the rotation of the end-effector.

Keywords Boltzmann-Hamel equations · Lower-mobility parallel manipulators

1 Introduction

An explicit set of differential equations describing the dynamics in terms of the actuation variables (i.e. the set of generalized coordinates), the mass properties and the actuation forces, is necessary for the implementation of simulation and control on real time applications. Such set is called explicit or detailed form since inertia, gravitational and Coriolis terms are obtained separately and in closed form.

Dynamic equations in implicit form have been frequently obtained using Newton-Euler method and the principle of virtual work. Lagrangian formalism is a

O. Altuzarra (✉) · P.M. Eggers · F.J. Campa · C. Roldan-Paraponiaris · C. Pinto
University of the Basque Country UPV/EHU, Bilbao, Spain
e-mail: oscar.altuzarra@ehu.es

B. Corves et al. (eds.), *Mechanisms, Transmissions and Applications*,
Mechanisms and Machine Science 31, DOI 10.1007/978-3-319-17067-1_17

method reported to be very efficient in obtaining an explicit form of such equations in open chain mechanisms, but a prohibitive task for parallel manipulators because of the kinematic constraints due to closed loops [1]. However, exploiting less known capacities of Analytical Mechanics, such as quasi-velocities, Boltzmann-Hamel equations and the principle of energy equivalence, it is possible to find the aforementioned dynamic equations with a good computational efficiency in parallel mechanisms with full-mobility, i.e. six degrees of freedom (DOFs) [2– 4].

In this paper, authors aim to apply the same principles to the finding of explicit dynamic equations in lower-mobility parallel mechanisms, pointing out the effect of considering their constraints equations in the methodology.

2 Dynamic Modeling of the 3PRS

Approaching the aforementioned dynamic modeling in terms of input variables $\mathbf{q}_\rho = [\rho_1, \rho_2, \rho_3]^T$ using Lagrange equations means that we should be able to get the Lagrange function $\mathscr{L}_{mech}$ of every link of the mechanism in terms these variables to apply the so called Euler operator $\mathscr{E}_{\mathbf{q}_\rho}$ [5]:

$$\mathscr{E}_{\mathbf{q}_\rho}(\mathscr{L}_{mech}) = \frac{d}{dt}\frac{\partial \mathscr{L}_{mech}}{\partial \dot{\mathbf{q}}_\rho} - \frac{\partial \mathscr{L}_{mech}}{\partial \mathbf{q}_\rho} = \mathbf{f}_\rho = \mathbf{t}_\rho + \mathbf{Q}_\rho \tag{1}$$

so that we can get the generalized forces $\mathbf{f}_\rho$, that include the input torques $\mathbf{t}_\rho$ in terms of the input variables for known mass and inertia of the mechanism, and the effect of output wrenches represented by the generalized forces $\mathbf{Q}_\rho$. Such an explicit equation can be applied to control the input currents depending on the motion variables measured at the encoders.

However, parallel manipulators, as the 3PRS in Fig. 1, are characterized by a complex direct kinematics so finding such form of $\mathscr{L}_{mech}$ is often impossible. Nevertheless, Analytical Mechanics offers some alternatives.

The use of dependent coordinates can simplify the issue of the nonlinear nature of the direct problem. Essentially the use of such redundant coordinates means that we can analyze subsystems of the mechanism as if we had disassembled the closed chained mechanism, as long as we add later on the conditions of the assembled motion. This is done when using the Lagrange multipliers λ, but adding in the analysis the finding of those λ's. The Principle of energy equivalence is an alternative to this that avoids such additional calculus. The 3PRS parallel mechanism will be analyzed by taking apart the end-effector and the limbs.

Spatial rotations add to the complexity of $\mathscr{L}_{mech}$ the fact that the kinetic energy results in terms of the derivatives of angular coordinates multiplied by trigonometric functions. The use of quasi-velocities as generalized coordinates simplifies

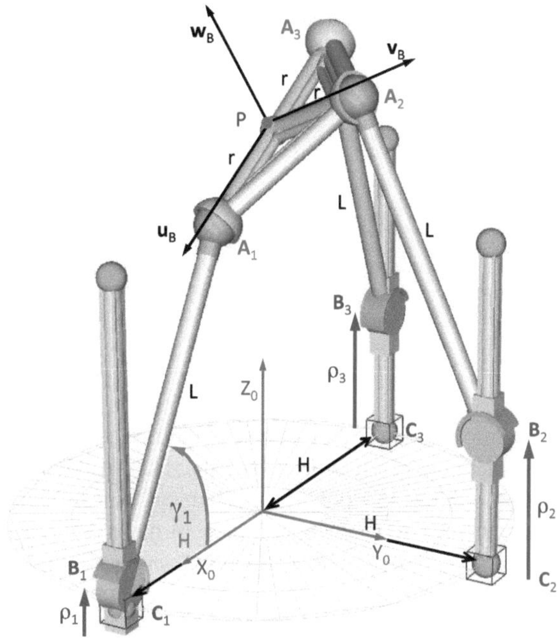

Fig. 1 Dynamic modelling of the 3PRS mechanism

such expression requiring the use of the Boltzmann-Hamel equations instead of Euler's operator. The 3PRS's end-effector will be analyzed with this approach, while Euler's operator will be used for the limbs moving on fixed vertical planes.

2.1 Principle of Energy Equivalence

The mechanism is split up into N free-body subsystems represented by their generalized coordinates $\mathbf{q}_{b_i}$. The condition that individuals subsystems b_i move as they were part of the assembled mechanism implies that $\mathbf{q}_b$, the set of all the generalized coordinates $\mathbf{q}_{b_i}$ pertaining to every subsystem b_i, is a function of the generalized coordinates chosen for the assembled mechanism, in our case the input variables $\mathbf{q}_\rho$. Furthermore, the virtual displacements are related by:

$$\delta\mathbf{q}_b = \frac{\partial \mathbf{q}_b}{\partial \mathbf{q}_\rho}\,\delta\mathbf{q}_\rho = \mathbf{J}\,\delta\mathbf{q}_\rho; \quad \delta\mathbf{q}_{b_i} = \mathbf{J}_{b_i}\delta\mathbf{q}_\rho \quad i = 1,\ldots,N \tag{2}$$

where the Jacobians $\mathbf{J}$ and $\mathbf{J}_{b_i}$ are found with the Kinematic analysis, showing the interdependence of kinematic parameters in the assembled mechanism.

The virtual work done by the assembled system and the set of subsystems is the same:

$$\delta W_{\rho} = \delta W_{b} = \sum_{i=1}^{N} \delta W_{b_i} \tag{3}$$

$$\delta \mathbf{q}_{\rho}^{T} \mathbf{f}_{\rho} = \delta \mathbf{q}_{b}^{T} \mathbf{f}_{b} = \sum_{i=1}^{N} \delta \mathbf{q}_{b_i}^{T} \mathbf{f}_{b_i} \tag{4}$$

$$\mathbf{f}_{\rho} = \mathbf{J}^{T} \mathbf{f}_{b} = \sum_{i=1}^{N} \mathbf{J}_{b_i}^{T} \mathbf{f}_{b_i} \tag{5}$$

what means that we can find the dynamic equations for each subsystem apart and then add them up. When considering the disassembled subsystems, joint forces appear at points where we split the mechanism. Such forces produce a virtual work on a subsystem that is cancelled out with the work produced in the adjacent subsystem when adding all dynamic equations, and hence there is no need to finding them.

2.2 Lagrange Equations

Let's define subsystems $i = 1, \ldots, 3$ with every limb $B_i A_i$ as free bodies moving on the corresponding vertical planes, being their local generalized coordinates $\mathbf{q}_{b_i} = \left[x_{b_i}, y_{b_i}, \gamma_{b_i}\right]^T$, position of the centre of mass and the orientation of the body defined with a local reference system Σ_{b_i}. Upon application of Euler's operator $\mathscr{E}_{\mathbf{q}_{b_i}}(\mathscr{L}_{b_i})$ we get:

$$\mathbf{f}_{b_i} = \begin{Bmatrix} 0 \\ F_{b_i} \\ 0 \end{Bmatrix} = \mathscr{E}_{\mathbf{q}_{b_i}}(\mathscr{L}_{b_i}) = \begin{bmatrix} m_{b_i} & 0 & 0 \\ 0 & m_{b_i} & 0 \\ 0 & 0 & I_{b_i} \end{bmatrix} \ddot{\mathbf{q}}_{b_i} + \begin{Bmatrix} 0 \\ m_{b_i} g \\ 0 \end{Bmatrix} = \mathbf{M}_{b_i} \ddot{\mathbf{q}}_{b_i} + \mathbf{g}_{b_i} \tag{6}$$

Using the kinematic relationship $\dot{\mathbf{q}}_{b_i} = \mathbf{J}_{b_i} \dot{\mathbf{q}}_{\rho}$ we obtain:

$$\mathbf{f}_{b_i} = \mathbf{M}_{b_i} \mathbf{J}_{b_i} \ddot{\mathbf{q}}_{\rho} + \mathbf{M}_{b_i} \dot{\mathbf{J}}_{b_i} \dot{\mathbf{q}}_{\rho} + \mathbf{g}_{b_i} \tag{7}$$

And the contribution of subsystem i to the mechanism's dynamics in terms of the actuation variables is obtained with the projection in Eq. (5):

$$\mathbf{f}_{\rho_{b_i}} = \mathbf{J}_{b_i}^{T} \mathbf{f}_{b_i} = \mathbf{J}_{b_i}^{T} \mathbf{M}_{b_i} \mathbf{J}_{b_i} \ddot{\mathbf{q}}_{\rho} + \mathbf{J}_{b_i}^{T} \mathbf{M}_{b_i} \dot{\mathbf{J}}_{b_i} \dot{\mathbf{q}}_{\rho} + \mathbf{J}_{b_i}^{T} \mathbf{g}_{b_i} \tag{8}$$

where matrices are operated and renamed as:

$$\mathbf{f}_{\rho_{b_i}} = \mathbf{I}_{b_i}\ddot{\mathbf{q}}_\rho + \mathbf{c}_{b_i} + \bar{\mathbf{g}}_{b_i} \tag{9}$$

being the inertia matrix $\mathbf{I}_{b_i}$ and gravitational vector $\bar{\mathbf{g}}_{b_i}$ only functional of the position of the mechanism, while the second term on the right contains the quadratic terms in velocities $\mathbf{c}_{b_i}$. Upon addition of Eq. (9) for $i = 1,\ldots,3$ we get the contribution of the limbs to the Dynamics of the mechanism:

$$\mathbf{t}_\rho = \mathbf{I}_{Ls}\ddot{\mathbf{q}}_\rho + \mathbf{c}_{Ls} + \bar{\mathbf{g}}_{Ls} \tag{10}$$

2.3 Boltzmann-Hamel Equations

The end-effector has a constrained spatial motion that involves translation and rotation. The principle of energy equivalence allows us to consider separately both motions by using a virtual mass at the centre of mass of the mobile platform subjected to translation, and a virtual inertial body subjected to rotations.

The contribution of the first virtual mass to the Dynamics can be solved using the Lagrange equations mentioned above. Considering the coordinates of the centre of mass as generalized coordinates $\mathbf{q}_G = [x_G, y_G, z_G]^T$, an applied external force at that point $\mathbf{f}_G$, and the kinematic relationship with the input variables through the corresponding Jacobian $\mathbf{J}_G$ we can get:

$$\mathbf{J}_G^T\mathbf{f}_G = \mathbf{J}_G^T\mathbf{M}_G\mathbf{J}_G\ddot{\mathbf{q}}_\rho + \mathbf{J}_G^T\mathbf{M}_G\dot{\mathbf{J}}_G\dot{\mathbf{q}}_\rho + \mathbf{J}_G^T\mathbf{g}_G \tag{11}$$

$$\mathbf{J}_G^T\mathbf{f}_G = \mathbf{I}_G\ddot{\mathbf{q}}_\rho + \mathbf{c}_G + \bar{\mathbf{g}}_G \tag{12}$$

The contribution of the virtual inertial body can be approached more effectively using the components of the angular velocity $\mathbf{w}$ in the moving frame as quasi-velocities for the application of the Boltzmann-Hamel equations. Such quasi-velocities $\mathbf{w}$ are dependent on the Euler angles and their derivatives $\mathbf{w}(\dot{\mathbf{q}}_e, \mathbf{q}_e)$, with an expression as:

$$\mathbf{w} = \mathbf{W}\mathbf{q}_e = \begin{bmatrix} \cos\phi & \sin\theta\sin\phi & 0 \\ -\sin\phi & \sin\theta\cos\phi & 0 \\ 0 & \cos\theta & 1 \end{bmatrix} \begin{Bmatrix} \dot{\theta} \\ \dot{\psi} \\ \dot{\phi} \end{Bmatrix} \tag{13}$$

Also, loop-closure and constraint equations in the mechanism's Kinematics provide an expression of the angular velocity $\mathbf{w}$ in terms of the input variables through the corresponding Jacobian $\mathbf{w} = \mathbf{J}_R\dot{\mathbf{q}}_\rho$. Hence, a relationship can be found between the derivatives of Euler angles and the inputs $\dot{\mathbf{q}}_e = \mathbf{W}^{-1}\mathbf{J}_R\dot{\mathbf{q}}_\rho = \mathbf{J}\mathbf{q}_\rho$.

The Lagragian function is:

$$\mathscr{L}_{mp} = \frac{1}{2}\mathbf{w}^T\mathbf{I}_{mp}\mathbf{w} \tag{14}$$

and the Boltzmann-Hamel equation projected onto the input variables is:

$$\mathbf{J}_R^T\mathbf{m}_{mp} = \mathbf{J}^T\left[\left(\frac{\partial \mathbf{w}}{\partial \dot{\mathbf{q}}_e}\right)^T \cdot \left(\frac{d}{dt}\frac{\partial \mathscr{L}_{mp}}{\partial \mathbf{w}}\right) + \frac{d}{dt}\left(\frac{\partial \mathbf{w}}{\partial \dot{\mathbf{q}}_e}\right)^T \cdot \frac{\partial \mathscr{L}_{mp}}{\partial \mathbf{w}} - \left(\frac{\partial \mathbf{w}}{\partial \mathbf{q}_e}\right)^T \cdot \frac{\partial \mathcal{L}_{mp}}{\partial \mathbf{w}}\right] \tag{15}$$

where the first term contains the projection of the applied external moments expressed in the moving frame $\mathbf{m}_{mp}$ onto the jointspace.

Upon differentiation of Eq. (14) and substitution of Eq. (13) where appropriate we get:

$$\mathbf{J}_R^T\mathbf{m}_{mp} = \mathbf{J}^T\left[\mathbf{W}^T\mathbf{I}_{mp}\dot{\mathbf{w}} + \dot{\mathbf{W}}^T\mathbf{I}_{mp}\mathbf{w} - \left(\frac{\partial \mathbf{w}}{\partial \mathbf{q}_e}\right)^T\mathbf{I}_{mp}\mathbf{w}\right] \tag{16}$$

It can be demonstrated that:

$$\left(\frac{\partial \mathbf{w}}{\partial \mathbf{q}_e}\right) = \mathbf{A} - \dot{\mathbf{W}} \tag{17}$$

with

$$\mathbf{A} = \left(\frac{\partial \dot{\mathbf{w}}}{\partial \dot{\mathbf{q}}_e}\right) \tag{18}$$

and then Eq. (16) can be expressed as:

$$\mathbf{J}_R^T\mathbf{m}_{mp} = \mathbf{J}^T\left[\mathbf{W}^T\mathbf{I}_{mp}\dot{\mathbf{w}} + 2\dot{\mathbf{W}}^T\mathbf{I}_{mp}\mathbf{w} - \mathbf{A}^T\mathbf{I}_{mp}\mathbf{w}\right] \tag{19}$$

Considering Eq. (13) and its derivative with respect to time $\dot{\mathbf{w}} = \dot{\mathbf{W}}\dot{\mathbf{q}}_e + \mathbf{W}\ddot{\mathbf{q}}_e$ into Eq. (16), we get:

$$\mathbf{J}_R^T\mathbf{m}_{mp} = \mathbf{J}^T\left[\mathbf{W}^T\mathbf{I}_{mp}\mathbf{W}\right]\ddot{\mathbf{q}}_e + \mathbf{J}^T\left[\mathbf{W}^T\mathbf{I}_{mp}\dot{\mathbf{W}} + 2\dot{\mathbf{W}}^T\mathbf{I}_{mp}\mathbf{W} - \mathbf{A}^T\mathbf{I}_{mp}\mathbf{W}\right]\dot{\mathbf{q}}_e \tag{20}$$

and using $\dot{\mathbf{q}}_e = \mathbf{J}\dot{\mathbf{q}}_\rho$ and its derivative $\ddot{\mathbf{q}}_e = \mathbf{J}\ddot{\mathbf{q}}_\rho + \dot{\mathbf{J}}\dot{\mathbf{q}}_\rho$, it yields:

$$\begin{aligned}\mathbf{J}_R^T\mathbf{m}_{mp} = {} & \mathbf{J}^T\left[\mathbf{W}^T\mathbf{I}_{mp}\mathbf{W}\right]\mathbf{J}\ddot{\mathbf{q}}_\rho + \left[\mathbf{J}^T\left[\mathbf{W}^T\mathbf{I}_{mp}\mathbf{W}\right]\dot{\mathbf{J}}\right.\\ & \left. + \mathbf{J}^T\left[\mathbf{W}^T\mathbf{I}_{mp}\dot{\mathbf{W}} + 2\dot{\mathbf{W}}^T\mathbf{I}_{mp}\mathbf{W} - \mathbf{A}^T\mathbf{I}_{mp}\mathbf{W}\right]\mathbf{J}\right]\dot{\mathbf{q}}_\rho\end{aligned} \tag{21}$$

Some further simplifications and renaming produce the Dynamics equation contributed by the rotation of the mobile platform:

$$\mathbf{J}_R^T \mathbf{m}_{mp} = \mathbf{J}_R^T \mathbf{I}_{mp} \mathbf{J}_R \ddot{\mathbf{q}}_\rho + \mathbf{c}_{mp} = \bar{\mathbf{I}}_{mp} \ddot{\mathbf{q}}_\rho + \mathbf{c}_{mp} \tag{22}$$

Adding the Dynamic equations of the limbs Eq. (10), the translational part of the mobile platform Eq. (12), and the rotational part Eq. (22) we get the explicit Dynamics equation for the parallel manipulator:

$$\mathbf{t}_\rho + \mathbf{J}_G^T \mathbf{f}_G + \mathbf{J}_R^T \mathbf{m}_{mp} = \left[\mathbf{I}_{Ls} + \mathbf{I}_G + \bar{\mathbf{I}}_{mp}\right] \ddot{\mathbf{q}}_\rho + \left[\mathbf{c}_{Ls} + \mathbf{c}_G + \mathbf{c}_{mp}\right] + \left[\bar{\mathbf{g}}_{Ls} + \bar{\mathbf{g}}_G\right] \tag{23}$$

The advantage of this approach is that the Kinematic analysis provides the Jacobians $\mathbf{J}_b$, $\mathbf{J}_G$, and $\mathbf{J}_R$ from the application of loop-closure equations and constraint equations in a straightforward way. Also, matrices **W** and **A** are defined for an specific Euler convention and do not depend on the mechanism analyzed. Hence, the procedure is systematic and reduces the risk of errors in stating the particular Dynamic equations for any specific mechanism.

3 A Numerical Example

A numerical example is done with the parallel manipulator in Fig. 1, a 3*PRS* with orthoplanar limbs. Dimensions are H = 300 mm, L = 500 mm and r = 200 mm. Limbs have a mass of m = 1.2 kg and end-effector M = 1.7 kg homogeneously distributed. Motion required at the end-effector keeps Z value constant while rotation about the X and Y axes, θ and ψ respectively, is as shown in Fig. 2.

The application of Eq. (23), for such an output motion required, produces the values of input forces required at the actuators as shown in Fig. 3.

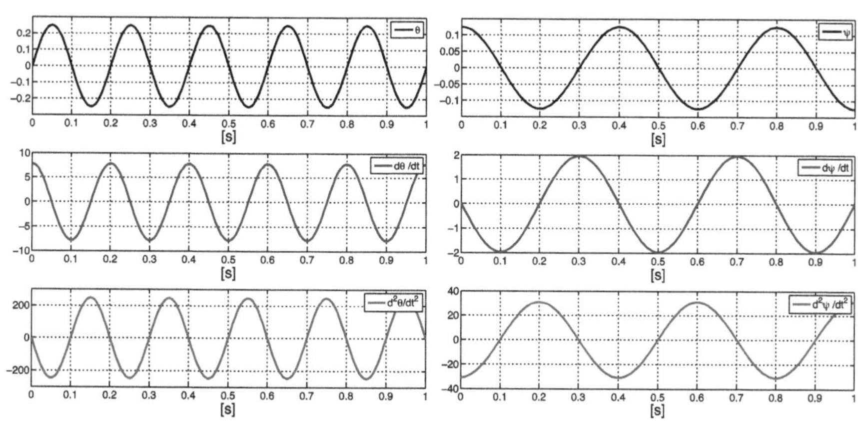

Fig. 2 Output motion of the 3PRS mechanism

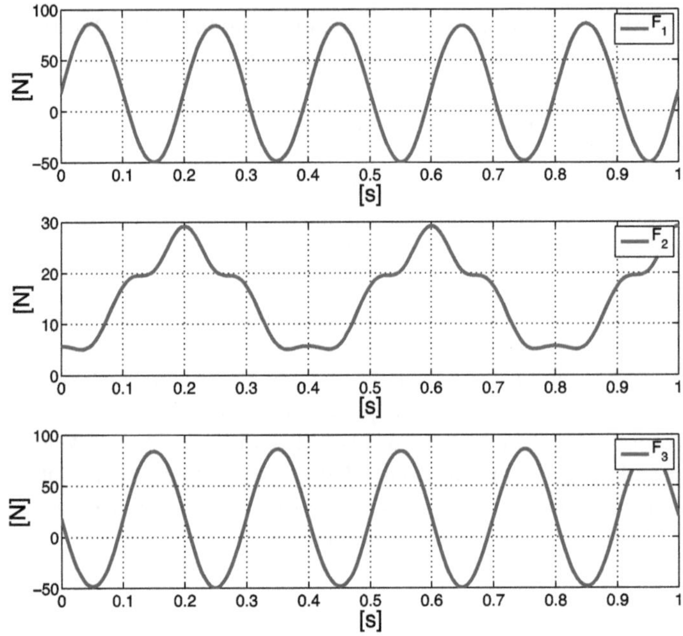

Fig. 3 Input forces of the 3PRS mechanism

4 Conclusions

The method used in this paper to develop the Dynamic equations of parallel manipulators with elements that move with spatial rotations shows a structure that is systematic and simple. For different manipulators, we should only change the mass and inertia terms, and find in the kinematic analysis the Jacobians needed.

The method provides the explicit dynamic equations in terms of the input variables without the need for solving other variables in the run. Therefore, they are very convenient for application into the control or simulation motors.

Acknowledgments The authors of this paper wish to acknowledge the finance received from the Spanish Government via the Ministerio de Educacion y Ciencia (Project DPI2011-22955), the ERDF of the European Union, the Government of the Basque Country (Project GIC07/78, IT445-10), and the University of the Basque Country (Project EHUA13/30).

References

1. Ebert-Uphoff, I., Kozak, K.: Review of the role of quasi-coordinates for the kinematic and dynamic modeling of parallel manipulators. In: Proceedings of the Workshop on Fundamental Issues and Future Research Directions for Parallel Mechanisms and Manipulators, Quebec, pp. 328–338 (2002)

2. Abdellatif, H., Heimann, B.: Computational efficient inverse dynamics of 6-DOF fully parallel manipulators by using the Lagragian formalism. Mech. Mach. Theory **44**, 192–207 (2009)
3. Le, A.Y., Mills, J.K., Benhabib, B.: Dynamic modeling and control design for six DOF meso-milling machine tool. Robotica **32**, 515–532 (2014)
4. Lurie, A.I.: Analytical mechanics. In: Series: Foundations of Engineering Mechanics. Springer, Berlin (2002)
5. Brecher, C., Ostermann, T., Friedrich, D.: Control concept for PKM considering the mechanical coupling between actuators, Chemnitz Parallel Kinematics Seminar, Chemnitz, Germany, pp. 413–427 (2006)

Part IV
Mechanical Transmissions

Gear Variator—Scientific Reality

Konstantin Ivanov, Almas Dinassylov and Ekaterina Yaroslavceva

Abstract Gear adaptive variator is the toothed transmission with the variable transfer ratio which is adapting to the variable loading. Unsuccessful attempts of creation of a gear variator were undertaken repeatedly. The main problem is—maintenance of constant engagement of cogwheels of a variator. The decision of this problem can be based on use of a kinematic chain with two degrees of freedom. Earlier it was proved that this kinematic chain with closed contour possesses the effect of force adaptation. The gear variator can be performed in the form of the closed differential mechanism. The external variable technological loading changes the transfer ratio itself without any control system. The gear variator opens essentially new concept of creation of adaptive engineering for machines with variable technological resistance (for example, adaptive gearbox of car). The theory of a gear variator is based on discovery «Effect of force adaptation in mechanics» which is published in leading editions of world press. In the paper the bases of the theory of a gear adaptive variator are presented.

Keywords Gear variator · Toothed engagement · Differential mechanism · Force adaptation

K. Ivanov · A. Dinassylov (✉)
Almaty University of Power Engineering and Telecommunications,
Almaty, Kazakhstan
e-mail: adinasylov@yandex.ru

K. Ivanov
e-mail: ivanovgreek@mail.ru

E. Yaroslavceva
Saint-Petersburg University of Technology and Design, Saint Petersburg, Russia
e-mail: yaroslavceva_elena@rambler.ru

B. Corves et al. (eds.), *Mechanisms, Transmissions and Applications*,
Mechanisms and Machine Science 31, DOI 10.1007/978-3-319-17067-1_18

1 Introduction

The variator is a mechanism for stepless regulating of the transfer ratio. At present there are only frictional variators.

The gear (toothed) variator is a wheelwork with constant engagement of toothed wheels and with the variable transfer ratio. The gear variator is dream of the designers creating drives of machines with variable technological resistance (for example, car gear boxes, drives of manipulators etc.). The gear variator unlike a frictional variator can procure reliable transfer of force and high efficiency. The gear variator can be performed as a toothed mechanical transmission with variable transfer ratio.

Earlier attempts of creation of continuously variable transmissions (CVT) were undertaken [1–3]. Ivanov [1] used a wheelwork as the kinematic chain between the mobile stator and a rotor of electric motor, Crockett [2] used a gear variator together with the hydraulic converter and Harris [3] used a brake for control of the mechanism. Patents of inventors did not contain authentic theoretical dependences for calculation of mechanisms.

The theoretical substantiation for creation of adaptive gear mechanisms with variable transfer ratio has been developed in works of Ivanov [4–6].

In Ivanov's patents [7–9] gear adaptive mechanisms with a backing, with a high range of transfer ratios and with the use of inertia properties are presented.

The purpose of the work is the short theory of the gear variator.

2 Description of Gear Variator

In present time the scientific researches of so-called adaptive mechanisms which procure self-adjustment to external loading are executed. The adaptive connecting gear provides transfer of motion from the engine of constant power on the tool with a speed, inversely proportional loading. Thus, it is possible to consider the adaptive mechanism as the mechanism self-adjusted to a variable load.

The adaptive wheelwork (gear variator) looks like the closed gear differential with two degrees of freedom (Fig. 1). It contains a frame 0, the carrier H_1, the closed contour, containing toothed wheels 1–2–3–6–5–4 and the carrier H_2. Solar wheels 1, 4 are joined in the block of wheels 1–4. Epicycle wheels 3, 6 are joined in the block of wheels 3–6. The kinematic chain has two degrees of freedom. The gear variator can work in a regime with two degrees of freedom and in a regime with one degree of freedom. Motion of the mechanism with two degrees of freedom occurs in operating conditions of motion with self-regulation. The motion principle is basing on the transform of input energy stream from carrier H_1 into output energy stream on carrier H_2 by closed contour 1–2–3–6–5–4 which has energy circulation. Motion with one degree of freedom occurs at variator start-up when the output carrier is stopped. Another case takes place when the moment of resistance on the output

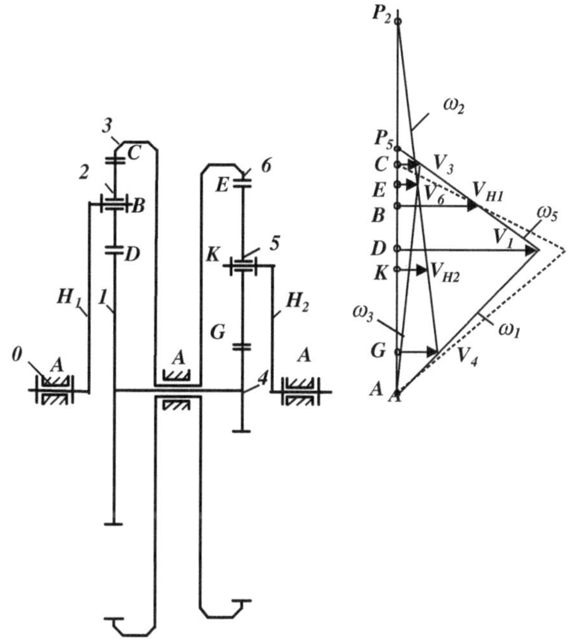

Fig. 1 The gear differential mechanism (gear variator) and a picture of its speeds

carrier exceeds the maximum value. Then the output carrier is stopped. The stop regime allows preventing a mechanism exit out of operation at overloading. After elimination of an overloading will occur the mechanism will continue work in motion operating condition.

3 Kinematic Analysis of Gear Variator

The kinematic analysis of the variator consists in definition of speeds of all points of the mechanism on the set speeds of two external links. It is convenient to make the kinematic analysis of the variator by means of a picture of speeds (Fig. 1).

In a picture of speeds linear speeds V_i points of the mechanism in the form of horizontal lines and angular velocities of links ω_i—in the form of inclined lines are presented. Angular velocities $\omega_1 = \omega_4, \omega_3 = \omega_6$ intermediate links 1–4 and 3–6 are defined in known angular velocities of external links, водил ω_{H1}, ω_{H2} and transfer ratios at the shut down carriers.

Transfer ratios of links of transfer we will define through numbers of teeth of wheels z_i $i = 1, 2, \ldots 6$. The interconnection of angular velocities of the mechanism is defined by equations

$$\frac{\omega_1 - \omega_{H1}}{\omega_3 - \omega_{H1}} = u_{13}^{(H1)}, \tag{1}$$

$$\frac{\omega_1 - \omega_{H2}}{\omega_3 - \omega_{H2}} = u_{46}^{(H2)}, \tag{2}$$

where $u_{13}^{(H1)} = -z_3/z_1$, $u_{46}^{(H2)} = -z_6/z_4$. From (1) and (2)

$$\omega_1 = u_{13}^{(H1)}(\omega_3 - \omega_{H1}) + \omega_{H1}. \tag{3}$$

$$\omega_1 = u_{46}^{(H2)}(\omega_3 - \omega_{H2}) + \omega_{H2}. \tag{4}$$

Let's subtract Eq. (4) of Eq. (3), we will receive

$$u_{13}^{(H1)}(\omega_3 - \omega_{H1}) + \omega_{H1} - u_{46}^{(H2)}(\omega_3 - \omega_{H2}) - \omega_{H2} = 0.$$

From here

$$(u_{13}^{(H1)} - u_{46}^{(H2)})\omega_3 - u_{13}^{(H1)}\omega_{H1} + u_{46}^{(H2)}\omega_{H2} = \omega_{H2} - \omega_{H1}.$$

From here

$$\omega_3 = \frac{\omega_{H2}(1 - u_{46}^{(H2)}) - \omega_{H1}(1 - u_{13}^{(H1)})}{u_{13}^{(H1)} - u_{46}^{(H2)}}. \tag{5}$$

Equations (4) and (5) define sequence of acts by definition angular velocities ω_3, ω_1 transfer links. Angular velocity of the satellite 2 is defined from a condition

$$\frac{\omega_2 - \omega_{H1}}{\omega_3 - \omega_{H1}} = u_{23}^{(H1)}, s \tag{6}$$

where $u_{23}^{(H1)} = z_3/z_2$. From here

$$\omega_2 = u_{23}^{(H1)}(\omega_3 - \omega_{H1}) + \omega_{H1}. \tag{7}$$

Angular speed of the satellite 5 is defined from a condition

$$\frac{\omega_5 - \omega_{H2}}{\omega_3 - \omega_{H2}} = u_{56}^{(H2)}, \tag{8}$$

where $u_{56}^{(H2)} = z_6/z_5$. From here

$$\omega_5 = u_{56}^{(H2)}(\omega_3 - \omega_{H2}) + \omega_{H2}. \tag{9}$$

It is necessary to note, that in the absence of mobility in a contour the kinematic chain will move in a condition with one degree of freedom. In this case angular velocities of all links are equal. Thus, all kinematic and power parameters are defined, and all mechanism has the kinematic and static definability.

4 Force Analysis of Gear Variator

Let's make the force analysis of the kinematic chain (Fig. 1) by the standard technique. The problem of the power analysis of the mechanism with two degrees of freedom consists in definition of reactions in kinematic pairs and in definition of the generalized forces on two external links.

Some features of act of forces occur for the considered kinematic chain. We will consider that on external links the generalized forces—moments M_{H1} and M_{H2} on carriers H_1 and H_2 are acting. On intermediate structural group of Assur the superposed forces do not act (a gravity of links and an inertial force of links are neglected because they are small in comparison with forces on external carriers).

The force analysis should be begun with consideration of structural group 1–2–3–6–5–4 in the form of the closed contour consisting of toothed wheels. The structural group contains the block of solar wheels 1–4, the satellite 2, the block epicycle wheels 3–6 and the satellite 5. Such structural group was never considered earlier in the power analysis of mechanisms. However for this structural group it is possible to make conditions of balance of statics which define interconnection internal and superposed forces. We will consider, that superposed forces for considered structural group is force F_{H1} transferred from the carrier H_1 on point *B*, and force F_{H2} transferred from the carrier H_2 on point *K*. Internal forces are reactions in kinematic pairs in points *C*, *E*, *D*, *G*.

The first feature of considered structural group consists that all internal forces R_{32}, R_{65}, R_{12}, R_{45} can be expressed on statics conditions through active forces F_{H1} and F_{H2}.

For links of a contour 2 and 5 we will express reactions in the kinematic парах *D*, *C*, *G*, *E* through external силы F_{H1}, F_{H2}, enclosed in точках *B*, *K*

$$R_{12} = R_{32} = 0.5F_{H1}. \tag{10}$$

$$R_{45} = R_{65} = 0.5F_{H2}. \tag{11}$$

Here $F_{H1} = M_{H1}/r_{H1}$, $R_{12} = M_{12}/r_1$, $R_{32} = M_{32}/r_3$, $F_{H2} = M_{H2}/r_{H2}$, $R_{45} = M_{45}/r_4$, $R_{65} = M_{65}/r_6$ M_{H1}, M_{H2}—moments on carriers, r_{H1}, r_{H2}—radiuses of carriers, M_{12}, M_{32}—moments, created on the satellite by 2 reactions R_{12}, R_{32} co of the party of toothed wheels 1 and 3, M_{45}, M_{65}—moments, created on the satellite by

5 reactions R_{45}, R_{65} co of the party of toothed wheels 4 and 6, $r_i (i = 1, 2...6)$—radiuses of wheels.

After substitution of values of forces in Eqs. (10) and (11) we will receive equations for definition of the internal moments through the external moments

$$M_{12} = 0.5 M_{H1} r_1 / r_{H1}, \tag{12}$$

$$M_{32} = 0.5 M_{H1} r_3 / r_{H1}, \tag{13}$$

$$M_{45} = 0.5 M_{H2} r_4 / r_{H2}, \tag{14}$$

$$M_{65} = 0.5 M_{H2} r_6 / r_{H2}. \tag{15}$$

Let's make for the satellite 2 and satellite 5 equilibrium equations in the form of the sum of the moments concerning its instant centers of speeds P_2, P_5.

After transformation with account the time taking into account $\omega_1 = \omega_4$, $\omega_3 = \omega_6$ we will receive

$$M_{12}\omega_1 + M_{32}\omega_3 = M_{H1}\omega_{H1}, \tag{16}$$

$$M_{45}\omega_1 + M_{65}\omega_3 = M_{H2}\omega_{H2}. \tag{17}$$

We will combine the made expressions. We will receive a condition of interacting of parameters of the kinematic chain as a whole

$$M_{12}\omega_1 + M_{32}\omega_3 + M_{45}\omega_1 + M_{65}\omega_3 = M_{H1}\omega_{H1} + M_{H2}\omega_{H2}. \tag{18}$$

In the left side of Eq. (18) the sum of powers (matching to the sum of works) contour internal forces occurs.

$$M_{12}\omega_1 + M_{32}\omega_3 + M_{45}\omega_1 + M_{65}\omega_3 = 0. \tag{19}$$

The right side of Eq. (18) represents the sum of powers (matching to the sum of works) contour superposed forces

$$M_{H1}\omega_{H1} + M_{H2}\omega_{H2} = 0. \tag{20}$$

From the Eq. (20) it follows

$$\omega_{H2} = M_{H1}\omega_{H1} / M_{H2}. \tag{21}$$

Additional constraint Eq. (21) provides effect of force adaptation to output loading: at the set constant parameters of aerial input M_{H1}, ω_{H1} and set output moment of resistance M_{H2} output angular speed ω_{H2} is in return proportional dependence on the variable output moment сопротивления M_{H2}.

From the Eq. (19) we will receive

$$(M_{12} - M_{45})\omega_1 + (M_{32} - M_{65})\omega_3 = 0. \tag{22}$$

The Eq. (22) represents the equation of works (powers) on intermediate links 1–4 and 3–6. The Eq. (22) means balance presence on intermediate links 1–4 and 3–6 simultaneously. In the mobile closed contour basic new situation occurs: balance in statics separately on each intermediate link is absent, but balance of intermediate links simultaneously on the move all contour occurs.

In the closed contour energy circulation occurs.

From Eq. (22) we will receive

$$-(M_{45} - M_{12})\omega_1 + (M_{32} - M_{65})\omega_3 = 0. \tag{23}$$

The Eq. (23) reflects an analytical form of circulation of energy unknown earlier in a contour during its motion. Thus, it is installed, that the gear variator represents the closed gear differential mechanism with two degrees of freedom and with one input.

5 Experimental Testing of the Gear Variator

Testing of force adaptation effect in the mechanism presented on Fig. 1 has been executed by M. Ceccarelli and G. Balbaev on the test-bed in LARM of university Cassino [10].

On Fig. 2 experimental tractive characteristic of the gear adaptive mechanism presented as function of output traction moment (or resistance moment) in Nm from output angular velocity in rpm at constant input angular velocity and engine torque.

The experimental tractive characteristic conforms to the theoretical one.

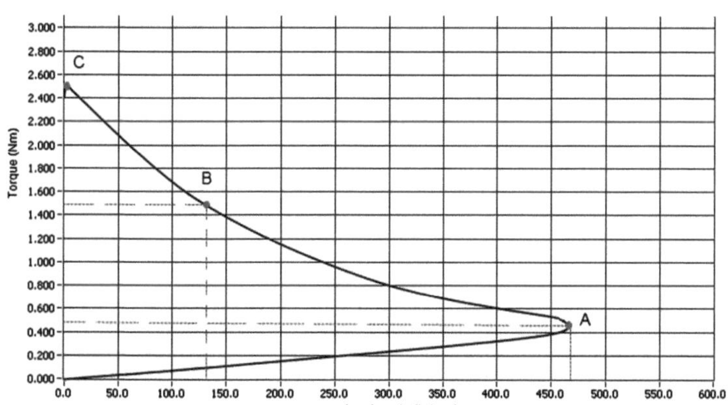

Fig. 2 Experimental tractive characteristic of gear variator

6 Conclusions

In the kinematic chain with two degrees of freedom the closed contour provides definiteness of motion. It allows considering the found regularity of the analysis of gear variator as the classical theory applied to mechanical systems with ideal constraints. The theory allows creating the gear variator possessing property of mechanical adaptation to variable technological loading. The adaptive gear variator provides possibility of motion of output link with a speed inversely proportional moment of resistance (torque) at a constant input power of engine.

The adaptive mechanical wheelwork uses the effect of power adaptation allowing automatically changing the transfer ratio. The gear variator ideally adapts for variable technological loading without use of any control systems. It allows simplifying a design, to reduce transmission sizes, to raise its reliability and efficiency.

References

1. Ivanov, K.S., Dmitrieva, N.A.: Non reactive engine. The copyright certificate of the USSR No. 769157, 10 p, 7 Sept 1980
2. Crockett, S.J.: Shiftless, continuously-aligning transmission. Patent of USA 4,932,928, Cl. F16H 47/08, U.S. Cl. 475/51; 475/47, 9 p, 1990
3. John, H.: Power transmission system comprising two sets of epicyclic gears. Patent of Great Britain GB2238090 (A), 11 p, 1991
4. Ivanov, K.S.: Theoretical of a Basis Gear Stepless Adjustable Transfer. The Theory of Mechanisms and Machines, V. 8, No. 2(16), pp. 36–48. SPb the State Polyengineeringal University, St.-Petersburg (2010)
5. Ivanov, K.S., Dinasylov, A.D., Yaroslavseva, E.K.: Adaptive-mechanical continuously variable transmission. In: Petuya, V., Pinto, C., Lovasz, E.C. (eds.) New Advances in Mechanisms, Transmissions and Applications. Proceedings of the Second Conference MeTrApp 2013. Mechanisms and Machine Science, vol. 17, pp. 83–90. Springer, Berlin (2014)
6. Ivanov, K.S.: Paradox of mechanics—a basis of creation CVT. In: Transactions of 2-d IFToMM Asian Conference on Mechanisms and Machines Science, pp. 245–264. Tokyo, Japan, 7–10 Nov 2012
7. Ivanov, K.S., Yaroslavtseva, E.K.: Way of automatic and continuous change of the twisting moment and speed of rotation of a output shaft depending on resistance to motion and the device for its realization. Patent of Russia RU No. 2398989, 10 p, 10 Sept 2010
8. Ivanov K.S.: Almaty KAZ—owner of the registered sample. The name—device of automatic and continuous change of a twisting moment—and changes of a corrected speed of output shaft depending on a tractive resistance. The deed on registration of the registered sample No. 20 2012 101 273.1. Day of Registration, 02 May 2012. The German patent and firm establishment, Federal Republic Germany, 12 p, 2012
9. Ivanov, K.S., Yaroslavtseva, E.K.: Device of a transmission of energy with continuously variable transfer ratio (variants). Patent of Kazakhstan No. 023907, 16 p, 23 Feb 2011
10. Ceccarelli, M., Balbayev, G.K.: Design and characterization of a new planetary gear box. In: Petuya, V., Pinto, C., Lovasz, E.C. (eds.) New Advances in Mechanics, Transmissions and Applications. Proceedings of the Second Conference MeTrApp 2013. Mechanism and Machine Science, vol. 17, pp. 91–98. Springer, Berlin (2014)

A Novel Approach for Conceptual Structural Design of Gearbox

Delun Wang, Huipeng Shen, Huimin Dong and Shudong Yu

Abstract The paper presents a novel approach for conceptual structural design of a complex machine part. The approach consists of the following four steps: (i) building a geometric-physical model, (ii) extracting a conceptual model, (iii) meeting the strength and stiffness requirements, and (iv) considering structural manufacturability and aesthetics. Due to the design constraints of a complex machine part in both geometrical and physical requirements, a geometric model is built as a simple geometric shape including all connection surfaces. And a physical model is a finite element model with loads and connection constraints incorporated. The conceptual model is a truss structure in accordance with the material distribution and load paths under different load cases. During designing conceptual structure, its skeleton is accomplished according to the optimal material distribution. The dimensions of the structural skeleton are determined based on the required structural strength and stiffness. Finally, by taking into account the manufacturability and aesthetic appearance, a conceptual structure with the required dimensions is accepted as the actual structure of the machine part. Using a gearbox design as an illustrative example, we demonstrated that the proposed approach is efficient and reliable. The new conceptual structure, produced using this approach for a gearbox, is superior to the empirical and analogical approaches because of lack of proven and reliable design methodology in the literature.

Keywords Structural skeleton · Conceptual design · Material distribution · Gearbox

---

D. Wang (✉) · H. Shen · H. Dong
Dalian University of Technology, Dalian, China
e-mail: dlunwang@dlut.edu.cn

H. Shen
e-mail: hpengshen@mail.dlut.edu.cn

S. Yu
Ryerson University, Toronto, Canada
e-mail: syu@ryerson.ca

B. Corves et al. (eds.), *Mechanisms, Transmissions and Applications*,
Mechanisms and Machine Science 31, DOI 10.1007/978-3-319-17067-1_19

1 Introduction

Some parts of a machine, which can provide support and determine the machine critical layout, are usually large, heavy, and difficult to manufacture. Those complex machine parts can take 70–90 % of the total machine weight. Examples include the chassis of a vehicle and the case of a gearbox and so on. Because of the lack of proven and reliable design methodology, it is difficult to design the geometric structures and determine their corresponding performances. Indeed, the traditional structural design adopts manual calculations and design by experience and analogy. As a result, the traditional design can be bulky, heavy and costly. This may reduce the dynamic performance of the part and the overall product competitiveness. Lightweight design [9, 12] and the appearance design have been a main research topic in green manufacturing technology.

With the development of the computational mechanics [2, 4], especially with the application of finite element method, the topology optimization method and calculation software, research on how to obtain a reasonable structure with excellent properties has been carried out in different industries [1, 3, 5, 11] and fields [6, 7, 14]. Kroll [10] discussed the general influence of lightweight design approaches on energy efficiency in machine tools and restrictions on the maximum mass reduction for structural components. Jang and Kim [8] developed a specialized integrated design for MEMS compliant mechanisms from topology to shape optimization. Remouchamps [13] used the topology optimization to design aircraft pylons. However, a general scientific method for the structural conceptual design of complex machine parts does not appear yet.

In this paper, we are interested in how to design a reasonable structure in connection with the performance requirements of a gearbox. The process of the conceptual structural design is introduced as follows. Firstly, the geometric model and the physical model are set up by analyzing load cases and constraint characteristics. And then, the structural conceptual model is represented according to the optimal material distribution based on the synthesized load paths. Meanwhile, the dimensions of the structure are determined by the required strength and stiffness. Finally, the structural manufacturability and good appearance are briefly introduced.

2 Geometric-Physical Model

The geometric model of a complex part is used to describe the shape and structural characteristics and show the spatial connection between the adjacent parts. The physical model is a finite element model containing loads (directions and magnitudes) and constraints. For a three-stage cylindrical gear reducer, its structure sketch is shown in Fig. 1. The installation on the input side (I-shaft) is connected by a screw shaft. The output shaft (IV-shaft) is supported by bearings.

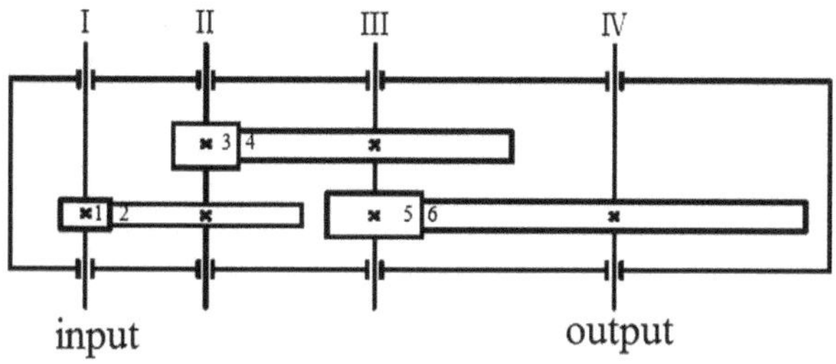

Fig. 1 Sketch of a three-stage cylindrical gear reducer

2.1 Geometric Model

Based on the main functions of parts, a geometric model is a simple shape including the connection surfaces. The shape and dimensions of the geometric model mainly show the connection surface dimensions, the position distribution, the movement situation. The process of building the geometric model can be expressed as follows. Firstly, in order to build the surface shape and structure for each connection, it is essential to ensure the relative spatial position and dimension between the connection surfaces. Secondly, a simple solid configuration is confirmed. Generally it is a simple geometric shape or combinations of shapes like cylinder, cuboid, etc. Thirdly, the internal and outside occupied space is determined in accordance with the required movement between adjacent parts, space and assembly. Then, the design domain is confirmed, where the structure, shape and dimensions can be redesigned. For the cylindrical gear reducer, a gearbox is indispensable to bearing and transferring loads to the foundation. The geometric model is shown in Fig. 2.

2.2 Physical Model

In building a physical model, we need to consider the loads, the connection constraints, and the finite elements. The first two factors are discussed in this section.

A machine part is subjected to static and dynamic interfacial loads at the connection surfaces, its own weight, the bearing weight, the inertia force, the thermal

Fig. 2 A geometric model

Fig. 3 A physical model

load, electro-magnetic forces, fluid forces, etc. The interfacial loads affect other parts via the connection surfaces. The calculation condition is usually chosen be to the most critical operating condition, or a typical operating condition with adequate load factors. The load is obtained by solving the mechanical model in the pre-set condition, and determining the domain of loads.

The connection constraint is limited by adjacent parts. The constraint type is determined by the connecting surface, the installation form and the number of degrees of freedom. In analyzing the performance and designing structure by the finite element method, the constraint must be implemented into the finite element model as the boundary conditions. There are three common types of constraints: displacement equivalent, spring element equivalent and contact element equivalent. A fixed connection surface can be modeled using the displacement equivalent constraint. Bolt and screw shaft connection surface is generally modeled using an equivalent spring element. The surface-to-surface contact between adjacent parts is modeled using the contact element.

Based on the meshed geometric model, a physical model with applied load and constraints is established. The load cases contain the clockwise and counterclockwise conditions. The connection surfaces with bearings and screw shaft are modeled using the spring elements in a finite element analysis. Figure 3 shows the physical model of the gearbox.

3 Conceptual Structural Design

In order to obtain a reasonable structure, the conceptual model is built according to the material distribution and load paths under different load cases. We then design the local details and come up with a preliminary structural scheme.

3.1 Structural Skeleton

When the design of a structure is accomplished according to the optimal material distribution, we consider the design as reasonable. The next question is how to find a structure of optimal material distribution. In this section, we propose a method for describing the material distribution by a space vector diagram. This method includes two steps: the topology optimization of a continuum structure, and the outputting load path with a single load case.

The detailed topology optimization of continuum structures can be found in the literature. In this paper, we only give a brief discussion of the objective functions and constraints. An objective function generally falls in three categories: minimum weighted compliance (maximum stiffness), minimum weighted reciprocal eigenvalue (maximum fundamental frequency), and minimum compliance index (the mixture of the first two). Constraints include performance constraints and manufacturing constraints, etc.

① The objective function for minimum weighted compliance is

$$\min C_w(X) = \frac{1}{2}\sum w_i u_i(X)^T K u_i(X) \tag{1}$$

where C_w is structural compliance; w_i is a corresponding weighting coefficient of every load case; u_i is the vector of structural displacements, K is a structural global stiffness matrix.

② The objective function for minimum weighted reciprocal eigenvalue is

$$\min f_w(X) = \sum w_i/\lambda_i(X) \tag{2}$$

where f_w is weighted reciprocal eigenvalue; w_i is a corresponding weighting coefficient of every eigenvalue; λ_i is structural eigenvalue.

③ The objective function for minimum compliance index is

$$\min S(X) = \sum w_i C_i(X) + NORM \frac{\sum w_j/\lambda_j(X)}{\sum w_j} \tag{3}$$

where S is structural compliance index; NORM is correction coefficient.

As mentioned above, the weighted compliance can be applied as objective function when the static stiffness is a critical influence of the structure. The weighted reciprocal eigenvalue can be used considering the impact of machine vibration on the structure. And the compliance index can be selected balancing the effect of the first two. For the gearbox, the last objective function is adopted.

Based on the physical model along with the related optimization parameters, a topology optimization model can be achieved. The material distribution can be obtained by the well-known algorithms available in the commercial finite element software. The results are shown in Fig. 4 for the clockwise condition, and Fig. 5 for the counterclockwise condition. According to the material utilization, a boundary identification rule is created. The load path is output by the space vector diagram. The structural skeleton form is the type shown in Fig. 6.

Fig. 4 Clockwise condition

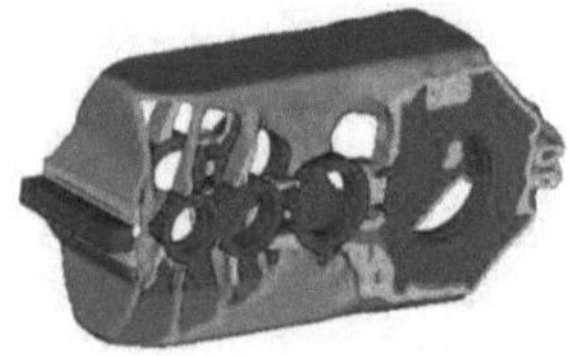

Fig. 5 Counterclockwise condition

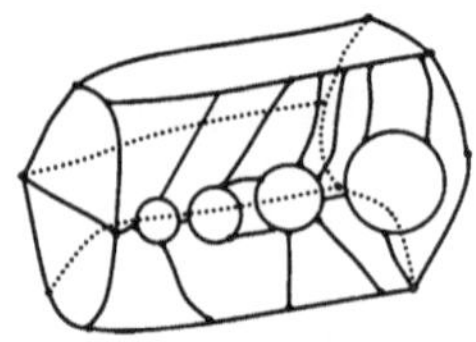

Fig. 6 Space vector diagram of load path

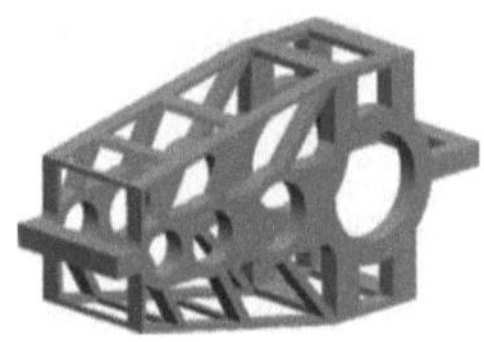

Fig. 7 Conceptual model

3.2 Integrated Skeleton

Considering the bearing capacity of various parts under complex load cases, the optimal path can be obtained by synthesizing space vectors for material distributions under different conditions. We can use the Boolean operations to synthesize space vectors. Then we build the conceptual model according to the optimal path.

It can be seen that the conceptual model is a truss structure integrated and regularized the space vector diagram of the optimal load path by synthesizing the material distributions under different load cases. Figure 6 shows the integrated space vector diagram. The conceptual model of the gearbox is shown in Fig. 7.

4 Strength and Stiffness Design

In order to meet the strength and stiffness requirements, the wall and local structure with proper type and scale should be added. The strength scale is taken as the maximum working stress as an index. The working stress is the maximum principal stress by incorporating the bending stress, the tensile stress, the compressive stress,

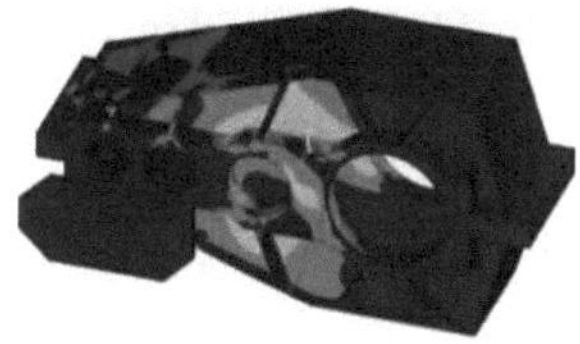

Fig. 8 Stress nephogram

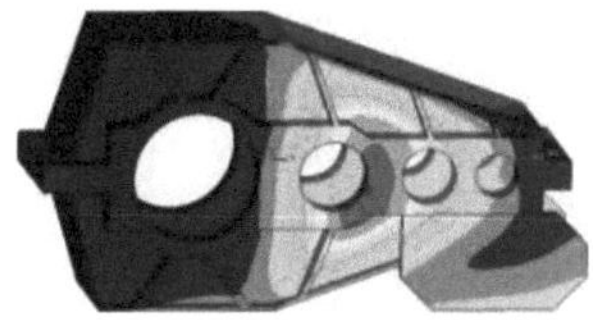

Fig. 9 Displacement nephogram

and the shear stress. And the stress concentration everywhere should be kept as small as possible. The stiffness scale is taken as the displacement in the most critical operating condition as an index. It should be pointed out that the greater the strength and stiffness of the structure, the heavier the weight. Overdesign results in poor dynamic performance, heavy product, material waste, and low efficiency. So the structure should be as light as possible provided that the strength and stiffness requirements are satisfied.

The overall performance is the result of the wall structure combined with local structure. Meanwhile, the type and scale of local structure vary depending on the main wall. Therefore, design for strength and stiffness must take the scale of the structure and the weight of the part into account.

For this gearbox, it is designed to seal the entire structure as bearings are considered to be lubricated by splashing oil inside the gearbox. Then a wall in the conceptual model is added. Of course, the conceptual model can be inside or outside the gearbox. This paper adopts the outside form considering the usage requirements. The whole structure scale is determined by size optimization. For walls of 8 mm in thickness, and the ribs of 10 mm thickness, the stress and displacement nephograms are shown in Figs. 8 and 9 by the commercial finite element software. The key results are given in Table 1.

Table 1 Design result of the gearbox

Performance parameters	Mass (Kg)	Static displacement (m)	Static strength (MP)
Objective	≤650	8.0×10^{-5}	≤160
Clockwise result	610	7.6×10^{-5}	38.3
Counterclockwise result	610	7.5×10^{-5}	41.5

Fig. 10 The first scheme

Fig. 11 The second scheme

5 Design for Structural Manufacturability

As mentioned above, the structure must meet not only the requirements of strength and stiffness, but also manufacturability and fine aesthetic appearance. We will discuss the two points in this section.

The manufacturability mainly reflects the geometric machinability of a structure. According to the above method of the conceptual design, the geometric structure shape is ideal. The manufacturability and costing must be considered when the structure of parts is designed. The manufacturability process includes the selection of material, processing and assembling, hoisting and transportation, etc.

After completing the strength, stiffness and process design, the overall structure needs to carry out aesthetics design for constructing a greater sense of stability. This kind of design method provides a different perspective to design structure shape and color to form a design style based on the scientific principle.

For the above structure of the gearbox, the arrangement of ribs is further designed considering the structural manufacturability and aesthetics. It can be seen that the arrangement of the well-distributed ribs makes the system, shown in Figs. 10 and 11, aesthetically pleasant as well as satisfactory in meeting the performance requirements.

6 Conclusions

1. This paper presents a novel approach for conceptual structural design, which consists of four steps: building a geometric-physical model, extracting a conceptual model, meeting the strength and stiffness requirements, and considering structural manufacturability and aesthetics.

2. The approach lays a scientific foundation of structure design compared to the traditional empirical and analogical approaches. The proposed approach has been successfully used to design gearbox. The outcome indicates that the approach is reasonable and effective in design for lightweight.
3. Using the proposed design methodology, we produced a satisfying gearbox that meets all the requirements and constraints.

Acknowledgments This paper acknowledges the support by National Key Technology Research and Development Program of the Ministry of Science and Technology of China (No. 2012BAA01B05).

References

1. Antoine, N.E., Kroo, I.M.: Framework for aircraft conceptual design and environmental performance studies. AIAA J. **43**(10), 2100–2109 (2005)
2. Bendsøe, M.P.: Optimal shape design as a material distribution problem. Struct. Optim. **1**(4), 193–202 (1989)
3. Bureerat, S., Boonapan, A., Kunakote, T., Limtragool, J.: Design of compliance mechanisms using topology optimisation. In: The 19th Conference of Mechanical Engineering Network of Thailand, pp. 421–427. Phuket, Thailand (2005)
4. Cheng, G.D., Gu, Y.X., Wang, J.: A prospective review on research and application of optimal design of mechanical systems in China. J. Mech. Strength **17**(2), 68–74 (1995)
5. Cho, S.K., Kim, H.J., Chang, S.H.: The application of polymer composites to the table-top machine tool components for higher stiffness and reduced weight. Compos. Struct. **93**, 492–501 (2011)
6. Dong, H.M., et al.: On structure and performance for machine tool lighteweight design-structural and static characteristics of components in horizontal lathe. Modular Mach. Tool Autom. Manuf. Tech. **6**, 1–5 (2013)
7. Dong, H.M., et al.: Physical property of basic unit and database in machine tool lightweight design. Manuf. Technol. Mach. Tool **12**, 78–82 (2012)
8. Jang, G.W., Kim, K.J., Kim, Y.Y.: Integrated topology and shape optimization software for compliant MEMS mechanism design. Adv. Eng. Softw. **39**(1), 1–14 (2008)
9. Kleiner, M., Geiger, M., Klaus, A.: Manufacturing of lightweight components by metal forming. CIRP Annals-Manuf. Technol. **52**(2), 521–542 (2003)
10. Kroll, L., et al.: Lightweight components for energy-efficient machine tools. CIRP J. Manuf. Sci. Technol. **4**, 148–160 (2011)
11. Lee, D.G., et al.: Design and manufacture of composite high speed machine tool structures. Compos. Sci. Technol. **64**, 1523–1530 (2004)
12. Rusinko, C.A.: Green manufacturing: an evaluation of environmentally sustainable manufacturing practices and their impact on competitive outcomes. IEEE Trans. Eng. Manage. **54**(3), 445–454 (2007)
13. Remouchamps, A., Bruyneel, M., Fleury, C., Grihon, S.: Application of a bi-level scheme including topology optimization to the design of an aircraft pylon. Struct. Multi. Optim. **44**, 739–750 (2011)
14. Xing, Q.F., et al.: Structure unit characteristics and static performance in machine tool lightweight. Manuf. Technol. Mach. Tool **12**, 44–48 (2012)

Dynamic Modeling of Planetary Gear Train for Vibration Characteristic Analysis

Huimin Dong, Kai Zhang, Delun Wang, Yangyang Wu and Shaoping Bai

Abstract In this paper, a torsional-translational coupled nonlinear dynamic model for planetary gear train with three planet gears is developed to investigate its vibration characteristic. In this lumped-parameter model, gears and carrier all have three degrees of freedom: two translations and one rotation. The equations of motion of the dynamic model are built in consideration of gear elastic deformation and time-varying mesh stiffness. Gear elastic compatibility equations are developed to describe the relationship between displacements, elastic deformations and transmission errors. Thus, dynamic equations of the planetary gear train consist of equations of motion and gear elastic compatibility equations can be found and the effects of support stiffness and flexible planet pin on vibration characteristic have been addressed. The results show floating central gear and flexible planet pin can avoid critical operating conditions, which provides a theoretical guideline for the design of planetary gear train.

Keywords Planetary gear train · Dynamic model · Vibration characteristic · Equivalent mesh error

1 Introduction

Planetary gear train with three planet gears is widely used in many industries and fields, because it offers several advantages of large transmission ratio, simple construction and compactness. Generally, due to manufacture and assembly errors and time-varying mesh stiffness, noise and vibration would occur in gear

H. Dong (✉) · K. Zhang · D. Wang · Y. Wu
Dalian University of Technology, Dalian, China
e-mail: donghm@dlut.edu.cn

S. Bai
Aalborg University, Aalborg, Denmark
e-mail: shb@m-tech.aau.dk

B. Corves et al. (eds.), *Mechanisms, Transmissions and Applications*,
Mechanisms and Machine Science 31, DOI 10.1007/978-3-319-17067-1_20

transmission which damages the gear teeth seriously. Therefore, extensive studies on gear noise and vibration problems have been conducted analytically [1, 2]. The effects of manufacturing errors [3], unequal planet stiffness [4], planet position errors [5], eccentricity errors [6] and time-varying mesh stiffness [7] have been considered. August [8] found floating sun gear can improve loading condition at a low speed; Kahraman [9] discussed the effects of planet mesh phasing and support stiffness on the gear vibration. To summarize, there are mainly two vibration excitations in planetary gear train, which are time-varying mesh stiffness, manufacturing and assembly errors.

In this paper, a torsional-translational coupled nonlinear dynamic model is developed for a three-planet planetary gear train. The model considers the multi-degree-of-freedom (M-DOF) of the planetary gear train; moreover, the two vibration excitations, which are time-varying mesh stiffness, manufacturing and assembly errors; both are discussed in this paper. Based on this model, the influences of support stiffness and flexible planet gear pin on vibration characteristics are studied.

2 Dynamic Modeling

2.1 Coordinate Systems and Parameters

The planetary gear train considered in this study consists of a sun gear, a carrier and a ring gear (subscripts s, c and r respectively), which are coupled to each other by three pinions (subscript ni). Each component undergoes planar motion, which means two translations and one rotation are considered. Figure 1 illustrates the dynamic model of planetary gear train: component bearings are modeled by linear springs and dampers, gear mesh interactions are represented by gear mesh stiffness and dampers acting along the line of action. Following assumptions are made (i) the translational displacements of all parts in axial direction and non-axial torsional displacements are ignored; (ii) The meshing angle variations of gear pairs are negligible; (iii) The dynamic model is based on lumped-parameter theory, planet gears have the same structure, also the same physical and geometry errors.

In the dynamic model, coordinate systems are defined as shown in Fig. 1, the fixed coordinate system $\{O; X, Y\}$; moving coordinate systems $\{O_c; X_c, Y_c\}$ and $\{O_{ni}; X_{ni}, Y_{ni}\}$, $i = 1, 2, 3$, attached to the carrier and rotate at the constant carrier angular velocity ω_c. The planetary gear train converted into a conventional one in $\{O_c; X_c, Y_c\}$, and the line of action is fixed respect to this coordinate system. Translations x_j, y_j and torsional displacement u_j, $j = c, s, r$ are measured in moving coordinate system $\{O_c; X_c, Y_c\}$, $j = ni$ in the coordinate system$\{O_{ni}; X_{ni}, Y_{ni}\}$; x_j, y_j represent planar vibrations of each components. u_j, $j = s, r$ is measured along the tangent to base circles; for the carrier, u_c is along the tangent to equivalent circle

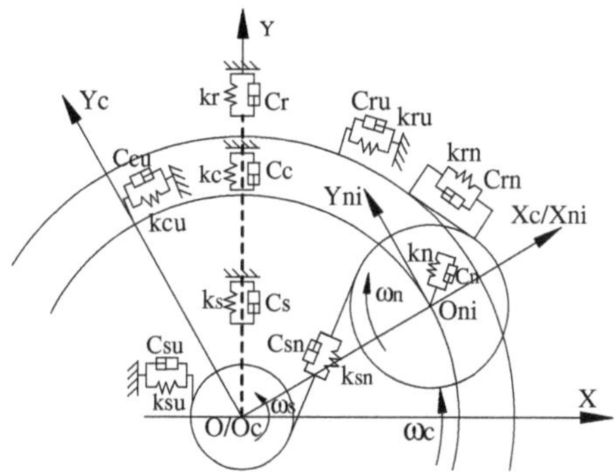

Fig. 1 Dynamic model of the 3-pinion planetary gear train

which radius is equal to distance between the pin and carrier centers; u_j is taken as positive when its direction is coincident with its angular velocity.

2.2 Equations of Motion

Since planar vibrations of each component x_j and y_j, $j = c, s, r, ni$, are measured in rotating coordinate systems, which is non-inertial; In order to describe the inertial forces in moving coordinate systems, the inertial accelerations are projected to moving coordinate. Acceleration in non-inertial frame is not only the second derivative of x_j and y_j, but also centrifugal acceleration and Coriolis acceleration, which means

$$\begin{cases} a_{jx} = \ddot{x}_j - 2\omega_c \dot{y}_j - \omega_c^2 x_j \\ a_{jy} = \ddot{y}_j + 2\omega_c \dot{x}_j - \omega_c^2 y_j \end{cases} \quad j = c, s, r, ni \tag{1}$$

Equations of Motion of Carrier. As shown in Fig. 1, the carrier is coupled with planet pin-bearing assembly via stiffness k_{cn} and damping C_{cn}. Moreover, the carrier support bearing is modeled by stiffness k_c and damping C_c. Thus, the carrier suffers interaction forces from carrier support bearing, planet-pin bearing and inertial force. In the fixed coordinate system $\{O; X, Y\}$, equations of motion of carrier in X_c and Y_c directions can be expressed as:

$$\begin{cases} m_c a_{cx} + C_{cx}\dot{x}_c + k_{cx}x_c - \Sigma(C_{cn}\dot{u}_c + k_{cn}u_c)\sin\psi_{ni} + \Sigma(C_{cnx}\dot{x}_{cni} + k_{cnx}x_{cni}) = 0 \\ m_c a_{cy} + C_{cy}\dot{y}_c + k_{cy}y_c + \Sigma(C_{cn}\dot{u}_c + k_{cn}u_c)\cos\psi_{ni} + \Sigma(C_{cny}\dot{y}_{cni} + k_{cny}y_{cni}) = 0 \end{cases} \tag{2}$$

where m_j, $j = c, s, r, ni$ is the mass of each component, ψ_{ni} is position angle of the ith planet gear, $\psi_{ni} = 2\pi(i - 1)/n$, x_{cni} and y_{cni} are the relative displacements of carrier respect to planet gears in X_c and Y_c directions respectively.

There are three torsional forces applied to carrier, which are interaction forces from carrier support bearing and pin-bearing assembly, and torsional inertial force. Thus, torsional equation of motion of carrier can be obtained:

$$(I_{zc}/r_c^2)\ddot{u}_c + (C_{cu}\dot{u}_c + k_{cu}u_c) - \Sigma(C_{cnx}\dot{x}_c + k_{cnx}x_c)\sin\psi_{ni} \\ + \Sigma(C_{cny}\dot{y}_c + k_{cny}y_c)\cos\psi_{ni} + \Sigma(C_{cnu}\dot{u}_c + k_{cnu}u_c) - \Sigma(C_{cny}\dot{y}_{ni} + k_{cny}y_{ni}) = -T_c/r_c \quad (3)$$

where T_c is the torque applied to carrier, r_c is radius of equivalent circle of carrier, which is equal to distance between the pin and carrier centers. I_{zj} represents moment of inertia about z-axis, $j = c, s, r, ni$.

Equations of Motion of Sun Gear and ring gear. The sun gear connects to planet gears along the line of action through gear mesh stiffness k_{jn} and damping C_{jn}. Moreover, the sun and ring gears coupled to the ground or shaft via support stiffness k_j and damping C_j, also a torsional stiffness k_{ju} and damping C_{ju}, $j = s, r$. There are three forces applied on central gears, which are meshing forces, interaction forces from support bearing and inertial force. Equations of motion of central gears with three-degree of freedom are given as follow:

$$\begin{cases} m_j a_{jx} + C_{jx}\dot{x}_j + k_{jx}x_j + \Sigma(C_{jn}\dot{\delta}_{jni} + k_{jni}\delta_{jni})\sin(\psi_{jni}) = 0 \\ m_j a_{jy} + C_{jy}\dot{y}_j + k_{jy}y_j - \Sigma(C_{jni}\dot{\delta}_{jni} + k_{jni}\delta_{jni})\cos(\psi_{jni}) = 0 \end{cases} \quad j = s, r \quad (4)$$

where $\psi_{sni} = \psi_{ni} - \alpha_{sni}$, $\psi_{rni} = \psi_{ni} + \alpha_{rni}$, α_{sni} and α_{rni} are the pressure angle of external, internal gear pair, respectively; δ_{sni}, δ_{rni} is gear elastic deformation along the line of action between the *i*th planet gear and sun gear, ring gear, respectively. Similar to the carrier, the torsional equations of motion of central gears can be written as:

$$\begin{cases} (I_{zs}/r_{bs}^2)\ddot{u}_s + (C_{su}\dot{u}_s + k_{su}u_s) + \Sigma(C_{sn}\dot{\delta}_{sni} + k_{sn}\delta_{sni}) = T_s/r_{bs} \\ (I_{zr}/r_{br}^2)\ddot{u}_r + (C_{ru}\dot{u}_r + k_{ru}u_r) - \Sigma(C_{rni}\dot{\delta}_{rni} + k_{rni}\delta_{rni}) = 0 \end{cases} \quad (5)$$

where r_{bj}, $j = s, r, ni$ are radius of base circle of sun gear, ring gear and the *i*th planet gear, respectively.

Equations of Motion of Planet Gear. For the *i*th planet gear, it is positioned at an angle ψ_{ni} respected to the first planet gear, meshing with the sun gear and ring gear, each planet gear jointed to the carrier through a pin-bearing assembly. Similar to carrier, planet gear subjects interaction force from pin-bearing assembly modeled by translational stiffness k_{cni} and damping C_{cni}, torsional stiffness k_{nu} and damping C_{nu}, internal and external meshing forces and inertial force.

$$\begin{cases} m_{ni}a_{nix} + C_{cni}\dot{x}_{cniX} + k_{cni}x_{cniX} - (C_{sni}\dot{\delta}_{sni} + k_{sni}\delta_{sni})\sin(\alpha_{sni}) \\ +(C_{rni}\dot{\delta}_{rni} + k_{rn}\delta_{rni})\sin(\psi_{rni}) = 0 \\ m_{ni}a_{niy} + C_{cni}\dot{y}_{cniY} + k_{cni}y_{cniY} - (C_{sni}\dot{\delta}_{sni} + k_{sni}\delta_{sni})\cos(\psi_{sni}) \\ -(C_{rni}\dot{\delta}_{rni} + k_{rni}\delta_{rni})\cos(\psi_{rni}) = 0 \\ (I_{zni}/r_{bni}^2)\ddot{u}_{ni} + (C_{nui}\dot{u}_{ni} + k_{nui}u_{ni}) - (C_{sni}\dot{\delta}_{sni} + k_{sni}\delta_{sni}) + (C_{rni}\dot{\delta}_{rni} + k_{rni}\delta_{rni}) = 0 \end{cases} \tag{6}$$

where x_{cniX}, y_{cniY} are relative displacements of planet gear with respect to the carrier in X_{ni} and Y_{ni} directions respectively.

Equations (2)–(6) stands for the equations of motion of the 3-pinion planetary gear train, expanding $i = 1, 2, 3$ and $j = s, r$, totally 18 equations. The 18 equations include translational and torsional displacements, totally 18 variables and 6 gear elastic deformations (δ_{sni} and δ_{rni}, $i = 1, 2, 3$). In other words, there are totally 24 variables in 18 equations, in order to determine the 24 variables, additional 6 supplementary equations are needed, which are presented by follow section.

3 Gear Elastic Deformations Compatibility

In order to solve the dynamic equations, supplementary equations are needed to describe the relationship between vibration displacements, equivalent mesh error and gear elastic deformations. According to the contact condition of tooth pair, the elastic and geometric motion involving manufacturing and assembly errors will be compatible along the line of action of gear pair.

According to the contact condition of engaged tooth pair, gear pair elastic deformation should be equal to sum of the equivalent displacements of gears and equivalent meshing error. That is referenced as gear elastic compatibility equation, can be written as:

$$\begin{cases} \delta_{sni} = p_{sni} \cdot H_{cont} \\ \delta_{rni} = p_{rni} \cdot H_{cont} \end{cases}, \quad H_{cont} = \begin{cases} 1 & p_{sni}, p_{rni} \geq 0 \\ 0 & p_{sni}, p_{rni} < 0 \end{cases} \tag{7}$$

where H_{cont} is defined as contact factor, p_{sni}, p_{rni} are the relative gear mesh displacement of sun-planet and ring-planet gear pair, respectively (shown in Fig. 2).

$$\begin{cases} p_{sni} = -x_s\sin(\psi_{sni}) + y_s\cos(\psi_{sni}) + u_s - x_{ni}\sin\alpha_{sni} - y_{ni}\cos\alpha_{sni} - u_{ni} + e_{sni} \\ p_{rni} = -x_r\sin(\psi_{rni}) + y_r\cos(\psi_{rni}) - u_r + x_{ni}\sin\alpha_{rni} - y_{ni}\cos\alpha_{rni} + u_{ni} + e_{rni} \end{cases} \tag{8}$$

where e_{sni} and e_{rni} are the equivalent mesh errors of sun-ith planet gear pair, ring-ith planet gear pair, respectively; and these errors are caused by manufacturing and assembly errors.

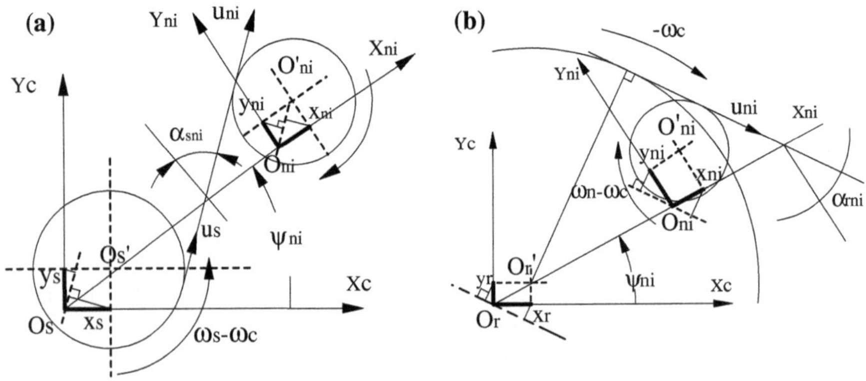

Fig. 2 Relative displacement of gear pairs: **a** external gear pair; **b** internal gear pair

4 Vibration Characteristics

The gear vibration characteristics of the 3-pinion planetary gear train are discussed in this section. Time-varying mesh stiffness and manufacturing and assembly errors are taken into account. The effects of support stiffness and flexible planet pin on vibration characteristic are quantified.

Time-varying mesh stiffness varies as the gears rotating, because meshing process always varies from one and two pairs of spur gear teeth in contact. The mesh stiffness k_{sni} and k_{rni} are time-dependent periodic functions over one mesh cycle; which can be approximated by the Fourier series:

$$k = k_{\min} + (k_{\max} - k_{\min})(\varepsilon - 1) + \sum_{s=1}^{\infty} \frac{k_{\max} - k_{\min}}{s\pi} \left[\sin 2\pi s(\frac{\Delta t}{T} + \varepsilon - 1 - \frac{t}{T}) - \sin 2\pi s(\frac{\Delta t}{T} - \frac{t}{T}) \right] \tag{9}$$

where $k_{\min}$ is minimum of time-varying mesh stiffness, and $k_{\max}$ is maximum one, ε represents contact ratio. The values of $k_{\min}$ and $k_{\max}$ can be determined through Ishikawa formula [10].

Effect of support stiffness. Using parameters in Table 1 and time-varying mesh stiffness, we solved the equations by Runge-Kutta numerical integral method. The translational and torsional vibration displacements, velocities and accelerations of gear centers are determined. As shown in Fig. 3, the planar vibration accelerations of sun gear center in coordinate system $\{O_c; X_c, Y_c\}$ and the accelerations of the first planet gear center in $\{O_{n1}; X_{n1}, Y_{n1}\}$ are presented.

Figure 4 shows the vibration accelerations of the system with sun gear support stiffness decreased 10 times to 1.0×10^5 N/mm, which is equivalent to floating sun gear condition. Compared with Fig. 3, the accelerations of sun gear and planet gear

Table 1 Parameters of the 3-pinion planetary gear train

	Sun	Planet	Ring	Carrier
Number of teeth	35	31	97	
Modulus (mm)	18	18	18	
Mass (kg)	642.8	474.5	742.8	2920.9
Pressure angle (deg)	20	20	20	
Tooth width (mm)	326	320	326	
Input speed (r/min)	545			
Input torque (N mm)	6.7×10^5			
Support stiffness (N/mm)	1.0×10^6	1.0×10^6	1.0×10^6	1.0×10^6
Torsional stiffness (N/mm)	3.44×10^7	2.68×10^7	1.0×10^6	1.78×10^7
Translational damping (N s/mm)	10	10	10	10
Torsional damping (N s/mm)	5	5	5	5
Eccentric errors (μm)	6	6	6	
Assembly error (μm)	6	6	6	

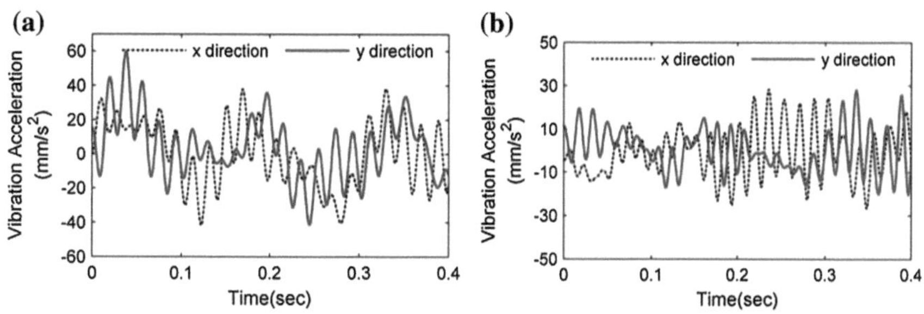

Fig. 3 Vibration accelerations of the 3-pinion planetary gear train: **a** sun gear, **b** planet gear

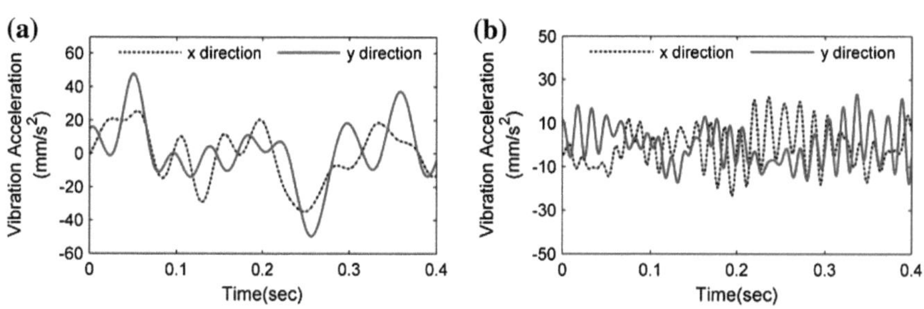

Fig. 4 Vibration accelerations of system with floating sun gear: **a** sun gear, **b** planet gear

fluctuate more steadily, and the maximum vibration acceleration value of sun gear decreased from 30 to 22 mm/s^2, 60 to 48 mm/s^2 in x and y direction, respectively, also the same to the acceleration of planet gear.

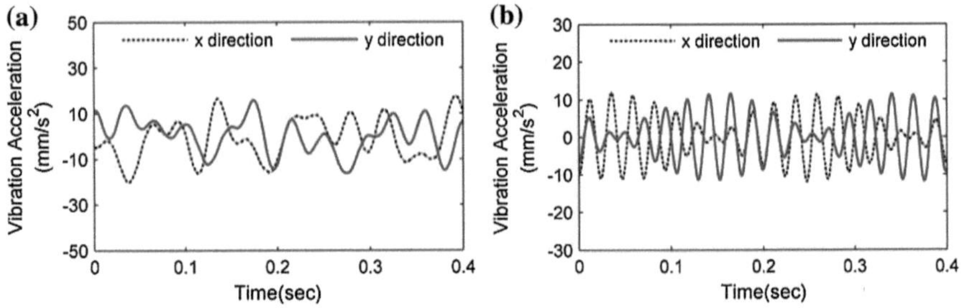

Fig. 5 Vibration acceleration of system with flexible planet pin: **a** planet gear, **b** ring gear

Effect of flexible planet pin. The influence of the support stiffness to the planet gears is investigated by a decrease value. In physical system, flexible planet pins are used, which allow the planet gears deflect independently in a circumferential direction, Fig. 5 shows the vibration acceleration when using flexible planet pin, for which the support stiffness of planet pins, is decreased to 1.67×10^5 N/mm. Comparing to Fig. 3b, the maximum vibration acceleration value of sun gear decreased from 28 to 12 mm/s^2, 30 to 12 mm/s^2 in x and y direction, respectively. The use of flexible planet bin improves the vibration among planet gears significantly.

5 Conclusions

In this paper, a new dynamic model for planetary gear train is developed by lumped-parameter method, incorporating time-varying mesh stiffness obtained through Ishikawa formula. The influences of support stiffness and flexible planet pin on vibration characteristic are simulated and discussed. It is shown that support stiffness has a significant influence on gear vibration. In particular, floating sun gear can improve the vibration obviously, and the use of a flexible pin can be helpful in improving planet vibration characteristics.

Acknowledgments The authors want to acknowledge the financial support from National Natural Science Foundation of China (No. 51375065), Liaoning Provincial Science and Technology Innovation major projects (No. 201303004-1).

References

1. Parker, R.-G., Vijayakar, S.-M., Imajo, T.: Non-linear dynamic response of a spur gear pair: modelling and experimental comparisons. J. Sound Vib. **237**(3), 435–455 (2000)
2. Kahraman, A.: Planetary gear train dynamics. J. Mech. Des. **116**(3), 713–720 (1994)

3. Detinko, F.-M.: Free vibration of a thick ring on multiple supports. Int. J. Eng. Sci. **27**(11), 1429–1438 (1989)
4. Singh, A.: Epicyclic load sharing map-development and validation. Mech. Mach. Theory **46** (5), 632–646 (2011)
5. Frater, J., August, R., Oswald, F.-B.: Vibration in planetary gear systems with unequal planet stiffness. NASA (TM-83428) (1982)
6. Guo, X., Velex, P.: On the dynamic simulation of eccentricity errors in planetary gears. Mech. Mach. Theory **61**, 14–29 (2013)
7. August, R., Kasuba, R.: Torsional vibrations and dynamic loads in a basic planetary gear system. J. Vib. Acoust. **108**(3), 348–353 (1986)
8. Zhang, D., Wang, S.: Parametric vibration of split gears induced by time-varying mesh stiffness. Proc. Inst. Mech. Eng. Part C J. Mech. Eng. Sci. **229**(1), 18–25 (2014)
9. Kahraman, A.: Natural modes of planetary gear trains. J. Sound Vib. **173**(1), 125–130 (1994)
10. Li, Y., Sun, W.: Study on the improved algorithm of the time-varying meshing stiffness of gear. J. Mech. Transm. **33**(5), 22–26 (2010)

Kinematics and Dynamics of Compound and Complex Gear System

Jósef Drewniak, Jerzy Kopeć and Stanislaw Zawiślak

Abstract Analysis of planetary gear coupled with additional friction gear is analyzed in the present paper. The parts of the system are a compound planetary gear and a friction or hydro-kinetic gear. The kinematical and dynamical analyzes are described. Some variants of the system are considered. The standard methods e.g. Willis formulas and graphical-analytical methods were utilized.

Keywords Mixed graphs · Force and moment analysis · Variants of gears

1 Introduction

Planetary gears are widely used due to high ratios, relatively low weight and compactness. However design of such gears requires a full and detailed analysis of velocities, torques, efficiencies, etc. The conceptual phase of design [1] consists of searching for alternative design solution and introductory calculations. The simulation is one of available approaches for analysis of virtual designs. The authors had analyzed a compound planetary gear in work [2] by means of traditional and graph-based approaches. Versatile graph modeling methods were applied by the authors in considerations of different gears [2–6]. Here the proposed design solution consists in aggregation of the earlier considered planetary gear with an additional friction gear. In work [2], due to the fact that the DOF of the single gear was 2, the case of two inputs and two outputs was considered, respectively. In other papers of the authors [3, 4] the reduction of mobility is achieved via system of clutches and

J. Drewniak · J. Kopeć · S. Zawiślak (✉)
University of Bielsko-Biala, Bielsko-Biala, Poland
e-mail: szawislak@ath.bielsko.pl

J. Drewniak
e-mail: jdrewniak@ath.bielsko.pl

J. Kopeć
e-mail: jkopec@ath.bielsko.pl

B. Corves et al. (eds.), *Mechanisms, Transmissions and Applications*,
Mechanisms and Machine Science 31, DOI 10.1007/978-3-319-17067-1_21

brakes aiming for reduction of the DOF to 1 and therefore precise calculations of the ratio and the output velocity are possible. This approach to design is very characteristic for the automotive automatic gear boxes where different drives (gears—ahead and reverse) are achieved via adequate rearrangements of control elements. Finally, efficiency analysis and optimization have to be performed [7].

In this paper, another approach is discussed i.e. an aggregation of two gears—it is the main idea, then full kinematical analysis as well as some bond graph modeling attempts have been done. Upon matching two gears, we obtain a possibility of mechanical control of the obtained system. An aim of this paper is just to propose and to analyze of the compound and complex gear system.

2 References Review

Book [8] encloses versatile gear, gearbox and automotive automatic gears schemes, their comparisons, modeling techniques, calculations, technical specifications etc. It is a very good practical manual for those who would be interested in gear design. There were at least two additions of this book. The modeling techniques for gear design and simulations are very different, but graphs are used very frequently [2–6, 9–15]. Versatile graph approaches are utilized. The review of the graph-based methods and their descriptions can be found in book [6] and paper [18]. The following graphs are used: mixed graphs [2, 6], contour graphs [2–5, 19], linear graphs [14, 15] as well as bond graphs [4, 9, 11–13]. The theoretical basics of bond graphs modeling can be found in books [16, 17] and for other graphs in thesis [6]. Searching for design solution of gear can be performed via search through the well-known general schemes [8] and modification or via dedicated synthesis procedure [18] or other approaches described in Pahl et al. [1]. The approach consisting in looking through variants is discussed in Drewniak and Zawiślak [5]. An aggregation of gears was proposed in Hsieh and Tang [19]. The tasks performed by means of graphs are as follows: synthesis [18], analysis [10], generation of atlases of design solutions [5], efficiency analysis, reverse problems, modeling [14] and some others.

Bond graphs seem to be the most powerful tool due to the fact that they can model several different subsystems simultaneously i.e. mechanical, electrical, pneumatic etc. Recently, bond graphs were used for drive systems of wind turbines [6, 9, 11, 12] and automobile gearboxes [13], but especially in hybrid systems.

3 Analysis of Designs of Complex Gears

The compound planetary gear can co-operate with another gear creating a complex gear system. Variants of such artifact are given in Figs. 1 and 2. The system could be applied e.g. for wind turbine drive system. Due to variable directions and power of wind blow the turbine gear system should be controllable. Besides mechatronic

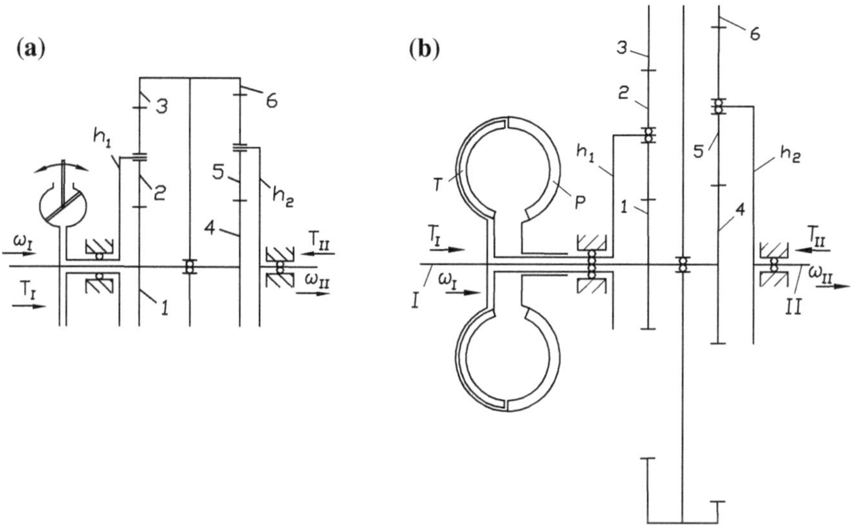

Fig. 1 Schemas of complex gear enclosing: **a** frictional gear, **b** hydro-kinetic gear

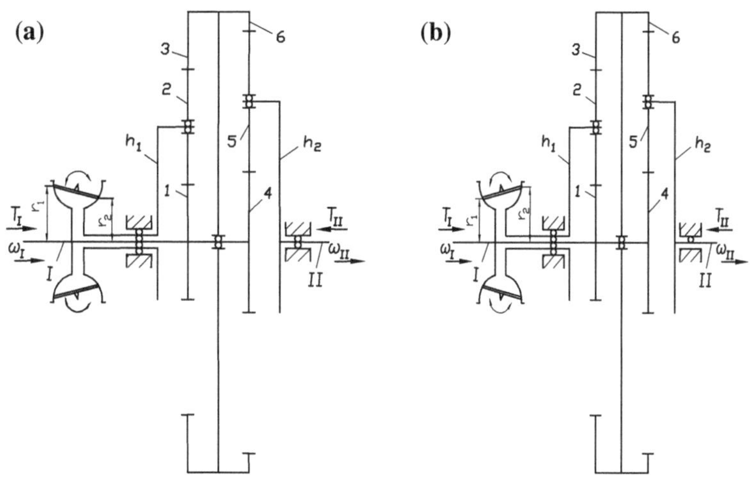

Fig. 2 Schemas of complex gear system enclosing friction gear (second variant): **a** reductor, **b** multiplicator

control systems, some mechanical subsystem allowing for variability of characteristics, are valuable. The variant of complex gear system differs in the introductory gear, when planetary gear remains the same. However, it itself can be modified via changes of teeth numbers and wheel dimensions.

The compound gear itself consists of two rows of elementary planetary gears connected mutually. The sun wheels are mounted together on the common shaft rotating along the main axis of symmetry (1 and 4). Planetary geared wheels (planets) 2 and 5 have carriers h_1 and h_2, respectively.

External drum has two internal toothings (teeth rings) 3 and 6. The gear without any clutch or brake has 2 DOF, therefore its performance could not be described without mutual dependence of parameters. The gear has interesting properties i.e. changes of dimensions of gear parts of these two mentioned rows allows for control of sign and value of the partial ratios and in consequence of the whole gear. However, precise determination of the ratio is needed, therefore it should be also done for the system of gears.

4 Kinematical and Dynamical Analysis of Gears

We start the considerations from the compound gear. The compound planetary gear is analyzed taking into account velocities and forces. Like it was mentioned before, there are two rows of elementary planetary gears. The simplified graphical analyses of velocities for both rows are presented in Fig. 3 and Fig. 4 for planets (planetary wheels) and sun wheels.

The rule of lack of losses is applied, therefore relative linear velocities are equal. The notions applied are as follows: v—linear velocities and ω—angular velocities. The goal of the current considerations is determination of all kinematical characteristic quantities. The second variant of dimensions is shown in Fig. 4. It is very interesting that increasing the diameter of the sun wheel *4*, we can change not only the value of kinematical ratio of the gear but also the direction of reverses of the

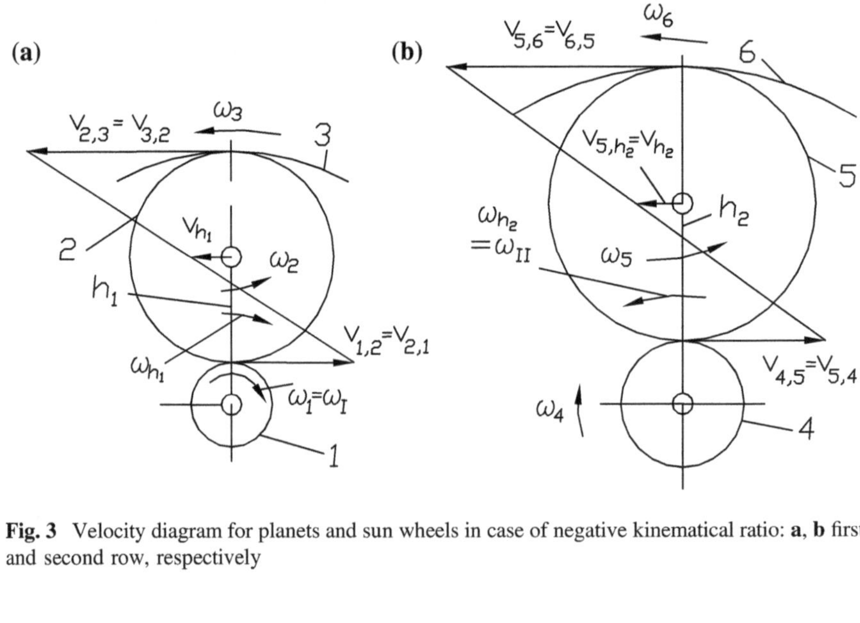

Fig. 3 Velocity diagram for planets and sun wheels in case of negative kinematical ratio: **a**, **b** first and second row, respectively

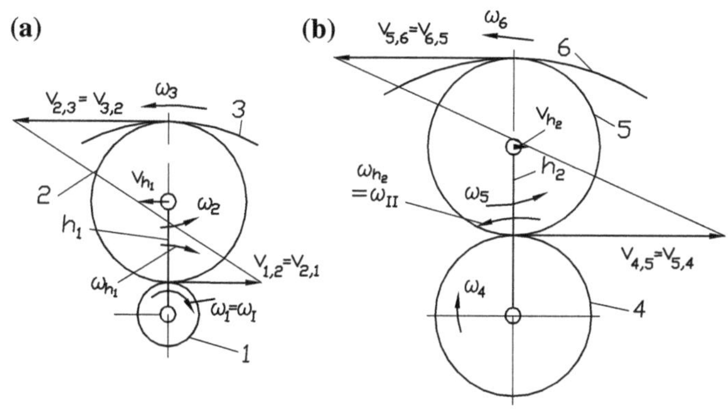

Fig. 4 Velocity diagram for planets and sun wheels in case of positive kinematical ratio

output shaft II as well as the carrier h_2. In Fig. 5, so called, free body diagrams of all three wheels for two rows are presented. Due to this approach, the forces analyses can be effectively performed based on the 3-rd Newton law.

In Fig. 5, the analysis of forces is graphically and schematically presented for two gear rows, abstracting from dimensions. Here, it was assumed that there are three planetary wheels in every case, however other assumption was done in the calculations presented underneath.

5 Calculations

The DOF for the gear system, consisting of two gears: friction and planetary can be calculated based on the following formula:

$$W = 3 \cdot n - 2 \cdot p_5 - p_4 = 3 \cdot 7 - 2 \cdot 7 - 6 = 1 \tag{1}$$

We can calculate the ration of the compound planetary gear using the Willis formula (taking into account the I-st and II-nd row of the planetary gear). The calculations were performed for the following data—(I row): $z_1 = z_2 = 24$, $z_3 = -72$, $s_I = 3$ number of planetary wheels; (II row): $s_{II} = 3$—number of planetary wheels, $z_4 = 24$, $z_5 = 18$, $z_6 = -60$ as well as $m_I = m_{II} = 2$—modules for I-st and II-nd row, respectively. We have also relationships connected with the design of the gear: $\omega_I = \omega_1 = \omega_4$ and $\omega_3 = \omega_6$ (Fig. 5 and Fig. 6).

The rotational velocity and moment (torque) are equal: $n_I = n_1 = n_4 = 900\ \frac{\text{obr}}{\text{min}}$, $T_I = 300\ \text{N} \cdot \text{m}$.

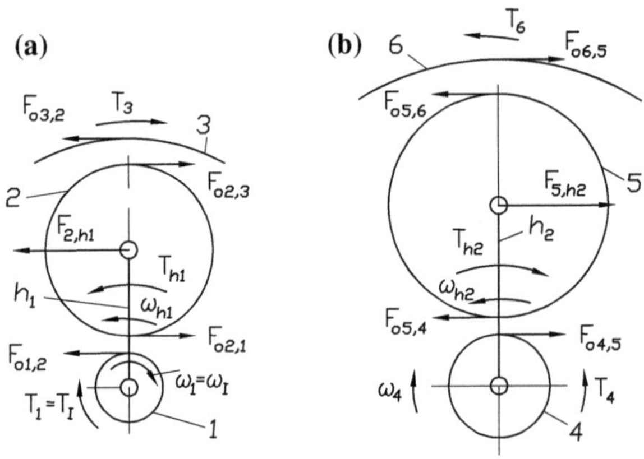

Fig. 5 Free body diagram of meshed wheels—analysis of tangent forces; in case of negative ratio (radial forces are omitted)

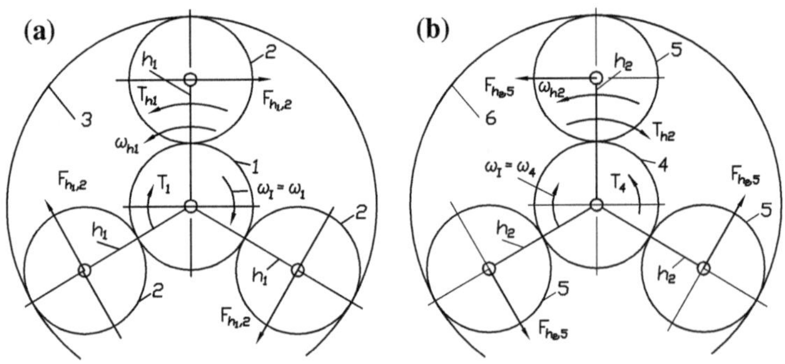

Fig. 6 Distribution of forces and torques acting on: **a** active carrier h_1; **b** passive carrier h_2

The ratio of friction gear is negative because the friction gear changes direction of rotations—Fig. 6a:

$$i_f = \frac{r_2}{r_1 \cdot (1 - \varepsilon)} = \frac{\omega_I}{\omega_{h1}} \tag{2}$$

It was assumed: $r_1 = 100$, $r_2 = 150$, $\varepsilon = 0,01$—slip coefficient (neglected in calculations). The carrier velocity is equal to:

$$\omega_{h1} = \frac{\omega_I}{i_f}, \quad \text{where } \omega_1 = \frac{\pi \cdot n_1}{30} \tag{3}$$

The direction of reverses of the carrier h_1 is opposite in comparison to the direction of reverses of the input shaft No I.

Ratio of friction gear is negative because of change of rotation direction. It can be written by means of the formula:

$$i_f = \frac{\omega_I}{\omega_{h1}} = -\frac{r_2}{r_1 \cdot (1-\varepsilon)} = -\frac{150}{100} = -1.5 \tag{4}$$

Therefore carrier velocity is equal:

$$\omega_{h1} = \frac{\omega_I}{i_f} = \frac{94.248}{-1.5} = -62.832\,\frac{\text{rad}}{\text{s}}, \tag{5}$$

where: $\omega_1 = \omega_4 = \frac{\pi \cdot n_1}{30} = \frac{\pi \cdot 900}{30} = 94.248\,\frac{\text{rad}}{\text{s}}$,

Willis formula for the I-st row of the planetary gear (set of wheels 1, 2 and 3) let us calculate the base ratio of the I-st row of the planetary gear:

$$i_{1,3}^{h1} = \frac{\omega_1 - \omega_{h1}}{\omega_3 - \omega_{h1}} = \frac{z_3}{z_1} = \frac{-72}{24} = -3 \tag{6}$$

As above, the Willis formula is used for the II-nd row of the planetary gear, and the basic ratio for the II-nd row of planetary gear is equal:

$$i_{4,6}^{h2} = \frac{\omega_4 - \omega_{h2}}{\omega_6 - \omega_{h2}} = \frac{z_6}{z_4} = \frac{-60}{24} = -2.5 \tag{7}$$

Angular velocity of drum wheels 3 and 6 was obtained from Willis formula either:

$$\omega_3 = \omega_6 = \frac{\omega_1 \cdot \left[1 + \frac{1}{i_f} \cdot \left(i_{1,3}^{h1} - 1\right)\right]}{i_{1,3}^{h1}} = -115.192\,\frac{\text{rad}}{\text{s}} \tag{8}$$

Angular velocity of the carrier h_2:

$$\omega_{h2} = \frac{\omega_3 \cdot i_{4,6}^{h1} - \omega_1}{\left(i_{4,6}^{h1} - 1\right)} = \frac{-115.192 \cdot (-2.5) - 94.248}{-2.5 - 1} = -55.352\,\frac{\text{rad}}{\text{s}} \tag{9}$$

In another way:

$$\omega_{h2} = \frac{\omega_1 \cdot \left[\frac{\left[1 + \frac{1}{i_f} \cdot \left(i_{1,3}^{h1} - 1\right)\right] \cdot i_{4,6}^{h1}}{i_{1,3}^{h1}} - 1\right]}{\left(i_{4,6}^{h1} - 1\right)} = -55.352\,\frac{\text{rad}}{\text{s}} \tag{10}$$

Kinematical ratio of the gear is equal to:

$$i_{I,h2} = \frac{\omega_I}{\omega_{h2}} = \frac{i_{1,3}^{h1} \cdot \left(i_{4,6}^{h1} - 1\right)}{-\left[1 + \frac{1}{i_f} \cdot \left(i_{1,3}^{h1} - 1\right)\right] \cdot i_{4,6}^{h1} - i_{1,3}^{h1}} = 1.7027 \tag{11}$$

The graphical-analytical method was applied for obtaining angular and linear velocities of wheels and carriers. For example the linear tangent velocity of sun wheel and satellite wheel in meshing point 2 is equal:

$$v_{1,2} = \omega_1 \cdot \frac{d_1}{2} = 94.248 \cdot \frac{48}{2} \cdot 10^{-3} = 2.262\,\frac{\mathrm{m}}{\mathrm{s}} \tag{12}$$

where:

$$\omega_1 = \frac{\pi \cdot n_1}{30} = \frac{\pi \cdot 900}{30} = 94.248\,\frac{\mathrm{rad}}{\mathrm{s}}$$

For obtaining linear velocity of wheel 2, meshed in point with wheel 3 we can write down the equation (Fig. 6a):

$$\frac{v_{2,3} - v_{h1}}{\frac{d_2}{2}} = \frac{v_{2,1} + v_{h1}}{\frac{d_2}{2}} \tag{13}$$

It should be noticed that: $v_{2,1} = v_{1,2}$. Thus we obtain:

$$v_{2,3} = v_{2,1} + 2 \cdot v_{h1} = 2.262 + 2 \cdot 3.016 = 8.294\,\frac{\mathrm{m}}{\mathrm{s}} \tag{14}$$

where:

$$v_{h1} = \omega_{h1} \cdot r_{h1} = 62.832 \cdot 48 \cdot 10^{-3} = 3{,}016\,\frac{\mathrm{m}}{\mathrm{s}}, \tag{15}$$

$$n_{h1} = \frac{n_I}{i_f} = \frac{900}{-1{,}5} = -600\,\frac{\mathrm{obr}}{\mathrm{min}}, \tag{16}$$

$$\omega_{h1} = \frac{\pi \cdot |n_{h1}|}{30} = \frac{\pi \cdot 600}{30} = 62.832\,\frac{\mathrm{rad}}{\mathrm{s}}, \tag{17}$$

$$r_{h1} = \frac{d_1 + d_2}{2} = \frac{m_I \cdot (z_1 + z_2)}{3} = \frac{2 \cdot (24 + 24)}{2} = 48. \tag{18}$$

Next we can calculate the velocity of drum wheel no 3:

$$\omega_3 = \frac{v_{3,2}}{\frac{|d_3|}{2}} = \frac{8.294 \cdot 10^3}{72} = 115.194 \, \frac{\text{rad}}{\text{s}}, \tag{19}$$

where $v_{3,2} = v_{2,3}$, and $d_3 = m_I \cdot z_3 = 2 \cdot (-72) = -144$.

The numerical values of angular velocity of drum wheels 6 and 3 are obviously the same ($\omega_6 = \omega_3$) and these angular velocities (ω_3 and ω_6) have the same direction.

For calculating the linear velocity $v_{h2,5}$ of carrier h_2 in rotating axis of satellite 5 ($v_{h2,5} = v_{5,h2} = v_{h2}$—linear velocity of center of item 5, Fig. 6b; we omit finally the index 5 as redundant) we use the following formula:

$$\frac{v_{5,6} - v_{h2}}{\frac{d_5}{2}} = \frac{v_{5,4} + v_{h2}}{\frac{d_5}{2}} \tag{20}$$

$$v_{h2} = 0,5 \cdot \left(v_{5,6} - v_{5,4}\right) = 0.5 \cdot (6.912 - 2.262) = 2.325 \, \frac{\text{m}}{\text{s}} \tag{21}$$

where:

$v_{5,6}$ linear velocity of wheel 5 in meshing point with drum wheel 6,

$v_{5,4}$ linear velocity of wheel 5 in meshing point with wheel

$$v_{5,6} = v_{6,5} = \omega_6 \cdot \left|\frac{d_6}{2}\right| = 115.194 \cdot 60 \cdot 10^{-3} = 6.912 \, \frac{\text{m}}{\text{s}}, \tag{22}$$

$$v_{4,5} = v_{5,4} = \omega_4 \cdot \frac{d_4}{2} = 94.248 \cdot \frac{48}{2} \cdot 10^{-3} = 2.262 \, \frac{\text{m}}{\text{s}}. \tag{23}$$

Angular velocity of carrier h_2:

$$\omega_{h2} = \frac{v_{h2}}{r_{h2}} = \frac{2.325 \cdot 10^3}{42} = 55.357 \, \frac{\text{rad}}{\text{s}}, \tag{24}$$

where:

$$r_{h2} = \frac{d_4 + d_5}{2} = \frac{m_{II} \cdot (z_4 + z_5)}{2} = \frac{2 \cdot (24 + 18)}{2} = 42.$$

Total gear ratio is

$$i_{I,h2} = -\frac{\omega_I}{\omega_{II}} = -\frac{\omega_I}{\omega_{h2}} = -\frac{94.248}{55.357} = -1.7025. \tag{25}$$

For comparison, second example was considered for the same planetary gear parameters. The only difference is that the ratio of friction gear is:

$$i_f = -\frac{r_2}{r_1 \cdot (1-\varepsilon)} = \frac{\omega_I}{\omega_{h1}} = -\frac{100}{100} = -1 \tag{26}$$

where $r_1 = 100$, $r_2 = 100$, $\varepsilon = 0,01$—slip coefficient (neglected in calculations).

The results of calculations were collected in Table 1.

Table 1 Collected values of velocities analysis

Variable	Value	Description
n_{h1}	$900\ \frac{\text{obr}}{\text{min}}$	
ω_{h1}	$94.248\ \frac{\text{rad}}{\text{s}}$	Angular velocity. The absolute value is considered, the direction of carrier velocity is compatible to direction assumed in Fig. 6
r_{h1}	48 mm	Radius of carrier h_1
ω_1	$94.248\ \frac{\text{rad}}{\text{s}}$	
$i_{1,3}^{h1}$	−3	Base ratio of I-st row of planetary gear
$i_{4,6}^{h2}$	−2.5	Base ratio of II-nd row of planetary gear
ω_3	$-157.080\ \frac{\text{rad}}{\text{s}}$	Angular velocity of drum wheel 3
d_3	−144 mm	
ω_{h2}	$-85.272\ \frac{\text{rad}}{\text{s}}$	
$i_{I,h2}$	1.1053	
$v_{1,2}$	$2.262\ \frac{\text{m}}{\text{s}}$	Row of planetary gear in meshing point with the satellite no 2
$v_{2,3}$	$11.310\ \frac{\text{m}}{\text{s}}$	Row of planetary gear 2 in meshing point with wheel 3
$v_{3,2} = v_{2,3}$	$11.310\ \frac{\text{m}}{\text{s}}$	
v_{h1}	$4524\ \frac{\text{m}}{\text{s}}$	
v_{h2}	$3.5815\ \frac{\text{m}}{\text{s}}$	Velocity of carrier h_2, axis of satellite 5
$v_{5,6}$	$9.425\ \frac{\text{m}}{\text{s}}$	Velocity of wheel 5 in the meshing point with wheel no 6; remark: $v_{5,6} = v_{6,5}$
$v_{4,5}$	$2.262\ \frac{\text{m}}{\text{s}}$	Velocity of 5 in meshing point with 4; remark: $v_{4,5} = v_{5,4}$, $\omega_4 = \omega_1$
ω_{h2}	$85.274\ \frac{\text{rad}}{\text{s}}$	Angular velocity of carrier h_2
r_{h2}	42 mm	Carrier radius h_2
$i_{I,h2}$	−1.1052	Total ratio

6 Conclusions

The complex gear systems were analyzed. The system consists of the planetary compound planetary gear and the friction gear. Kinematical and dynamical analyses were performed showing a range of achieved parameters, bond graph calculations are not entered but they confirm correctness of analyses. The calculations were performed based upon traditional approach. The immanent property of the system is possibility of essential modification via changing design parameters.

References

1. Pahl, G., Beitz, W., Feldhusen, J., Grote, K.-H.: Engineering design: A systematic approach, 3rd edn. Berlin, Springer (2007)
2. Drewniak, J., Zawiślak, S.: Linear graph and contour graph-based models of planetary gears. JTAM **48**(2), 415–433 (2010)
3. Drewniak, J., Zawiślak, S.: Graph-based models of compound planetary gear boxes. Solid State Phenom. **199**, 143–148 (2013)
4. Drewniak, J., Kopeć, J., Zawiślak, S.: Graph based analysis of planetary gears. ACC J. XVII **1**, 15–23 (2011)
5. Drewniak, J., Zawiślak, S.: Analysis and modification of planetary gears based upon graph-theoretical models. Transactions of the Universities of Košice, vol. 2, pp. 84–87 (2009)
6. Zawislak, S.: The graph-based methodology as an artificial intelligence aid for mechanical engineering design. Habilitation Thesis, University of Bielsko-Biala, Poland (2010)
7. Zhao, N., Jia, Q.J.: The efficiency analysis and optimization of gear trains. Adv. Manuf. Technol. **156–157**, 1000–1005 (2010)
8. Looman, J.: Zahnradgetriebe. Springer, Berlin (1988)
9. Agarwal, S., et al: Bond graph model of wind turbine blade. In: Bond Graph Modeling. Theory and Practice, MathMod, Vienna (2012)
10. Ardte, D., Lam, K.-T.: Analysis of kinematics of a gear steering mechanism for the tracked vehicles. J. Sci. Innov. **2**(3), 183–188 (2012)
11. Badoud, A.E., et al.: Bond graph algorithms for fault detection and isolation in wind energy conversion. Arab. J. Sci. Eng. **39**, 4057–4076 (2014)
12. Bakka, T., Karimi, H.R.: Bond graph modeling and simulation of wind turbine systems. J. Mech. Sci. Technol. **27**(6), 1843–1852 (2013)
13. Coudert, N., Dauphin-Tanguy, G.: Mechatronic design of an automatic gear box using bond graphs. In: Proceedings of the Conference on Systems, Man and Cybernetics, vol. 2, pp. 216–221 (1993)
14. Lang, S.Y.T.: Graph-theoretic modeling of epicyclic gear system. Mech. Mach. Theory **40**, 511–529 (2005)
15. Wojnarowski, J., Kopec, J., Zawislak, S.: Gears and graphs. J. Theor. Appl. Mech. **44**, 139–162 (2006)
16. Brown, F.T.: Engineering systems dynamics: A unified graph-centered approach. Marcel Dekker, New York (2001)
17. Merzouki, R., Samataray, A.E., et al.: Intelligent mechatronic systems. Springer, London (2013)
18. He, R., et al.: The topological synthesis design of hybrid electrically variable transmission. Appl. Mech. Mater. **229**, 38–43 (2013)
19. Hsieh, L-Ch., Tang, H-Ch.: The kinematic design of 2 K-2H planetary gear reducers with high reduction ratio. Appl. Mech. Mater. **319**, 610–615 (2013)

Part V
Robotics

Step Design of a Cassino Tripod Leg Mechanism

Mingfeng Wang and Marco Ceccarelli

Abstract In this paper, we present the mechanical design of a tripod leg mechanism for a Cassino biped locomotor. Kinematic properties are investigated to select the proper stroke of the linear actuators and the radius difference between the moving and base plates. Workspace optimization of the adopted 3-UPU manipulator is computed by considering the Global Condition Index and universal joints constraints. Step design of the foot plate is presented and analyzed to improve the performance by changing the initial length and reassembling of the three limbs in the leg mechanism.

Keywords Biped locomotors · Workspace analysis · Tripod leg mechanisms · Step design

1 Introduction

Biped locomotors, as biped robots, have attracted interests of many research communities in the past decade, and a lot of prototypes have been built in the laboratories and even for specific application tasks [1, 2].

However, most of the existing biped locomotors are based on leg designs with human-like architectures by using serial chain solutions, such as ASIMO, NAO, HUBO, HRP-4 and so on, while WL-16 (Waseda Leg-No. 16) is a design that achieved world first dynamic human-carrying biped walking as based on leg designs with Gough-Stewart parallel mechanisms [3]. Ota et al. [4] and Sugahara et al. [5] have also proposed to use Gough-Stewart parallel mechanisms for leg

M. Wang (✉) · M. Ceccarelli
LARM: Laboratory of Robotics and Mechatronics,
DICeM-University of Cassino and South Latium, Cassino, Italy
e-mail: wang@unicas.it

M. Ceccarelli
e-mail: ceccarelli@unicas.it

B. Corves et al. (eds.), *Mechanisms, Transmissions and Applications*,
Mechanisms and Machine Science 31, DOI 10.1007/978-3-319-17067-1_22

modules. Nevertheless, the potentiality of parallel manipulators (PMs) for leg mechanisms has not been fully investigated, since the typical 6-DOF PMs also suffer from some disadvantages, e.g., reduced workspace, difficult mechanical design, complex direct kinematics, and accurate control algorithms. To overcome the above disadvantages, PMs with fewer than six DOFs, namely reduced DOF manipulators, have been widely attempted both in industry and academy. In the field of leg designs for biped robots, Ceccarelli and Carbone [6] have investigated the possibility of using PMs with less than six degrees of freedom for leg design as inspired from the human leg muscular system. Pan and Gao [7] have presented a new kind of hexapod walking robot as based on 3-DOF parallel mechanisms for applications when nuclear disaster happens. Wang et al. [8] has proposed a quadruped/biped reconfigurable walking locomotor as based on four 3-UPU PMs during quadruped walking and converting into two 6-SPU PMs during biped working. Indeed, 3-DOF PMs could be enough for reducing the total cost and operations of leg mechanisms for biped locomotors, since that fewer actuators are required.

At the Laboratory of Robotics and Mechatronics (LARM), University of Cassino and South Latium, one research line is devoted to design and analysis of parallel mechanisms with reduced DOFs for multiple purposes, and several prototypes of parallel mechanisms have been built with low-cost and easy-operation characteristics. Research activities have been carried out on both theoretical aspects and application implementations. A 3-DOF PM has been designed and used as a novel design solution of a tripod leg mechanism for a Cassino biped locomotor [9]. To improve the experimental performance of the built prototype of the tripod leg mechanism [10], the kinematic optimization process is presented in this paper by taking into account reachable workspace, kinematic isotropy and step design.

In this paper, the built prototype of the tripod leg mechanism is presented as based on a 3-UPU parallel manipulator. In order to obtain the well-conditioned workspace, kinematic analysis and optimization progress are proceeded by considering the Global Condition Index and angular limits of universal joints. The step design of the foot plate is presented and analyzed by comparing two configurations of three limbs in the leg mechanism.

2 The Cassino Tripod Leg Mechanism

A first prototype of the proposed tripod leg mechanism, as shown in Fig. 1a, has been built and verified for prescribed movements [10]. Its kinematic scheme of a 3-UPU (U and P stand for universal and prismatic pair, respectively) parallel manipulator is shown in Fig. 1b.

The leg mechanism consists of a waist plate, a moving foot plate, and three identical limbs with linear actuators and U-joints at each end. By properly orienting the axes of U-joints, the moving foot plate can only perform purely translational motion, as pointed out in [11].

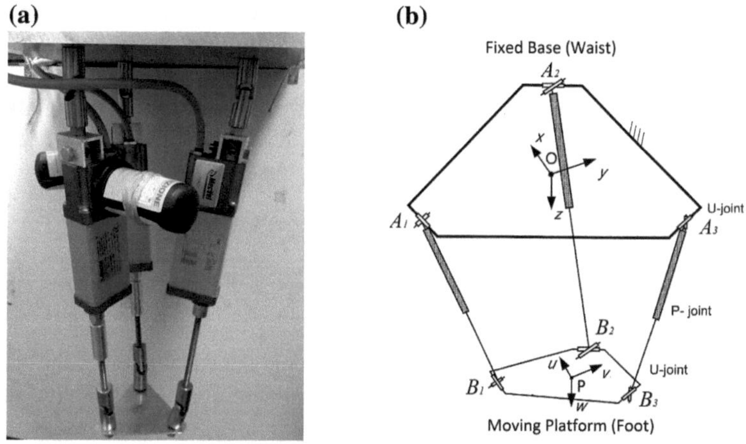

Fig. 1 The Cassino tripod leg mechanism: **a** a prototype; **b** a kinematic scheme of 3-UPU PM

The main specifications and details of mechanical design parameters of the tripod leg mechanism has been indicated in the previous work [10], as referring to dimension parameters in Fig. 2, where L, W, and H are length, width and height of the mechanism, respectively; S_L and S_H are step length and height; step trajectory starts from point S-1 and follows the sequences of S_1-2-3-4-1. H_w and H_f are the thicknesses of waist and foot plates; L_f and W_f are length and width of the foot,

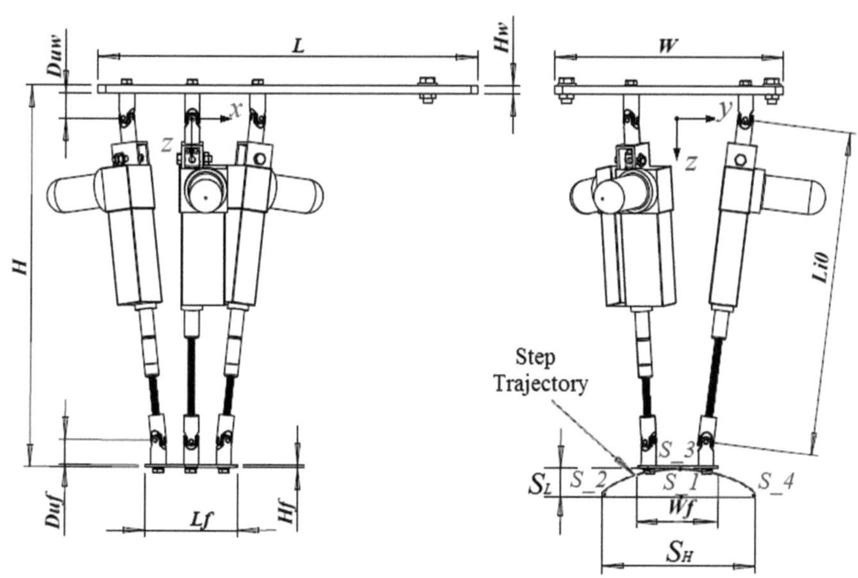

Fig. 2 Dimension parameters of the prototype in Fig. 1a

respectively. In each limb, the distance between the rotation center of upper U-joint and waist plate is equal to that between the rotation center of lower U-joint and foot plate, i.e. the half-length of the universal joint, is noted as $D_{uw} = D_{uf}$, the distance between the two rotation centers of upper and lower U-joints, i.e. the length of each limb, is noted as l_i (i = 1, 2, 3), and the initial value of l_i is set as L_{i0}, which determines the initial height of the proposed leg mechanism.

3 Kinematic Analysis

In this section, the direct and inverse kinematics and the calculation of the Jacobian matrix of the proposed 3-UPU parallel manipulator are based on the solution methodology as presented in [9, 11].

In Fig. 1b, the position vectors of points A_i and B_i (i = 1,…,3) with respect to base coordinate frame A: O-xyz and moving coordinate frame B: P-uvw (the origins O and P placed at the corresponding triangle's center) respectively, can be written as

$$^{A}\mathbf{a_i} = \left[a_{ix}, a_{iy}, 0\right]^T, \quad \text{and} \quad ^{B}\mathbf{b_i} = {}^{A}\mathbf{b_i} = \left[b_{ix}, b_{iy}, 0\right]^T \tag{1}$$

where a superscript indicates the reference coordinate frame in which a vector is expressed. The vector **p** determines the position of point P in the base coordinate frame. Furthermore, since vectors $\mathbf{a_i}$ and $\mathbf{b_i}$ are constants, the difference of them, which is defined as vector $\mathbf{c_i}$, is a constant too. Hence, a vector-loop equation can be obtained for each limb as

$$\mathbf{l_i} = \mathbf{p} + \mathbf{b_i} - \mathbf{a_i} = \mathbf{p} - \mathbf{c_i} \tag{2}$$

For the direct kinematics, the position vector $\mathbf{p} = [p_x, p_y, p_z]^T$ of the moving platform can be computed according to the given length of each limb l_i. The solution can be obtained by expanding the Eq. (2) in the form

$$\mathbf{p}^2 - 2\mathbf{p}\cdot\mathbf{c_i} + \mathbf{c_i^2} = l_i^2 \quad \text{for i} = 1, 2, 3. \tag{3}$$

For the inverse kinematics, the length of each limb l_i can be computed according to the given position vector **p**. The solution can be obtained by dot-multiplying Eq. (2) with itself and taking the square root as

$$l_i = \pm\sqrt{\left(p_x - c_{ix}\right)^2 + \left(p_y - c_{iy}\right)^2 + p_z^2} \quad \text{for i} = 1, 2, 3. \tag{4}$$

where only positive root can be obtained in the presented leg mechanism, while negative root cannot be obtained unless reassembling the mechanism.

For the Jacobian matrix **J**, it relates the velocities of moving platform and linear actuator and can be written as a 3 × 3 matrix as

$$\mathbf{J}\dot{\mathbf{p}} = \dot{\mathbf{l}}_\mathbf{i} \tag{5}$$

$$\mathbf{J} = \left[\mathbf{u}_{13}{}^{\mathrm{T}}, \mathbf{u}_{23}{}^{\mathrm{T}}, \mathbf{u}_{33}{}^{\mathrm{T}}\right]^{\mathrm{T}} \tag{6}$$

where $\mathbf{u}_{i3}$ are the unit vectors along the axes of the prismatic joint axis for limb *i*.

Due to the direct kinematic analysis above, the platforms workspace depends on two geometric parameters: the magnitude c_i of vector $\mathbf{c}_\mathbf{i}$ and the stroke s_i of the linear actuator which is directly related to each limb's length l_i. Since the proposed 3-UPU platform has equilateral triangular base and moving plates, and three identical linear actuators, therefore, c_i and s_i have the same value in each limb, so that they are two free parameters to be defined in the kinematic design, which are denoted by *C* and *S* respectively.

4 Kinematic Optimization

In this section, the kinematic tool used to formulate the workspace optimization problem is based on the Global Condition Index which is introduced in [12] and used in [11, 13], as defined in Eq. (7)

$$\eta = \frac{\int_W \frac{1}{\kappa} dW}{\int_W dW} \tag{7}$$

where *W* is the workspace and · is the condition number of the manipulator Jacobian matrix **J**

$$\kappa = \|\mathbf{J}\| \|\mathbf{J}\|^{-1} \tag{8}$$

The objective of the well-conditioned workspace is to determine the values of the tripod leg mechanism design parameters that result in the best kinematic isotropy by considering the allowable angular limits of the universal joints. The range of values of the actuators' stroke, *S*, is considered to be 0.2 and 0.8 of the minimum length of the limb, i.e., *Li*0, which are based on the technical specifications of the majority of commercially available linear actuators. The range of values of the radius difference of two plates, *C*, is assumed between 0.27 and 0.645 of the maximum length of the limb, i.e., *Li*0 + *S*, which is indicated in the workspace optimization work as indicated in Ref. [13].

By using the Matlab optimization toolbox, the maximum workspace is produced for values of design parameters *C* around 0.3 and *S* as large as possible, i.e., around 0.8 when 1,000,000 points are randomly selected inside the determined enclosure

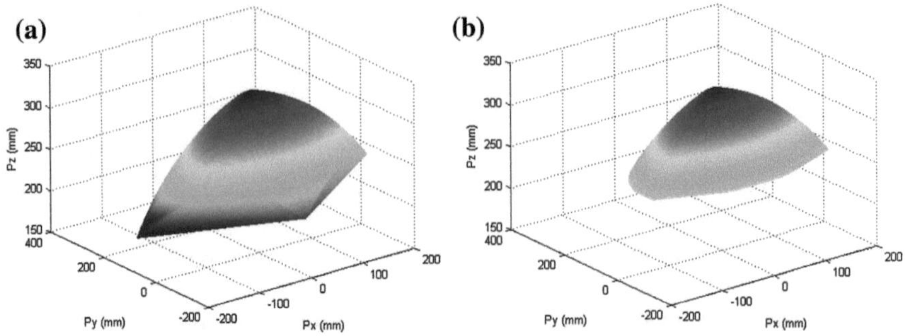

Fig. 3 Optimized workspace of tripod leg mechanim for C = 120 mm and S = 200 mm: **a** without U-joints angular constraints; **b** with U-joints angular constraints

space by Monte Carlo method. Under considering the limitation of the built prototype, namely the stroke of the linear actuator is fixed as S = 100 mm, in order to obtain the optimal workspace, the minimum value of $Li0$ is selected as 250 mm and then the value of C is calculated as 100 mm. The constraints of U-joints are considered by treating them as two revolute joints at the base. For each point within the obtained workspace, if all of the revolute joints fall between the allowable angular limits, then the point is within the well-conditioned workspace. Specifically, the angular limits of the revolute joints are chosen with the condition number remains below 10, as pointed out in [13].

Figure 3a, b present the workspace of the manipulator for the optimized values without and with angular constrains of U-joints, respectively. The color of the points represents the value of the Global Condition Index, the dark red points have large values and they are generally found toward the center of the workspace, and the dark blue points have small values and they are generally found toward the bottom edges of the workspace. It can be noted that by imposing constraints on the revolution range of the universal joints, the workspace isotropy is improved while the workspace volume is slightly decreased.

5 Step Design

Based on the well-conditioned workspace of the tripod leg mechanism, the initial length of three limbs can be determined as $Li0$ = 250 mm, hence, the objective of step design is proceeded to optimize the step size by assembling three limbs in the leg mechanism.

During the biped walking gait, the human-like foot step trajectory is an ovoid curve, as shown in Fig. 2, where the straight line segment is related to the supporting phase and the curve segment is related to the swinging phase. During the foot motion cycle, the step trajectory will start from point S-1 and follows the sequences of S_1-2-3-4-1. The step size can be characterized by the step length S_L and step height S_H.

By assuming the direction of the biped walking gait as forward, the tripod leg mechanism was assembled with one limb ahead and the other two rear in the previous design [9], while in this paper the one is assembled with one on the right

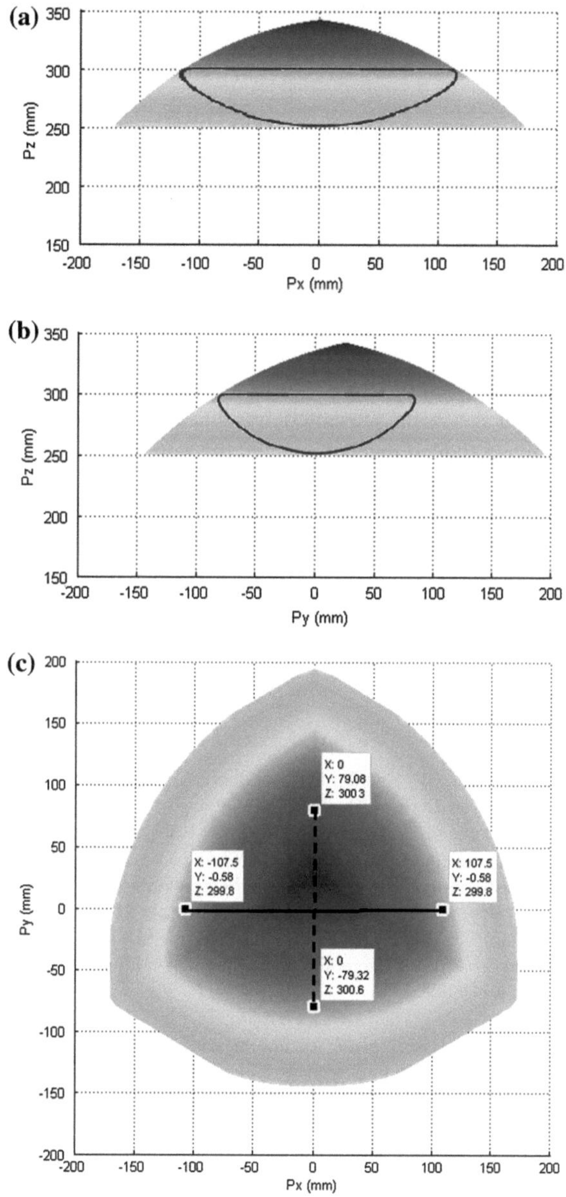

Fig. 4 Cross-sections of the workspace with the step trajectory inside for forward walking: **a** in *O-xz* plane, for the foot configuration of one limb on the *right* and the other two on the *left*; **b** in *O-yz* plane, for the foot configuration of one *ahead* and the other two *rear*; **c** in *O-xy* plane

and the other two on the left. In order to compare the step sizes in different assemblies, Fig. 4a–c present three cross-sections of the optimized workspace of the manipulator in *O-xz*, *O-yz* and *O-xy* planes, respectively. Assumed that the step trajectory starts from the initial position P_0 = [0, 0, 300] with step height S_H = 50 mm, the obtained maximum step size are shown in Fig. 4a, b, and the corresponding step lengths can be obtained from position data in Fig. 4c, which are computed as S_{La} = 215 mm and S_{Lb} = 158.4 mm, respectively. It can been seen that the step size is improved about 35.7 % by assembling three limbs with one right and the other two left.

6 Conclusions

In this paper we have presented the mechanical design and kinematic analysis of a tripod leg mechanism for a Cassino biped locomotor. The well-conditioned workspace has been obtained by taking into account the Global Condition Index and angular limits of universal joints. Considering the constraints of the built prototype, the minimum value of *Li*0 is selected as 250 mm and the value of *C* is calculated as 100 mm. Based on the obtained well-conditioned workspace, step design has been presented by changing the initial length and assembly of three limbs in the leg mechanism and the step length can been improved 35.7 % based on the same step height.

Acknowledgments The first author would like to acknowledge China Scholarship Council (CSC) for supporting his PhD study and research at the Laboratory of Robotics and Mechatronics (LARM) in the University of Cassino and South Latium, Italy, for the years 2013–2015.

References

1. Siciliano, B, Khatib, O.: Handbook of Robotics, Part G, Legged Robots, pp. 361–390. Springer, Berlin (2008)
2. Carbone, G., Ceccarelli, M.: Legged Robotic Systems, Cutting Edge Robotics, pp. 553–576. ARS Scientific Book, Vienna (2005)
3. Hashimoto, K., et al.: Biped landing pattern modification method and walking experiments in outdoor environment. J. Robot. Mechatron. **20**(5), 775–784 (2009)
4. Ota, Y., et al.: Research on a six-legged walking robot with parallel mechanism. In: Proceedings of International Conference on Intelligent Robots and Systems, Victoria, pp. 241–248 (1998)
5. Sugahara, Y., Sugahara, Y., Endo, T., Lim, H.O., Takanishi, A..: Design of a battery-powered multi-purpose bipedal locomotor with parallel mechanism. In: Proceedings of International Conference on Intelligent Robots and Systems, Lausanne, vol. 3, pp. 2658–2663 (2002)
6. Ceccarelli, M., Carbone, G.: A new leg design with parallel mechanism architecture. In: Proceedings of International Conference on Advanced Intelligent Mechatronics, Singapore, pp. 1447–1452 (2009)

7. Pan, Y., Gao, F.: Payload capability analysis of a new kind of parallel leg hexapod walking robot. In: Proceedings of International Conference on Advanced Mechatronic System, Luoyang, pp. 541–544 (2013)
8. Wang, H.B., et al.: Application of parallel leg mechanisms in quadruped/biped reconfigurable walking robot. J. Mech. Eng. **45**(8), 24–30 (2009)
9. Wang, M.F., Ceccarelli. M.: Design and simulation of walking operation of a Cassino biped locomotor. In: New Trend in Mechanism and Machine Science, vol. 24, pp. 613–621. Springer, Berlin (2015)
10. Wang, M.F., Ceccarelli, M., Carbone, G.: Experimental experiences with a LARM tripod leg mechanism. In: Proceedings of 10th IEEE/ASME International Conference on Mechatronic and Embedded Systems and Applications (MESA), Senigallia, pp. 1–6 (2014)
11. Tsai, L.W., Joshi, S.A.: Kinematics and optimization of a Spatial3-UPU parallel manipulator. ASME J. Mech. Des. **122**, 439–446 (2000)
12. Gosselin, C., Angeles, J.: A global performance index for the kinematic optimization of robotic manipulators. ASME J. Mech. Design **113**(3), 220–226 (1991)
13. Badescu, M., Morman, J., Mavroidis, C.: Workspace optimization of 3-UPU parallel platforms with joint constraints. In: Proceedings of IEEE International Conference on Robotics and Automation (ICRA), vol. 4, pp. 3678–3683 (2002)

Self-crossing Motion Analysis of a Novel Inpipe Parallel Robot with Two Foldable Platforms

Wan Ding and Yan-an Yao

Abstract A novel pipeline robot that can realize both inchworm gait and a new concept of *Self-crossing motion* (SCM) gait is proposed. The robot is composed of two identical spatial foldable platforms and three identical RPR (revolute-prismatic-revolute) kinematic chains. The foldable platform that is essentially an improved plane-symmetric Bricard linkage with special twist angle can achieve a continuous turning inside out. By taking advantage of the locomotion cooperation among two platforms and three kinematic chains, the robot possesses large deformation capabilities (adapt to different pipe size), different locomotion modes (SCM, inchworm), and foldable and expandable capabilities (fold into flat or prism shape). Two different SCM strategies are described respectively. As displayed in dynamic simulation with integrated locomotion gaits, the SCM movement performs a larger step length and fast speed than general inchworm locomotion.

Keywords Inpipe parallel robot · Spatial folding platform · Self-crossing motion · Inchworm motion

1 Introduction

Parallel mechanism consisted of two platforms and several kinematic chains can be improved to design inchworm pipeline robot. The construction characteristics of these robots include an extendable body for moving forward and two deformable platforms for supporting [1]. Bekhit et al. [2], presented an inchworm robot based

W. Ding · Y. Yao (✉)
Beijing Jiaotong University, Beijing, China
e-mail: yayao@bjtu.edu.cn

W. Ding
e-mail: dingwan@bjtu.edu.cn

B. Corves et al. (eds.), *Mechanisms, Transmissions and Applications*,
Mechanisms and Machine Science 31, DOI 10.1007/978-3-319-17067-1_23

on the Gough-Stewart (G-S) parallel mechanism with two improved 3-DoF (degree of freedom) triangular platforms. The platform was constructed by replacing each edge with linear actuator and each vertex with revolute joint accordingly. Aracil et al. [3] proposed a climbing robot based on G-S parallel mechanism with two ring platforms. Each platform was designed with four grip devices arranged equally around the ring for supporting. Jeon et al. [4] designed a steerable inchworm type inpipe robot that was consisted of two 1-DoF four crank-slider mechanisms constructed extendable platforms and one pinion-rack driven extensor mechanism. Similarly, Kim et al. [5] developed an inpipe robot that was composed of two sets of scissor-type mechanisms on both ends and a single pneumatic cylinder on center.

However, for these robots, the step length and inpipe radius adaptability were limited by their deformation capability and physical size of the deformable platform. What's more, the foldable ability for most inchworm pipeline robot was restricted by their planar configurations. Thus, in order to acquire the large deformation and high mobility, the robots should be equipped with special fabricated large extension ratio telescopic rod [6] or more complex structure [7], which would definitely increase the difficulty of design, manufacture and control system.

In totally different way, by adopting the folding and single DoF properties of the Bricard linkage acting as the deforming platform and by using three RPR kinematic chains to achieve moving forward, we propose a novel inpipe robot. The platform possesses large deformation capability. What's more, a novel locomotion mode called *Self-crossing Motion* is proposed to achieve large step length and fast locomotion speed. And, the robot can easily fold into flat or prism shape.

2 Mechanism Design and SCM Description

2.1 Mechanism Design

Figure 1a shows the simplified 3D model of the robot. It contains two identical deforming platforms (platforms A and B) and three RPR kinematic chains. See Fig. 1b, each platform is a single DoF plane-symmetric Bricard linkage with six revolute joints and six links [8]. See Fig. 1c, two platforms are connected with three RPR kinematic chains, which form a 3-RPR parallel manipulator. Figure 1d illustrates the partial view of vertex mechanism that aims at keeping the plane of circular ring always coplanar with the angular bisector, namely, $\theta_1 = \theta_2$. Figure 1e is the sketch diagram of the vertex mechanism that is a single DoF symmetric double planar slider-rocker mechanism calculated by G-K criterion [9].

The revolute property of the two links of Bricard linkage at A_{12} keeps unchanged. For the plane-symmetric Bricard linkage with special given parameters illustrated in what follows, the two triangular planes composed of three interval vertices, such as $\Delta A_1A_3A_5$ and $\Delta A_2A_4A_6$, always parallel to each other and keep equilateral triangle shapes [10]. Since each platform has single DoF, the platform

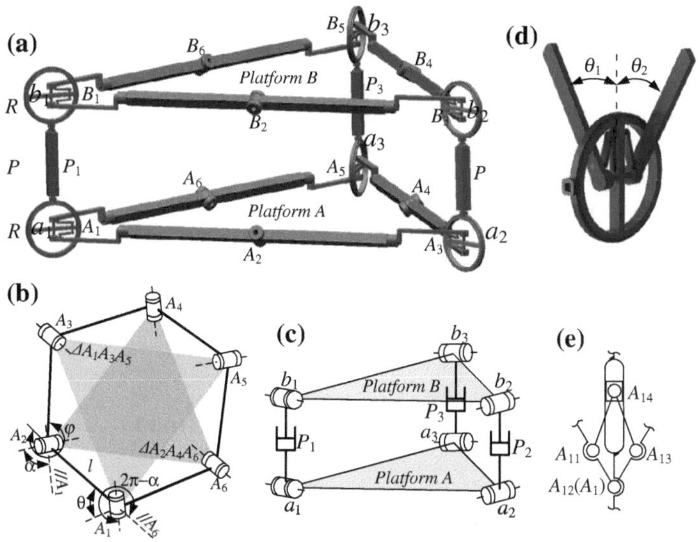

Fig. 1 The 3D model and mechanism sketch of the robot

will be locked into single link if one of the six revolute joints of the Bricard linkage is fixed. And then, the robot will be degenerated to a single DoF 3-RPR parallel manipulator (PM) with only translational motion if two deforming platforms are locked simultaneously. Therefore, the robot has three DoFs totally.

2.2 SCM Description

See Fig. 2a, the robot has three states between platforms A and B: State I, below state; State II, coplanar state; State III, above state. By properly controlling two platforms, three states can transform orderly, namely, $\Delta a_1a_2a_3$ can cross $\Delta b_1b_2b_3$, or reverse. Thus, we define this special motion as: *Self-crossing motion.*

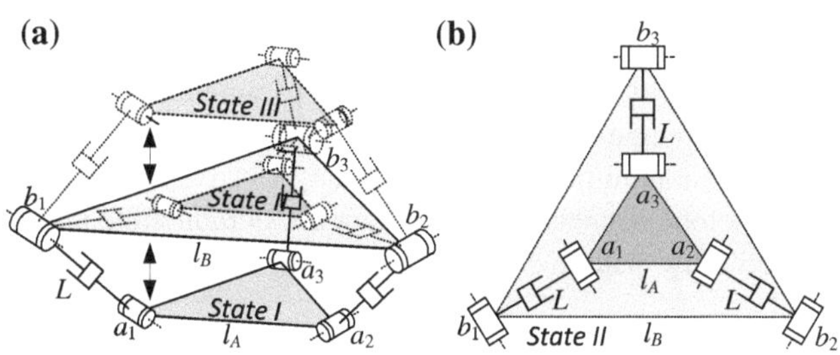

Fig. 2 SCM description: **a** three states, **b** constraint singularity position

However, at state II (Fig. 2b), the robot reaches the constraint singularity position (CSP) as two triangles are coplanar. The characteristics of instantaneous (infinitesimal) degrees of freedom changed, which changes into two independent expandable platforms and one instantaneous translation that is vertical to the triangle plane [11]. As labelled in Fig. 2, parameters l_A, l_B, and L represent the edge length of $\Delta a_1a_2a_3$ and $\Delta b_1b_2b_3$, and length of a_ib_i (i = 1, 2, 3). According to the relation of coplanar, the geometry condition satisfies:

$$L = \frac{2}{\sqrt{3}}|l_A - l_B| \quad (1)$$

3 Motion Analysis of the Deformation Platform

According to the Ref. [12], for the plane-symmetric Bricard linkage shown in Fig. 1 (b), the output angle φ can be represented by input angle θ and twist angle α:

$$\varphi_i = 2\arctan\frac{2\cos\alpha\sin\theta \pm \sqrt{2\cos^2\alpha + 1 + 2\sin^2\alpha\cos\theta}}{\cos^2\alpha - \sin^2\alpha - 2\cos^2\alpha\cos\theta}, \quad (i = 1, 2; \varphi_1 \text{uses '+'}, \varphi_2 \text{uses '−'}) \quad (2)$$

In Ref. [12], we know that the Bricard linkage keeps moving continuously when twist angle satisfies $60° \leq \alpha \leq 120°$, and, the linkage with twist angle α behaves the same way as one whose twist angle set as $180° - \alpha$. Meanwhile, the six linkages of the Bricard linkage with twist angle $\alpha = 60°$ or 120° will coincide together as both θ and φ_1 reach to 180°, which is impossible to reach this point of a physical prototype due to interference. What's more, the input-output curve called compatibility path forms a closed loop as $60° < \alpha < 120°$.

As shown in Fig. 3a, the variation of edge lengths $\|A_1A_3\|$ of $\Delta A_1A_3A_5$ with respect to variation of twist angle $\alpha \in [60°, 120°]$, input angle $\theta \in [\theta_{\min}, \theta_{\max}]$, and length of each linkage l = 620 mm are displayed.

See Fig. 3b, the compatibility path of $\alpha = 62°$ compared with $\alpha = 60°$ is adopted as an example to show new features of the Bricard linkage. The length $\|A_1A_3\|$ varies from 241 to 1240 mm. We can see that both of two twist angles are pass through four fixed state (S) points, S_2, S_3, S_6, and S_7. The compatibility path S_1S_5 (from S_1 to S_5) generate by choosing input-output parameters (θ, φ_1). As we can see, the variation of input angle θ changes from $\theta \in [-157.5°, 157.5°]$, and the output angle φ_1 reaches to the minimum value at point S_4(134.1°, 157.5°). Meanwhile, the S_1S_4 obeys to monotone decreasing, while S_4S_5 obeys monotone increasing. On the contrary, compatibility path S_5S_1 (from S_5 back to S_1) acts as in symmetry locomotion.

Interestingly, we find that the Bricard linkage can overcome two extreme states (S_1 and S_5) and achieve a locomotion cycle (from S_1 back to S_1) by exchanging the

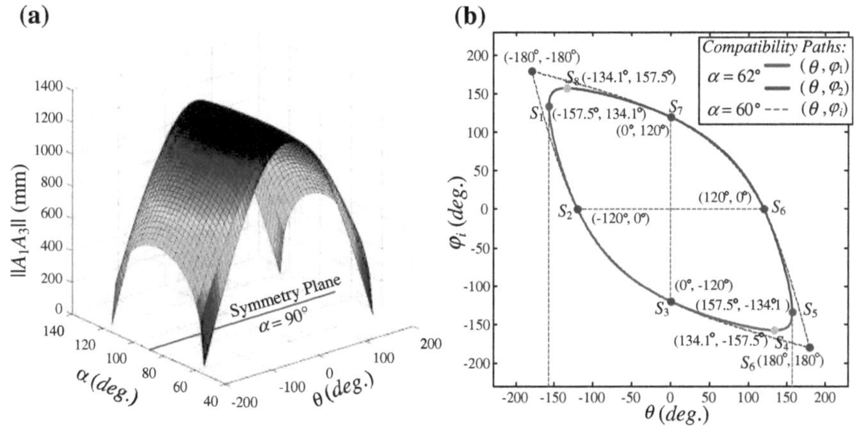

Fig. 3 Variation of parameters: **a** edge length $||A_1A_3||$, **b** compatibility paths

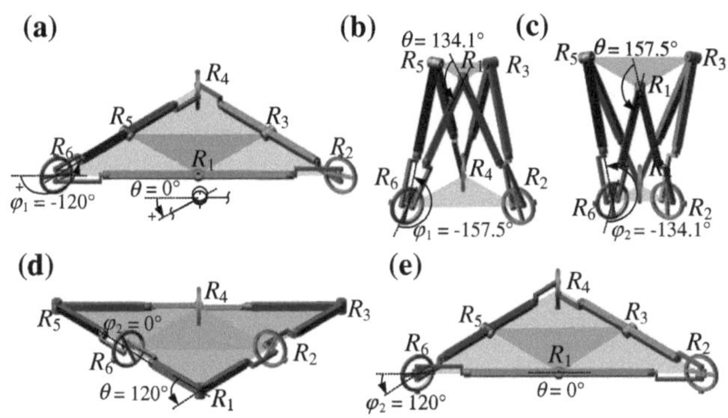

Fig. 4 Five special states of locomotion as $\alpha = 62°$

input and output at states S_1 and S_5. Therefore, in order to show this transformation and deformation, in Fig. 4, we choose five states (S_3 to S_7) that have symmetry motion with five states (from S_7 back to S_3) are shown accordingly. Especially, the output angle φ_2 can be enable after the minimum value at S_4 shown in Fig. 3b, c.

4 SCM Analysis

As discussed in part 2.2, in order to deal with the DoF variation, the instantaneous translation motion can be controlled by adding one actuation at one of the six revolute joints (a_i, b_i) shown in Fig. 2b. Additionally, we find an interesting gait of the SCM movement without additional actuation by using the contact force between two deforming platforms to overcome the CSP. The following part will illustrate the two different locomotion modes by using dynamic simulation.

4.1 SCM with Additional Actuation

Figure 5a is the initial position; input parameters are set as $\theta_A = \theta_B = 0°$, $L = 360$ mm; L keeps unchanged in whole locomotion cycle. Figure 5b shows that two platforms do not have interference as $\theta_A = 52°$, $\theta_B = 20°$. See Fig. 5c, $\theta_A = 137°$ and $\theta_B = 80°$. Figure 5d, disable θ_A and enable θ_P at the beginning of this step; the robot reaches CSP as $\theta_P = 130°$. Figure 5e, by further increasing $\theta_P = 220°$, platform A moves above platform B. Figure 5f, the robot recovers to its original configuration as $\theta_A = 0°$.

Additionally, the disable or enable process can be replaced by redundant actuation, namely, all of the six parameters are mounted with actuations.

4.2 SCM Without Additional Actuation

The omitted initial and final states have the same state as Fig. 5a, f. Platform B keeps unchanged in the cycle. Figure 6b, disable L, increase θ_A to extreme position $\theta_A = 157.5°$. Figure 6c, disable θ_A and enable φ_A; increase $\varphi_A = -70°$ so as to let

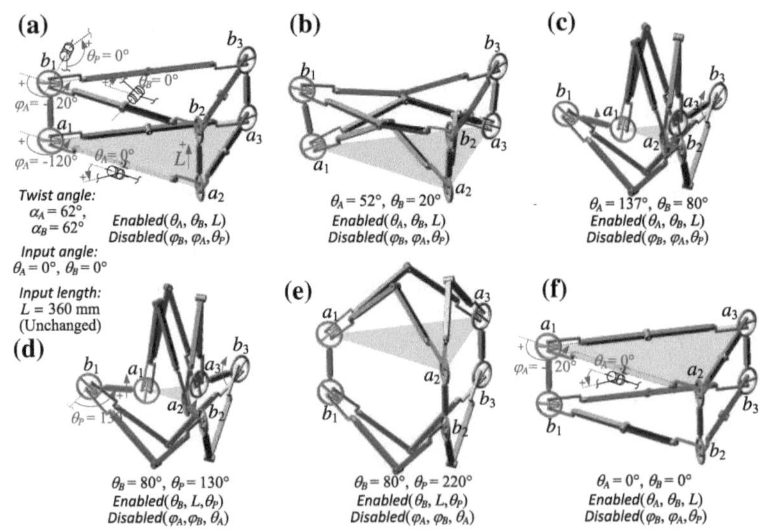

Fig. 5 Gait I: SCM with additional actuation (θ_P)

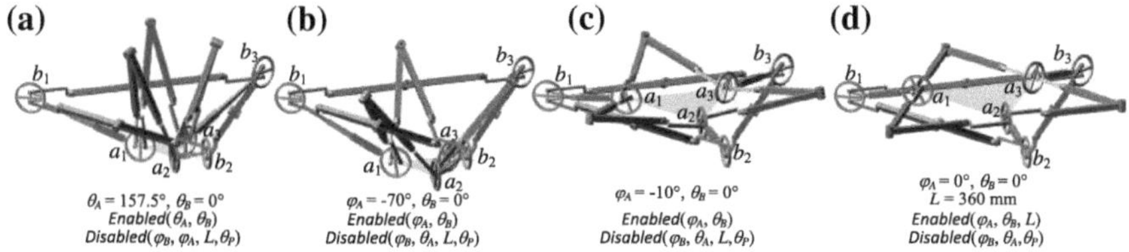

Fig. 6 Gait II: SCM without additional actuation (θ_P)

platform A contact with platform B. Figure 6d, the robot reaches the CSP, and platform A takes advantage of the contact force to "extrude" from platform B. Figure 6e, by further increasing $\varphi_A = 0°$, the robot utilize contact force and folding motion to overcome and pass the CSP, and then, L is enabled, $\Delta a_1 a_2 a_3$ is above $\Delta b_1 b_2 b_3$. Figure 6f, the robot recovers to its original configuration as $\varphi_A = 120°$.

4.3 Combined Locomotion Gaits in Pipeline with a Gap

Figure 7 shows that the robot takes advantage of the three gaits together to overcome a gap in a straight line inpipe (side view), which include two SCM gaits shown in Figs. 5 and 6 and one inchworm gait. The gaits shown from Fig. 7a–c are gait I. The gaits shown from Fig. 7c–e are inchworm gaits. The gaits shown from Fig. 7e–g are gait II. The enabled and disabled parameters are illustrated in each step. Figure 7g shows that the robot moves a distance upward.

By comparing the two locomotion modes, it can be seen that, SCM has a larger step length than inchworm motion in a cycle, while inchworm motion can pass

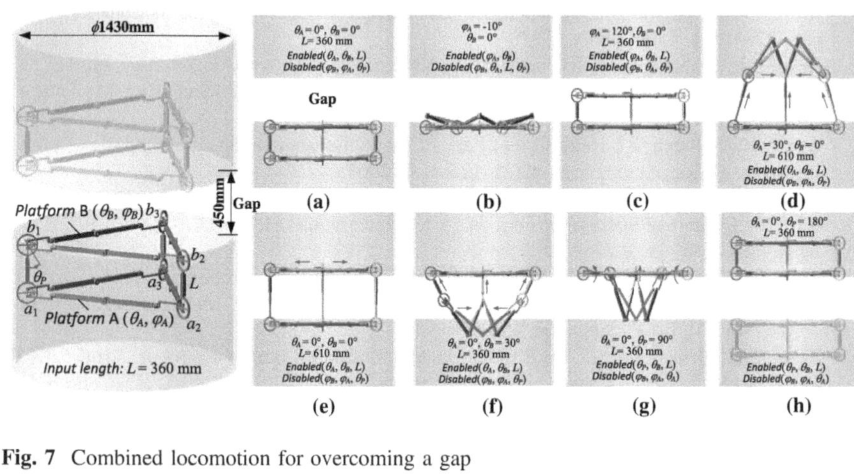

Fig. 7 Combined locomotion for overcoming a gap

through a narrower inpipe size than SCM. However, as mentioned above, inchworm motion and SCM can exchange into each other. And also, both of them are possible to give a fast and stable locomotion in pipeline.

5 Conclusions

This paper presented a pipeline robot composed of two identical improved plane-symmetric Bricard linkage with twist angle 62° acting as spatial foldable platform and three identical RPR kinematic chains. The two Bricard linkages of which continuous locomotion of turning inside out was illustrated in details were used to achieve a novel locomotion type called SCM. And then, the condition of realizing SCM and two different SCM strategies according to different locomotion modes to overcome CSP were shown and analyzed.

The SCM gait by using its own contact force to overcome the CSP was a novel feature for parallel robots. The dynamic simulations of SCM gaits and combined locomotion verified the locomotion feasibility of moving inpipe. What's more, the combined locomotion gaits illustrated that the SCM gait got a larger step length and faster speed than inchworm gait. The new type of pipeline locomotion, *Self-crossing Motion*, provided a new locomotion type for inpipe robot family.

Acknowledgments This work was supported by National Natural Science Foundation of China (51175030), Research Fund for the Doctoral Program of Higher Education (20130009110030).

References

1. Bertetto, A.M., Ruggiu, M.: Inpipe inch-worm pneumatic flexible robot. In: Proceedings IEEE/ASME International Conference on Advanced Intelligent Mechatronics, vol. 2, pp. 1226–1231. Como, Italy (2001)
2. Bekhit, A., Dehghani, A., Richardson, R.: Kinematic analysis and locomotion strategy of a pipe inspection robot concept for operation in active pipelines. Int. J. Mech. Eng. Mechatron. **1** (2), 15–27 (2012)
3. Aracil, R., Saltarén, R., Reinoso, O.: Parallel robots for autonomous climbing along tubular structures. Robot. Auton. Syst. **42**(2), 125–134 (2003)
4. Jeon, W., Park, J., Kim, I., et al.: Development of high mobility inpipe inspection robot. In: IEEE/SICE International Symposium on System Integration (SII), pp. 479–484. IEEE, Kyoto, Japan (2011)
5. Kim, Y.J., Yoon, K.H., Park, Y.W.: Development of the inpipe robot for various sizes. In: IEEE/ASME International Conference on Advanced Intelligent Mechatronics AIM2009, pp. 1745–1749. IEEE, Singapore (2009)
6. Zagal, J.C., Armstrong, C., Li, S.: Deformable octahedron burrowing robot. Artif. Life **13**, 431–438 (2012)
7. Clark, P.E., Curtis, S.A., Rilee, M.L.: A new paradigm for robotic rovers. Phys. Procedia **20**, 308–318 (2011)
8. Bricard, R.: Leçons des cinematique, vol. 2, pp. 7–12. Gauthier-Villars et cie, Paris (1972)

9. Gogu, G.: mobility of mechanisms: acritical review. Mech. Mach. Theory **40**(9), 1068–1097 (2005)
10. Racila, L., Dahan, M.: Bricard mechanism used as translator. In: Proceedings of 12th world congress IFToMM, vol. 3, pp. 337–341 (2007)
11. Zlatanov, D., Bonev, I. A., Gosselin, C. M.: Constraint singularities of parallel mechanisms. In: Proceeding of IEEE International Conference on Robotics and Automation, vol. 1, pp. 496–502. ICRA'02, Washington, DC (2002)
12. Chen, Y., You, Z., Tarnai, T.: Threefold-symmetric bricard linkages for deployable structures. Int. J. Solids Struct. **42**(8), 2287–2301 (2005)

A Novel Skid-Steering Walking Vehicle with Dual Single-Driven Quadruped Mechanism

Jianxu Wu, Qiang Ruan, Yan-an Yao and Meili Zhai

Abstract A novel legged-vehicle called dual quadruped vehicle (DQV) is proposed. The DQV contains two identical single-driven quadruped mechanisms which are composed of single degree-of-freedom (DoF) planar mechanisms. Based on the functional requirements, the construction of leg mechanism is presented. The arrangement of overall vehicle is described, and the steering analysis is discussed. Finally, the dynamic simulation and experimental system are carried out to verify the validity of the theoretical analysis and the maneuverability of the prototype. By decreasing the DoFs, it is useful for the reduction of the cost and simplification of the control algorithm. What's more, the reliability and practicality of the walking vehicle can also be enhanced.

Keywords Walking vehicle · Closed chain · Mechanism analysis · Skid steering

1 Introduction

Compared with the tracked and wheeled vehicles, the legged machines are more flexible, efficient and versatile when operated on off-road terrains. In the past decades, plenty of the prototypes have been widely used in various fields. For

J. Wu · Q. Ruan · Y. Yao (✉)
Beijing Jiaotong University, Beijing, China
e-mail: yayao@bjtu.edu.cn

J. Wu
e-mail: 13116343@bjtu.edu.cn

Q. Ruan
e-mail: 12116320@bjtu.edu.cn

M. Zhai
The High School Affiliated to Beijing Normal University, Beijing, China
e-mail: 11121443@bjtu.edu.cn

B. Corves et al. (eds.), *Mechanisms, Transmissions and Applications*,
Mechanisms and Machine Science 31, DOI 10.1007/978-3-319-17067-1_24

example, the Bigdog [1] has been used as a delivery vehicle in the military tasks; the WL-16 and Hyperion 4 [2, 3] have been used as the transportation for injured individuals; the ASV and DANTE II [4, 5] have been used for supervision task and volcanoes exploration; the Walking Harvester [6] has been used for the forestry and agriculture tasks; the ASIMO [7] has been used in daily life.

From the above literature reviews, it can be seen that the researchers around the world have devoted themselves to the studies on legged machines. The crucial mechanical design of these robots has been focused on the leg mechanisms, which can be divided into open-chain linkage and close-chain linkage. In the open-chain mechanism, actuators were installed on the hip, knee and ankle. Nevertheless, a certain amount of legged robots were restricted to the laboratory for the following reasons: (1) The larger number of actuators increases the kerb weight and energy consumption of the robot. (2) If all of the DoFs need to be simultaneously actuated while walking on rough terrain, the control algorithm will become complex. This may make the robot less reliable in a severe environment. (3) Not only the increment in actuators and sensors, but also the intricate control could result in the expensive design and maintenance.

In order to improve the practicability for a legged robot, the enhancement of mechanical design with lower DoFs is urgently required. Therefore, the single-DoF planar linkages with close-chain have been implemented in different prototypes, such as: Chebyshev mechanism, Walking Vehicle [8], Walking Apparatus [9], and also, the reconstruction study for ancient machinery by Chiu, Hwang and Chen, et al. [10–12]. However, the above robots not only keep balance by adding supporting wheels, but also fail to steer. Different from these, the Klann mechanism [13] and the Theo Jansen mechanism [14] improved the mobility by adding more leg modules. With the aim of building an easy-operation and low-cost walking vehicle, this paper deals with the DQV that is composed by close-chain linkages on the basis of skid-steering theory.

2 Design of the Leg Mechanism

In order to be capable of passing various terrains, the closed kinematic chain used as the leg mechanism should be subject to the design constraints. Meanwhile, the foot trajectory, generally divided into the "support phase" and the "transfer phase", should meet the functional requirements, which can be summarized as follows. The desirable foot trajectory is shown in the Fig. 1:

1. The trajectory must be a closed symmetrical curve without intersections.
2. The support phase of the foot-point is approximately a straight line.
3. The height of the stride H in the transfer phase is as high as possible, and the length L in the support phase is as long as possible to ensure the walking ability.

Based on the functional requirements, the 1-DoF planar leg mechanism with the close-chain full-pivot feature is presented in Fig. 2. The linkages in *Watt* chain [15]

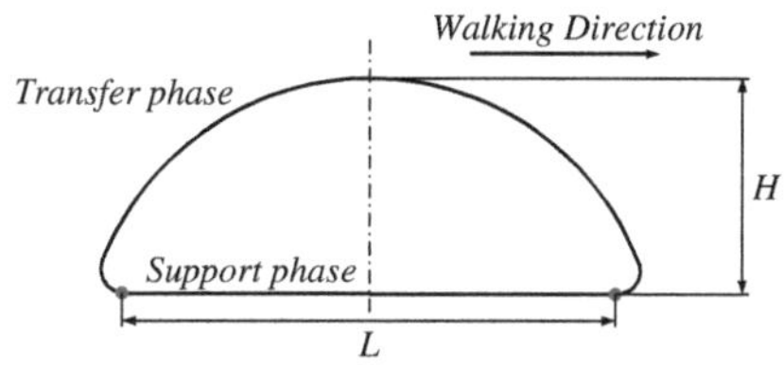

Fig. 1 The foot trajectory

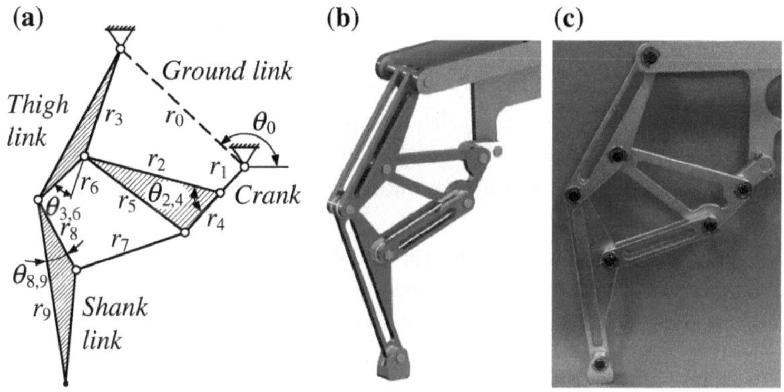

Fig. 2 The leg mechanism construction: **a** sketch diagram, **b** 3D module, **c** prototype

Table 1 Design parameters of the six-bar mechanism

r_0 (mm)	r_1 (mm)	r_2 (mm)	r_3 (mm)	r_4 (mm)	r_5 (mm)	r_6 (mm)
120.93	25	86.4	72	36	97.2	43.2
r_7 (mm)	r_8 (mm)	r_9 (mm)	θ_0 (°)	$\theta_{3,6}$ (°)	$\theta_{2,4}$ (°)	$\theta_{8,9}$ (°)
79.2	55.8	139.46	147.90	29.69	62.06	20

are designed as ground link, crank, thigh link, and shank link, respectively. In Fig. 2a, the parameters are described as: r_i ($i = 1, 2, \ldots, 9$), the length of linages; $\theta_{i,j}$ ($i = 1, 2,\ldots,9$; $j = 1, 2, \ldots,9$), the angle between the linkages i and j. The design parameters are listed in Table 1.

3 Dual Quadruped Vehicle

3.1 Arrangement of Overall Vehicle

The arrangement of the DQV is shown in Fig. 3a. The whole mechanism consists of two identical quadruped mechanisms on each side of the frame. In each quadruped

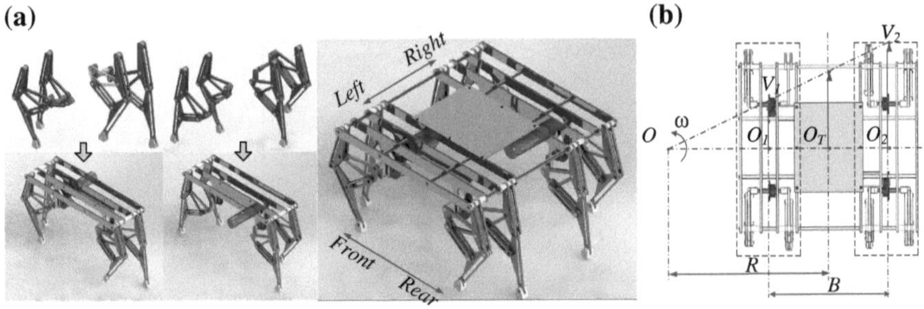

Fig. 3 The description and analysis of DQV: **a** arrangement, **b** steering model

mechanism, the cranks arranged in the front and rear legs on the same side have a zero radian phase difference, and the cranks arranged in the left and right legs have a 180° phase difference. The front and the rear legs follow the feature of moving in the same plane.

In order to simplify the motion control algorithm and reduce the number of motors further, the cranks of the left and right legs share the public driving shaft, and the front and rear shafts are driven by the motor through the chain transmission. Consequently, each quadruped robot is driven independently by a single motor.

3.2 Skid-Steering Analysis

Following the arrangement described above, the two quadruped mechanisms are equivalent to the two sides pedrail system, and they achieve different speed by changing the driving forces. Figure 3b depicts the no-load full-vehicle steering model. O and R are the steering center and the steering radius respectively, and B represents the distance between the longitudinal symmetry planes of the two quadruped robots. O_T is the projection of point O in the longitudinal symmetry plane of the vehicle, and V_T represents the average speed of the legged-vehicle. Similarly, V_1 and V_2 are the speed of the two quadruped mechanisms. Therefore, the steering parameters can be obtained as:

$$\omega_T = \frac{v_1}{R - 0.5B} = \frac{v_2}{R + 0.5B} = \frac{v_2 - v_1}{B} \tag{1}$$

$$R = \frac{v_2 - 0.5B\omega_T}{\omega_T} = \frac{v_1 + 0.5B\omega_T}{\omega_T} = \frac{v_1 + v_2}{2\omega_T} = \frac{B(v_1 + v_2)}{2(v_2 - v_1)} \tag{2}$$

4 Dynamic Simulations

Figure 4 displays a cycle of walking simulation by ADAMS^{TM}, the motor speed is 30r/min. After a cycle of walking simulation, it takes 4 s and the moving distance is 390 mm. The foot point trajectory shown in Fig. 4a is a closed ovoidal curve, the height and length of the stride are 40 and 200 mm respectively.

The fluctuation of the platform in DQV is shown in Fig. 5a, and the amplitude is 6 mm which meets the application requirements. The torque on the crank is shown in Fig. 5b.

Figure 6a displays a cycle of steering simulation. And, the motor speed of the two quadruped mechanisms is set as 15 and 45r/min, respectively. Besides, Fig. 6b illustrates the lateral force on the leg mechanism which results in loosing connection and increasing friction between linkages. Therefore, the detailed mechanical design strategy of a revolute joint shown in Fig. 6c is adopted in each leg.

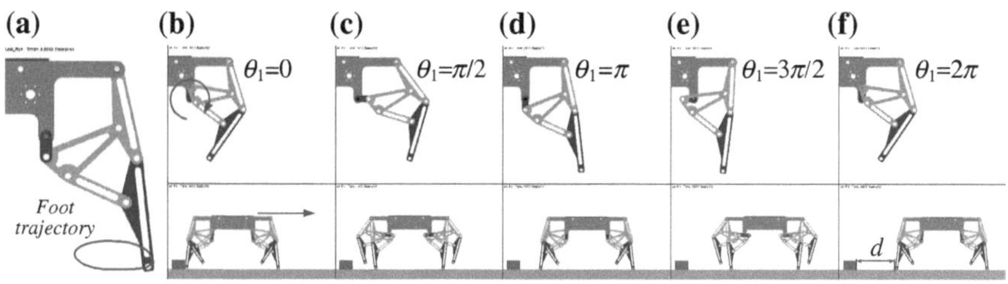

Fig. 4 Walking simulation

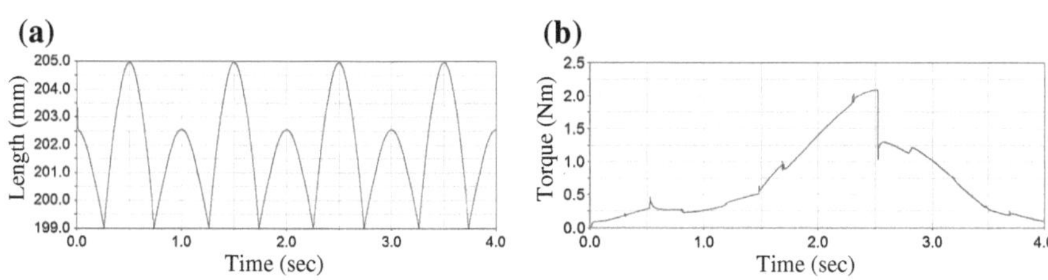

Fig. 5 Simulation parameters analysis: **a** fluctuation of vehicle **b** torque on the crank

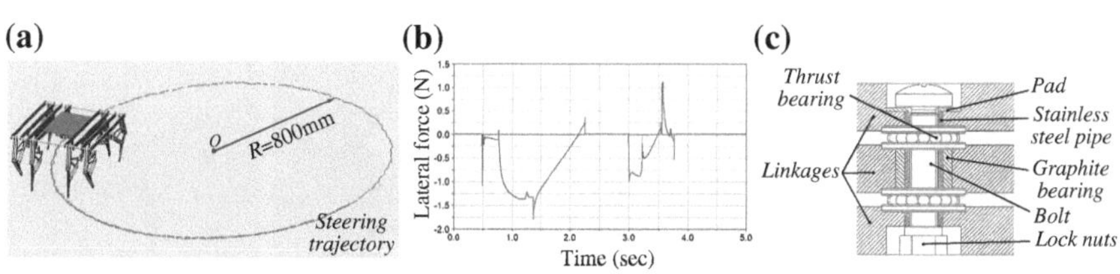

Fig. 6 Steering simulation: **a** steering trajectory **b** lateral force **c** rotational joint

5 Prototype and Experiments

As shown in Fig. 7, the experimental system composed of walking system, a control unit, a power supply, an operation unit is built (see Table 2 for a summary of the specifications of the DQV).

See Fig. 8, the turning experiments are carried out to validate the feasibility of the experimental system and the locomotion of constructed prototype. The experiment illustrated in Fig. 9 testifies the zero radius steering ability of the DQV.

Figure 10 shows the vertical obstacle passing experiments. The obstacle with the height of 20 mm is easy to overcome for the DQV. Furthermore, the experiment succeeds finally when one or two legs step on the obstacle which is 35 mm in height.

Fig. 7 The experimental system

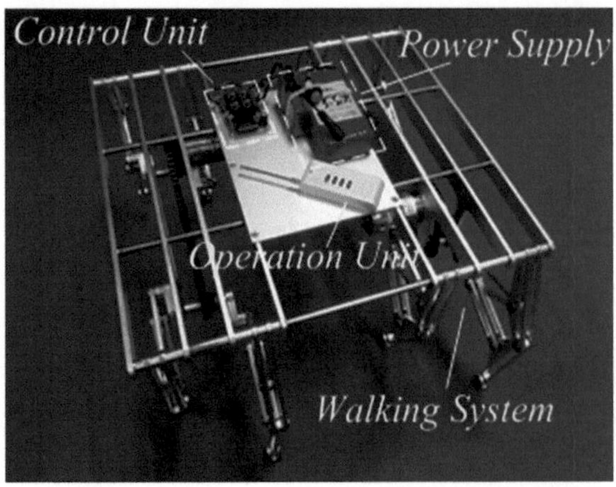

Table 2 Summary of the specifications for DQV

Curb weight (kg)	Size			Mobility			
	Length (mm)	Width (mm)	Height (mm)	Vertical obstacles(mm)	Slops (°)	Payload (kg)	Speed (mm/s)
4.2	459	428	258	350	30	6	300

Fig. 8 Steering with the Radius (R = 300 mm)

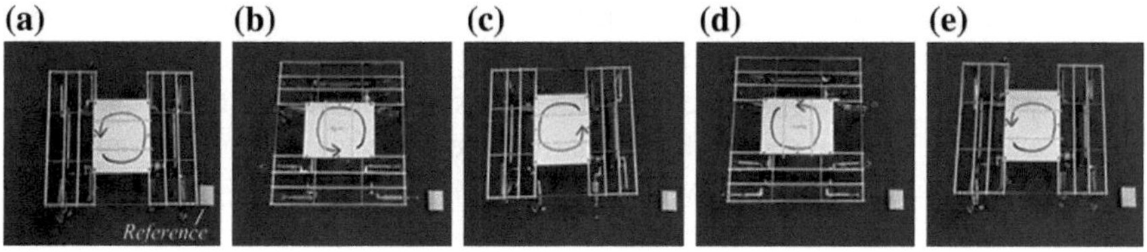

Fig. 9 Pivot steering experiment

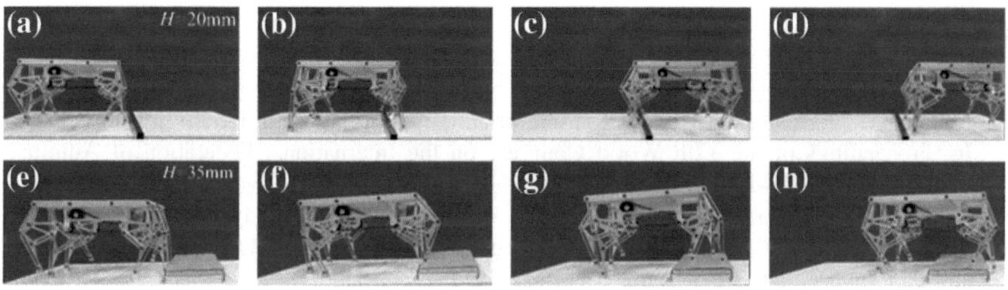

Fig. 10 Vertical obstacle passing experiments

Fig. 11 Slope and rugged terrain experiments

The climbing ability of the DQV is verified by the experiment shown in Fig. 11a, and the DQV is able to keep stability by taking full advantage of multiple legs when operated on the rugged terrain, as seen in Fig. 11b.

6 Conclusions

This paper proposed a novel walking vehicle called DQV. The mechanism design was presented according to the three functional requirements. The arrangement of the overall vehicle was described, and the turning mode was analyzed. Furthermore, dynamic walking motions were simulated in ADAMS^{TM}, and the results showed the locomotion feasibility of the mechanism design. The experiments were carried out successfully to testify the theoretical analysis and the reliability of the experimental walking system. Generally speaking, the DQV with simple control and flexible steering is able to cope with the different environment.

Acknowledgments This work was supported by Scientific Foundation of Beijing Jiaotong University (2014YJS112).

References

1. Raibert, M., Blankespoor, K., Nelson, G., et al.: Bigdog, the rough-terrain quadruped robot. In: Proceedings of the 17th World Congress on the International Federation of Automatic Control, pp. 10823–10825, Korea (2008)
2. Sugahara, Y., Hosobata, T., Mikuriya, Y., et al.: Realization of dynamic human-carrying walking by a biped locomotor. In: Proceedings of IEEE International Conference on Robotics and Automation, pp. 3055–3060, America (2004)
3. Warren, R.: Hyperion4 Ridable Robot Tiptoes Around, Haunts Dreams. http://www.switched.com/2010/06/20/hyperion-4-ridable-robot-tiptoes-around-haunts-dreams/
4. Waldron, K.J., McGhee, R.B.: The adaptive suspension vehicle. IEEE Control Syst. Mag. **6**, 7–12 (1986)
5. Bares, J.E., Wettergreen, D.S.: Dante II: technical description, results, and lessons learned. Int. J. Robot. Res. **18**, 621–649 (1999)
6. Paakkunainen, M.: Method for adjusting supply pressure. U.S. Patent 6,305,163 (2001)
7. Sakagami, Y., Watanabe, R., Aoyama, C., et al.: the intelligent Asimo: system overview and integration. Proc. IEEE/RSJ Int. Conf. Intelligent Robot. Syst. Switz. **3**, 2478–2483 (2002)
8. Ikeda, K., Matsumoto, S., Nozaki, T., and Taguchi, K.: Walking vehicle. U.S. Patent 3,850,259 (1974)
9. Shkolnik, N.: Walking apparatus. U.S. Patent 4,462,476 (1984)
10. Chiu, C.P.: On the design of a wave gait walking horse. Master thesis, Department of Mechanical Engineering, National Cheng Kung University, Tainan, Taiwan (1996)
11. Hwang, K.: On the design of an optimal 8-link type walking horse. Master thesis, Department of Mechanical Engineering, National Cheng Kung University, Tainan, Taiwan (1997)
12. Chen, Y.G., Yan, H.S.: On the design of a quadruped walking machine with a single actuator. Master thesis, Department of Mechanical Engineering, National Cheng Kung University, Tainan, Taiwan (2003)
13. Klann, R.D.: Jansen Linkage-Klann Linkage Comparison. http://www.mechanicalspider.com/comparison.html
14. Kim, S.W., Han, S.H., Kim, D.H.: Analysis of a crab robot based on Jansen mechanism. In: Proceedings of the 11th International Conference on Control, Automation and Systems, pp. 856–858, Korea (2011)
15. Yan, H.S.: Creative design of mechanical devices. Springer (1998)

Automatic Generation of Serial Manipulators to Be Used in a Combined Structural Geometrical Synthesis

Daniel Ramirez, Jens Kotlarski and Tobias Ortmaier

Abstract In the present paper, an approach to generate a minimal set of serial manipulators that provide the required motion directions without redundant solutions is introduced. Besides the architectures, the Denavit-Hartenberg parameters that can be used as optimization parameters in a posterior geometrical synthesis are extracted. The motion directions are defined through a required motion vector which is modified in order to consider the orientation of the robot base. All suitable architectures are generated from a set of discrete Denavit-Hartenberg parameters. The method removes all isomorphisms from the set of suitable architectures significantly reducing the number of solutions, i.e. the computational effort. In order to illustrate the advantage of the proposed method, the number of solutions generated with and without detection of isomorphisms are given. Finally, an example of architectures with four degrees of freedom is given.

Keywords Serial manipulators · Robot design · Structural synthesis · Isomorphisms · Denavit-Hartenberg parameters

D. Ramirez (✉) · J. Kotlarski · T. Ortmaier
Institute of Mechatronic Systems, Leibniz Universität Hannover, Hannover, Germany
e-mail: danielandres.ramirez@imes.uni-hannover.de; danielramirez@usantotomas.edu.co; andresr36@gmail.com

J. Kotlarski
e-mail: jens.kotlarski@imes.uni-hannover.de

T. Ortmaier
e-mail: tobias.ortmaier@imes.uni-hannover.de

D. Ramirez
Universidad Santo Tomás, Bogotá, Colombia

B. Corves et al. (eds.), *Mechanisms, Transmissions and Applications*,
Mechanisms and Machine Science 31, DOI 10.1007/978-3-319-17067-1_25

1 Introduction

The robot design process involves two different procedures: the structural synthesis and the geometrical synthesis. The first aims to find the number, type, and orientation of the robot joints in order to generate a manipulator with a given number of degrees of freedom (DOF). The second looks for the optimal dimensions of robot links to accomplish a given task [15]. However, in general the two procedures are carried out separately. Usually, an architecture is chosen a priori to perform the geometrical synthesis. Thus, the obtained manipulator does not necessarily represent the best mechanism. In order to perform a combined (structural and geometrical) synthesis, it is necessary to determine all suitable architectures that provide the motion directions of the required task as well as to identify the parameters that can be used in the optimization process maintaining the DOF at the end effector (EE) of the manipulator.

Most existing methods carrying out the structural synthesis of manipulators are based on the graph theory [2, 9], screw theory [8, 11, 12], groups of displacement [1, 10, 14], linear transformations [5–7], or position and orientation characterization [18, 19]. Most of these approaches have been applied in the design process of parallel manipulators. Only few works address serial manipulators: in [3] and [13], the rule-based design of serial kinematic chain is proposed. The derivation of the design rules are carried out using groups of the displacement and the screw theory, respectively. Additionally, in [4] 6 DOF serial robotic manipulators with only revolute joints are found. There, traveling coordinate systems are applied to describe the geometry of the manipulator and evolutionary morphology (EM) is used for generating new architectures.

Nonetheless, the rule-based design method does not provide the parameters and restrictions that should be subsequently used in a geometrical synthesis. Furthermore, the solutions generated by EM should be grouped before performing the geometrical synthesis in order to reduce the computational cost. Hence, a general approach to find serial architectures that fulfil a required motion is proposed. The architectures are described using the Denavit-Hartenberg (DH) parameters. The proposed method identifies the DH parameters that can be used as optimization parameters as well as the necessary restrictions that should be considered in the subsequent geometrical synthesis. With the goal of reducing the computational cost and the complexity of the optimization problem, the kinematic chains are grouped using detection of isomorphisms in the set of suitable architectures.

The paper is organized as follows. Firstly, the required motion vector is defined. Secondly, the general algorithm to generate the suitable architectures without redundant solutions is introduced. Thirdly, as part of this algorithm, the generation of all possible architectures that fulfil the required motion as well as the conditions for the detection of isomorphisms are presented. Then, the capability of the proposed approach is demonstrated comparing the number of architectures obtained both with and without detection of isomorphisms. Finally, two manipulators with 4 DOF are shown in order to clarify the obtained results.

2 Definition of the Required Motion

The first step of the structural synthesis is the definition of the required motion of the end effector, i.e. the required translational and rotational motion directions. The velocity $\boldsymbol{\xi}$ of the EE can be used for this purpose.

The velocity $\boldsymbol{\xi}$ of the end effector of a serial manipulator with n generalized coordinates can be described using the origin $\mathbf{O}_n$ of its reference frame $(\mathrm{RF})_n$ as a reference point [17]:

$$\boldsymbol{\xi} = \begin{pmatrix} \mathbf{v}_n \\ \boldsymbol{\omega}_n \end{pmatrix} = \mathbf{J}\dot{\mathbf{q}}, \tag{1}$$

where $\mathbf{v}_n$ and $\boldsymbol{\omega}_n$ are the linear and angular velocity of the EE, $\mathbf{J} = (\mathbf{j}_1, \mathbf{j}_2, \ldots, \mathbf{j}_n)$ is the Jacobian of the manipulator and the joint rates $\dot{\mathbf{q}} = (\dot{q}_1, \dot{q}_2, \ldots, \dot{q}_n)^{\mathrm{T}}$, corresponding to $\dot{\theta}_i$ in case of a revolute joint R and $\dot{d}_i$ in case of a prismatic joint P. The vector $\mathbf{j}_i$, with $i = 1, 2, \ldots, n$, is the ith column of the Jacobian and represents the effect of the ith joint rate $\dot{q}_i$ on the velocity $\boldsymbol{\xi}$ of the EE.

The required degrees of freedom of a given application are defined through the required motion vector $\boldsymbol{\xi}_{\mathrm{req}} \in \mathbb{R}^6$, corresponding to the vector $\boldsymbol{\xi} \in \mathbb{R}^6$. Elements of $\boldsymbol{\xi}_{\mathrm{req}} = \left(\mathbf{v}^{\mathrm{T}}_{\mathrm{req}_n}, \boldsymbol{\omega}^{\mathrm{T}}_{\mathrm{req}_n}\right)^{\mathrm{T}} = \left(\xi_{\mathrm{req}_1}, \xi_{\mathrm{req}_2}, \xi_{\mathrm{req}_3}, \xi_{\mathrm{req}_4}, \xi_{\mathrm{req}_5}, \xi_{\mathrm{req}_6}\right)^{\mathrm{T}}$ being zero ($\xi_{\mathrm{req}_i} = 0$) indicate that there is no required motion in this direction for any $\dot{\mathbf{q}}$. The required DOF of the manipulator are the amount of $\xi_{\mathrm{req}_i} \neq 0$. For instance, the vector $\boldsymbol{\xi}_{\mathrm{req}} = \left(\xi_{\mathrm{req}_1}, \xi_{\mathrm{req}_2}, \xi_{\mathrm{req}_3}, 0, 0, \xi_{\mathrm{req}_6}\right)^{\mathrm{T}}$ defines the motion of a manipulator with 4 DOF (1R3T) at the EE: three translational motions along the x-, y-, and z-axis (3T) and one rotational motion around the z-axis (1R).

By using the DH parameters to describe a serial manipulator, the axis of the first joint is always the z-axis and all configurations will have a motion along or around it. To avoid this problem, the required motion vector $\boldsymbol{\xi}_{\mathrm{req}}$ can be rotated as shown in [16]:

$$\begin{aligned} \boldsymbol{\xi}^{(z)}_{\mathrm{req}} = \boldsymbol{\xi}_{\mathrm{req}}, \quad \boldsymbol{\xi}^{(y)}_{\mathrm{req}} &= \begin{pmatrix} \mathbf{R}^{\mathrm{T}}_x(-\pi/2) & \mathbf{0}_{[3\times3]} \\ \mathbf{0}_{[3\times3]} & \mathbf{R}^{\mathrm{T}}_x(-\pi/2) \end{pmatrix} \boldsymbol{\xi}_{\mathrm{req}}, \\ \boldsymbol{\xi}^{(x)}_{\mathrm{req}} &= \begin{pmatrix} \mathbf{R}^{\mathrm{T}}_y(\pi/2) & \mathbf{0}_{[3\times3]} \\ \mathbf{0}_{[3\times3]} & \mathbf{R}^{\mathrm{T}}_y(\pi/2) \end{pmatrix} \boldsymbol{\xi}_{\mathrm{req}}, \end{aligned} \tag{2}$$

where $\mathbf{R}_x(-\pi/2)$ and $\mathbf{R}_y(\pi/2) \in \mathbb{R}^{3\times3}$ are the rotation matrices rotating $-\pi/2$ and $\pi/2$ around the x- and y-axis, respectively. $\boldsymbol{\xi}^{(z)}_{\mathrm{req}}$, $\boldsymbol{\xi}^{(y)}_{\mathrm{req}}$, and $\boldsymbol{\xi}^{(x)}_{\mathrm{req}}$ are the required motion vectors for a manipulator whose first axis is aligned with the z-, y-, and x-axis, respectively.

3 Structural Synthesis Algorithm

The procedure for the structural synthesis of serial manipulators is shown in Algorithm 1. At the beginning a set of suitable DH parameters for the first link is generated as pointed out in [16]. This set becomes the set of suitable architectures (*SoA*) for the first DOF of the mechanism. It is important to note that the resulting set includes several similar architectures with different DH parameters. These isomorphisms are found using the procedure proposed in Sect. 3.2. Further DOF are generated and added to every member of *SoA* until the DOF of the architectures in SoA is equal to the required DOF in ξ_{req}. The complete algorithm is carried out for the three motion constraint vectors $\xi^{(z)}_{\text{req}}$, $\xi^{(y)}_{\text{req}}$, and $\xi^{(x)}_{\text{req}}$.

3.1 Generation of Suitable Architectures

The proposed generation of all architectures fulfilling ξ_{req} is shown in [16]. These architectures are described by means of the DH parameters using prismatic (P) and revolute (R) joints. In the case of a P joint, d_i corresponds to the joint coordinate q_i and in case of a R joint $\theta_i = q_i$. Additionally, each DH parameter is limited to two possible values: $\theta_i = \{0, \frac{\pi}{2}\}, d_i = \{0, d_i^*\}, a_i = \{0, a_i^*\}, \alpha_i = \{0, \frac{\pi}{2}\}$. The values $d_i = 0$ and $a_i = 0$ mean that these parameters must be zero. In contrast, $d_i = d_i^*$ and $a_i = a_i^*$ mean that these parameters can take any value different from zero. The value $\alpha_i = 0$ indicates that two consecutive joint axes are parallel and $\alpha_i = \frac{\pi}{2}$ that these axes are perpendicular. All possible parameter combinations as well as further details can be found in [16].

Algorithm 1: Structural synthesis algorithm, the variable SoA corresponds to the set of suitable architectures

```
Input: ξ_req^(z)
ξ_req^(y) = R'_x ξ_req^(z);
ξ_req^(x) = R'_y ξ_req^(z);
foreach ξ_req in (ξ_req^(z), ξ_req^(y), ξ_req^(x)) do
    SoA = GenSuitableLinks(L, ξ_req);
    SoA = DetectIsomorphisms(SoA, ξ_req);
    for id = 2 to DOF do
        for Rob ∈ SoA do
            SoA* = GenSuitableArchitec(Rob, L, ξ_req);
            SoA* = DetectIsomorphisms(SoA*, ξ_req);
        SoA = SoA*
```

The suitability of an architecture is evaluated through two conditions. Firstly, the terms of ξ_{req} that are zero must also be zero in the robot motion vector ξ of the evaluated architecture. Secondly, the rank of the Jacobian $\mathbf{J}$ must be equal to the number of required DOF.

3.2 Detection of Isomorphisms

In the set of suitable architectures, there are configurations described with different DH parameters having similar kinematic behaviour (isomorphisms). In order to reduce the computational cost and the complexity of the optimization problem (combined structural and geometrical synthesis), these similar architectures are gathered in groups that can be represented by a common DH parameter matrix. Figure 1 shows an example of two architectures that are isomorphisms.

In the first case of Fig. 1, the parameters of the first link are $\theta_1 = 0$ and $a_1 = a_1^*$, while in the second case $\theta_1 = \pi/2$ and $a_1 = 0$. As can be seen, θ_1 and a_1 can take any value without affecting the motion directions of the EE. Hence, both manipulators can be considered particular cases of the same DH parameter matrix. Here, the notation θ_i, d_i, a_i, or α_i with $i = 1, 2, \ldots, n$ indicates that these parameters can take any value including zero.

The detection of isomorphisms is performed for every member of the set of suitable architectures. If two architectures can be described with the same DH parameter matrix, they are replaced in the set of suitable architectures by the new DH parameter matrix. Hence, the SoA contains the minimal set of possible manipulators for each new DOF (see Algorithm 1).

Table 1 summarizes the conditions to be evaluated for each DH parameter $(d_j, a_j, \alpha_j, \theta_j)$ of the jth link of a manipulator. ω_j represents the angular velocity generated by the first j links. $\mathbf{z}_{j-1}$ is the axis of the jth joint as shown in Fig. 2.

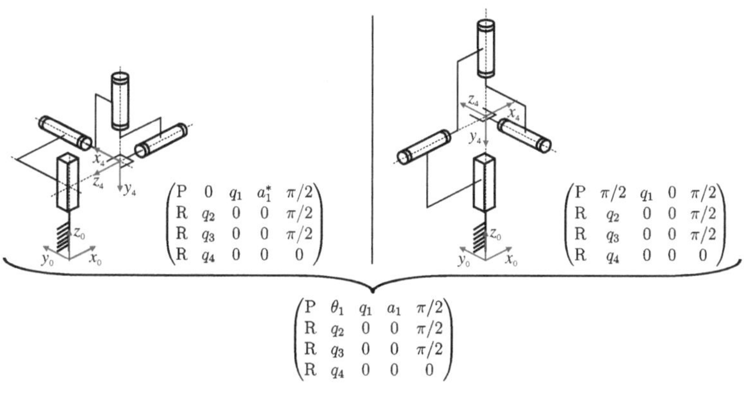

Fig. 1 Example of two isomorphisms with $\xi_{\text{req}} = (0, 0, \xi_{\text{req}_3}, \xi_{\text{req}_4}, \xi_{\text{req}_5}, \xi_{\text{req}_6})^{\text{T}}$

Table 1 Summary of the conditions to compare two architectures

Parameter	Joint	Condition	Possible values
d_j	R	$\boldsymbol{\omega}_j \times d_j\mathbf{z}_{j-1} = 0 \vee \boldsymbol{\omega}_j \times d_j\mathbf{z}_{j-1} \in \xi_{req}$	$d_j \in \mathbb{R}$
a_j	P	$\boldsymbol{\omega}_{j-1} \times a_j\mathbf{x}_j = 0 \vee \boldsymbol{\omega}_{j-1} \times a_j\mathbf{x}_j \in \xi_{req}$	$a_j \in \mathbb{R}$
a_j	R	$\text{rank}(\mathbf{J}_{j+1})_{a_j=0} < \text{rank}(\mathbf{J}_{j+1})_{a_j=a^*}$	$a_j \neq 0\,(a_j^*)$
	R	$\boldsymbol{\omega}_{j-1} \times a_j\mathbf{x}_j \in \xi_{req}$	$a_j \in \mathbb{R}$
α_j	R, P	$\text{rank}\left({}^j\mathbf{J}^{v}_{j+1,n}\right) = \{0,3\} \wedge \text{rank}\left({}^j\mathbf{J}^{\omega}_{j+1,n}\right) = \{0,3\}$	$\alpha_j \in \mathbb{R}$
θ_j	P	$\text{rank}\left({}^j\mathbf{J}^{v}_{j+1,n}\right) = \{0,3\} \wedge \text{rank}\left({}^j\mathbf{J}^{\omega}_{j+1,n}\right) = \{0,3\}$	$\theta_j \in \mathbb{R}$
	P	$\alpha_j = 0 \vee \boldsymbol{\omega}_{j-1} \times d_j z_{j-1} = 0$	

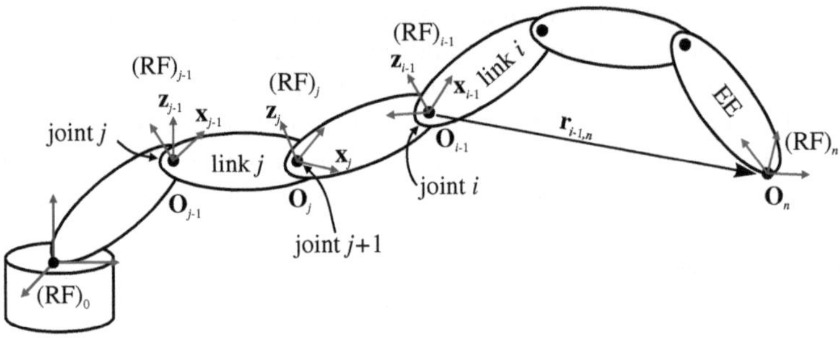

Fig. 2 Location of the reference frames of a serial manipulator

$(\mathbf{J}_{j+1})_{a_j=0}$ is the matrix composed by the first $j+1$ columns of the Jacobian matrix replacing $a_j = 0$. ${}^j\mathbf{J}^{v}_{j+1,n}$ represents the matrix formed by the first three rows of the last $n-j$ columns of the Jacobian matrix, i.e. the elements of the Jacobian that represent the effect of the last $n-j$ joints $(q_{j+1}, q_{j+2}, \ldots q_n)$ on the linear velocity described in the jth reference frame. ${}^j\mathbf{J}^{\omega}_{j+1,n}$ represents the matrix formed by the last three rows of the last $n-j$ columns of the Jacobian described in the jth reference frame.

It is important to note that the subindex j represents the link under consideration while i is used as the index to indicate any other link.

4 Results of the Proposed Approach

Based on the presented procedure (see Algorithm 1), the DH parameters of serial manipulators fulfilling a required motion vector ξ_{req} can be determined without any redundant solution. Table 2 compares the number of architectures generated using the proposed approach and the number of architectures reported in [16] (without

Table 2 Number of architectures generated with and without isomorphisms detection

ξ_{req}	Without isomorphisms detection				With isomorphisms detection			
	$N_s^{(z)}$	$N_s^{(y)}$	$N_s^{(x)}$	N_s	$N_s^{(z)}$	$N_s^{(y)}$	$N_s^{(x)}$	N_s
$(0, 0, \xi_{req_3}, 0, 0, \xi_{req_6})^T$	12	0	0	12	2	0	0	2
$(\xi_{req_1}, 0, 0, 0, 0, \xi_{req_6})^T$	0	0	4	4	0	0	1	1
$(\xi_{req_1}, \xi_{req_2}, 0, 0, 0, 0)^T$	0	8	8	16	0	1	1	2
$(0, \xi_{req_2}, \xi_{req_3}, 0, 0, \xi_{req_6})^T$	8	24	0	32	1	2	0	3
$(\xi_{req_1}, \xi_{req_2}, 0, 0, 0, \xi_{req_6})^T$	112	64	64	240	4	3	3	10
$(0, 0, 0, \xi_{req_4}, \xi_{req_5}, \xi_{req_6})^T$	2	2	2	6	1	1	1	3
$(\xi_{req_1}, \xi_{req_2}, \xi_{req_3}, 0, 0, 0)^T$	32	32	32	96	1	1	1	3
$(0, 0, \xi_{req_3}, \xi_{req_4}, \xi_{req_5}, \xi_{req_6})^T$	20	0	0	20	2	0	0	2
$(\xi_{req_1}, \xi_{req_2}, \xi_{req_3}, 0, 0, \xi_{req_6})^T$	1984	704	704	3392	17	9	9	35
$(0, \xi_{req_2}, \xi_{req_3}, \xi_{req_4}, \xi_{req_5}, \xi_{req_6})^T$	220	220	388	828	7	5	11	23

isomorphims detection). $N_s^{(z)}, N_s^{(y)}$, and $N_s^{(x)}$ represent the number of architectures with the first joint axis aligned with the *z*-, *y*-, and *x*-axis respectively. N_s is the total number of architectures found for each required motion vector. In comparison to [16], the number of architectures generated can be substantially reduced using the presented method. Therefore, the obtained architectures are able to be used in a subsequent efficient geometrical synthesis process. Hence, all possible architectures can be considered and the selection of the architecture can be included in the robot optimization process.

The proposed approach is capable for manipulators up to 6 DOF. In this case, the required motion vector ξ_{req} does not have to be rotated using Eq. (2). Without detection of isomorphisms, the number of architectures generated for a 6 DOF manipulator would increase up to a few millions. Such a set of suitable architectures could not be handled due to the very high computational cost. However, using the presented method, a total number of 326 architectures are obtained for a required motion vector $\xi_{req} = (\xi_{req_1}, \xi_{req_2}, \xi_{req_3}, \xi_{req_4}, \xi_{req_5}, \xi_{req_6})^T$.

As an example, Table 3 shows all possible architectures that can be used to obtain a manipulator with three rotational and one translational DOF along an axis parallel to the *z*-axis ($\xi_{req} = (0, 0, \xi_{req_3}, \xi_{req_4}, \xi_{req_5}, \xi_{req_6})^T$). The DH parameters of these architectures show which parameter has to be a certain value, which has to be different from zero, and which can take any value without changing the motion directions of the EE. In case of the PRRR manipulator on the left side, θ_1, a_1, α_1, and d_2 can take any value. Despite this, the EE maintains the direction of each DOF. Therefore, these parameters can be used as optimization parameters in case of a geometrical optimization process of this manipulator. In contrast, the parameters α_2 and α_3 must be $\pi/2$ and a_2, d_3, a_3, d_4, and a_4 must be zero.

Table 3 Architectures generated for a required motion vector $\xi_{req} = (0, 0, \xi_{req_3}, \xi_{req_4}, \xi_{req_5}, \xi_{req_6})^T$

R/P	θ_i	d_i	a_i	α_i
P	θ_1	q_1	a_1	α_1
R	q_2	d_2	0	$\frac{\pi}{2}$
R	q_3	0	0	$\frac{\pi}{2}$
R	q_4	0	0	α_4

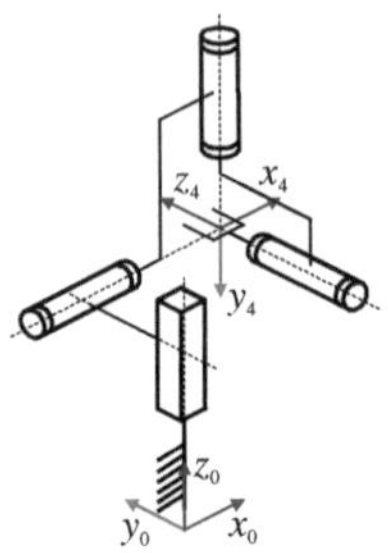

R/P	θ_i	d_i	a_i	α_i
R	q_1	d_1	0	0
P	$\frac{\pi}{2}$	q_2	0	$\frac{\pi}{2}$
R	q_3	0	0	$\frac{\pi}{2}$
R	q_4	0	0	α_4

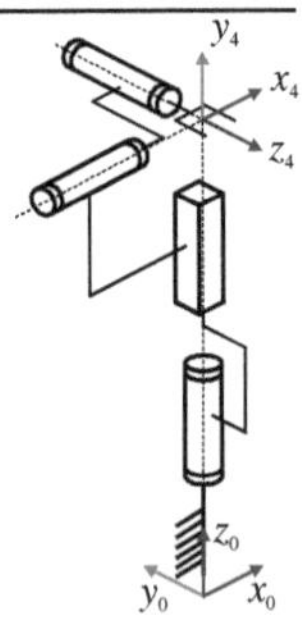

5 Conclusions

The paper presented an approach to generate serial architectures that fulfil a required motion vector. It provides the DH parameters that can be used as optimization parameters in a subsequent geometrical synthesis. A set of suitable architectures is generated from a modified required motion vector. The isomorphisms are detected and removed from this set in order to reduce significantly the computational cost and the complexity of the optimization problem. Hence, the obtained architectures can be efficiently used in a computer design process to carry out a combined structural geometrical synthesis, evaluating their performance regarding requirements as workspace, dynamics, etc. In comparison to the case without detection of isomorphisms, the number of architectures generated is reduced up to 99 % (for a 5 DOF manipulator) using the proposed approach. In order to clarify its effectiveness, the number of architectures generated by the presented method is compared to results previously published without the detection of isomorphisms.

Acknowledgments Daniel Ramirez would like to thank Universidad Santo Tomas and also the Administrative Department of Science, Technology, and Innovation of Colombia Colciencias for financial support through the Call 529 2011.

References

1. Angeles, J.: The qualitative synthesis of parallel manipulators. J. Mech. Des. **126**(4), 617–624 (2004)
2. Campos, A., Budde, C., Hesselbach, J.: A type synthesis method for hybrid robot structures. Mech. Mach. Theory **43**(8), 984–995 (2008)
3. Caro, S., Khan, W.A., Pasini, D., Angeles, J.: The rule-based conceptual design of the architecture of serial Schoenflies-motion generators. Mech. Mach. Theory **45**(2), 251–260 (2010)

4. Gogu, G.: Families of 6R orthogonal robotic manipulators with only isolated and pseudo-isolated singularities. Mech. Mach. Theory **37**(11), 1347–1375 (2002)
5. Gogu, G.: Fully-isotropic parallel manipulators with five degrees of freedom. In: Proceedings of the IEEE International Conference on Robotics and Automation, ICRA 2006, pp. 1141–1146 (2006)
6. Gogu, G.: Structural synthesis of fully-isotropic parallel robots with Schönflies motions via theory of linear transformations and evolutionary morphology. Eur. J. Mech. A. Solids **26**(2), 242–269 (2007)
7. Gogu, G.: Structural synthesis of parallel robots: part 1: methodology. Springer, New York (2007)
8. Gosselin, C.M., Kong, X.: Type synthesis of 4-DOF SP-equivalent parallel manipulators: a virtual chain approach. Mech. Mach. Theory **41**, 1306–1319 (2006)
9. Guo, Z., Wang, K., Xu, Z., Qi, H.: Topological design and genetic synthesis of the variable topology parallel mechanisms. In: ASME/IFToMM International Conference on Reconfigurable Mechanisms and Robots, ReMAR 2009, pp. 221–228 (2009)
10. Hervé, J.: The Lie group of rigid body displacements, a fundamental tool for mechanism design. Mech. Mach. Theory **34**, 719–730 (1999)
11. Kong, X.: Type synthesis of 3-DOF parallel manipulators with both a planar operation mode and a spatial translational operation mode. J. Mech. Robot. **5**(4), 041015 1–8 (2013)
12. Kong, X., Gosselin, C.M.: Type synthesis of 3-DOF spherical parallel manipulators based on screw theory. J. Mech. Des. **126**(1), 101–108 (2004)
13. Kuo, C., Dai, J.: Structural synthesis of serial robotic manipulators subject to specific motion constraints. In: Proceedings of the ASME 2010 International Design Engineering Technical Conferences and Computers and Information in Engineering Conference, Montreal (2010)
14. Li, Q., Huang, Z., Herve, J.: Type synthesis of 3R2T 5-DOF parallel mechanisms using the Lie group of displacements. IEEE Trans. Robot. Autom. **20**(2), 173–180 (2004)
15. Merlet, J.: Parallel Robots, 2nd edn. Springer, New York (2006)
16. Ramirez, D., Kotlarski, J., Ortmaier, T.: Automatic generation of serial robot architecture from required motion directions. In: Proceedings of the Symposium on Automated Systems and Technologies, pp. 55–64, Garbsen (2014)
17. Tsai, L.: Robot analysis: the mechanics of serial and parallel manipulators. Wiley, New York (1999)
18. Yang, T.L., Liu, A.X., Jin, Q., Luo, Y.F., Shen, H.P., Hang, L.B.: Position and orientation characteristic equation for topological design of robot mechanisms. J. Mech. Des. **131**(2), 17 (2009)
19. Yang, T.L., Liu, A.X., Shen, H.P., Luo, Y.F., Hang, L.B., Shi, Z.X.: On the correctness and strictness of the position and orientation characteristic equation for topological structure design of robot mechanisms. J. Mech. Robot. **5**(2), 18 (2013). doi:10.1115/1.4023871

Part VI
Biomechanics and Medical Engineering

Bioinspired Mechanism Synthesis for Flapping Flight with Unsteady Flow Effects

Hrishikesh Raste, Anupam Saxena, Roger Sauer and Burkhard Corves

Abstract Optimal 2D flapping wing kinematics and wing shape are obtained based on well-established unsteady flow characteristics to maximize the mean lift. Flow characteristics, e.g., creation and evolution (strength and position) of the attached leading and trailing edge vortex, vortex shedding and subsequent wake capture, Kramer's and transient effects, are modeled via the potential flow analyses performed on a finite set of cross flow planes. The stabilizing axial flow across the helical vortex is computed by considering the variation in span-wise vortex properties and centrifugal forces due to wing rotation. Post detachment, vortex decay is captured in the wake. The consequent variation in pressure around the wing chord is used to compute the mean lift per cycle.

Keywords Biomimetics · Unsteady flow mechanisms · Mechanism synthesis · Wing profile

1 Introduction

Flapping flight in insects has intrigued scientists for long. For a dragonfly, flow patterns associated with its aerial acrobatic feats are even more intricate. It flaps its wings in rowing motion along an inclined plane at a rate close to 40 Hz with *Re* in the range 3000–6000 [10]. Ellington [3] notes that the flow around the wing profile of a dragonfly is unsteady. Flapping flight is analyzed previously using quasi-steady

H. Raste (✉) · A. Saxena (✉)
Department of Mechanical Engineering, IIT Kanpur, Kanpur 208016, India
e-mail: hrishikesh.raste@gmail.com

A. Saxena
e-mail: anupams@iitk.ac.in

R. Sauer
AICES, RWTH Aachen-University, 52056 Aachen, Germany

B. Corves
IGM, RWTH Aachen-University, 52056 Aachen, Germany

B. Corves et al. (eds.), *Mechanisms, Transmissions and Applications*,
Mechanisms and Machine Science 31, DOI 10.1007/978-3-319-17067-1_26

models [2] wherein wing forces are computed treating fluid flow as steady at each time instant.

Sane [7] highlights the importance of two unsteady flow mechanisms around a flapping wing, namely the delayed stall with leading edge vortex and Kramer effect. It is known that a stably attached helical shaped leading edge vortex contributes significantly to the lift during translation while Kramer's effect accounts for the high peaks in lift during wing rotation. The wing shape can also influence the flow in different length scales. The temporally evolution of a leading edge vortex, and its convection and diffusion into the wake, makes flow around the flapping wing remarkably stable for Re ≥ 100 [7]. This vortex when shed encounters the flapping wing in the wake causing variations in lift.

Unsteady characteristics via the potential flow theory (Sect. 2) are considered to optimally synthesize a 2D mechanical linkage to simulate the wing kinematics (Sect. 4), and wing shape (Sect. 5) so that the lift is maximized. Creation and convection (position and strength) of an attached leading edge vortex is modeled as a singularity in potential flow (Sect. 2). Per Ellington's [4] observation, a single attached leading edge vortex is considered present at a time per cycle of wing motion. Viscous decay of the shed vortex is also accounted for in Sect. 2. Thereafter, for a candidate linkage and wing profile, hydrodynamic forces (per unit length) acting on the wing are computed analytically (Sect. 3). Flow computations are then coupled with mechanism synthesis to obtain results presented in Sect. 7.

2 The Potential Flow Model

The flow is assumed incompressible, inviscid and irrotational except for the presence of a concentrated vortex modeled as a flow singularity. The analysis is performed in a plate attached reference frame ($x''Ry''$, plane P) on a wing chord MN in Fig. 1b. Gradients of quantities along the wing span are assumed small, hence the 3D flow is modeled as a sequence of 2D flows on parallel cross planes (not shown in Fig. 1b). The wing is assumed thin, rigid, and uncumbered. The 2D flow is characterized using the following circulations—Γ around the plate that models the Kramer's effect, and Γ_j that is either attached to the leading/trailing edge, or free/shed in the wake. The model herein closely follows the analytical development in Minotti [6] but with salient differences.

2.1 Complex Potential at Far Field

This section describes the 2D flow in one of the cross-flow planes in Fig. 1b. The point R translates with a velocity $\mathbf{U}(t)$ and the chord rotates with an angular speed $\Omega(t)$ (ω_{BA}, Fig. 1b) about R. Defining a complex variable $z = x + iy$, the complex potential, following from Minotti's derivation [6], is written as,

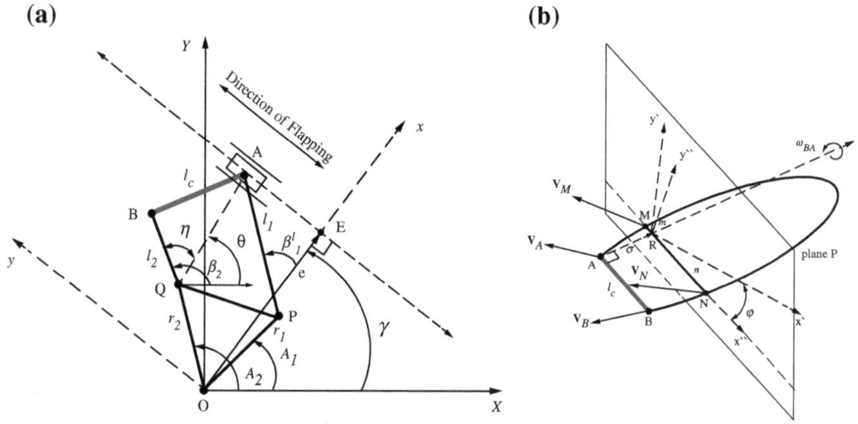

Fig. 1 **a** Schematic of the proposed six bar mechanism with a ternary link. Reference frames *XOY* (global) and *xOy* (local) are defined for simplifying kinematic calculations. Point *O* is fixed on the insect body. **b** A 3D view of the wing attached with the coupler link l_c. Chord *MN* of the wing is analyzed for lift using the 2D potential flow theory in the crossflow plane *P*

$$C(z) = -U^* z + \Omega i z^2/2 \tag{1}$$

where superscript star represents a complex conjugate. For ease of evaluating the flow around complex boundaries, Kutta-Joukowski (conformal) transformation is used. It maps the geometry of the wing chord from one complex plane $z = x + iy$ to another $\varsigma = \xi + i\eta$. The transformation $f : z \to \varsigma$ is given by

$$z = x_o + \varsigma + a^2/\varsigma \tag{2}$$

where a is the radius of cylinder in conformal plane. One now applies the Milne-Thomson theorem to model the surface of the cylinder as streamline, so that the flow does not penetrate it. Thus,

$$C_t(\varsigma) = C(\varsigma) + C^*(a^2/\varsigma^*) \tag{3}$$

2.2 Modeling an Attached/Shed Circulation and Its Center

Circulations, Γ centered at $\varsigma = 0$ and Γ_i centered at $\varsigma_i = a(p_i + iq_i)$ (p_i and q_i are scaling factors) in the conformal plane model the flow around the plate. In presence of these circulations $C(\varsigma)$ is modified to

$$C(\varsigma) = -U^*\varsigma + i\Omega(x_o + \varsigma)^2/2 + \Gamma \ln(\varsigma)/2\pi i + \sum_{i=1,2,3,4} \Gamma_i \ln(\varsigma - \varsigma_i)/2\pi i \tag{4}$$

Γ_1/Γ_2 (unknown variable) is a leading edge vortex at edge M/N, while Γ_3/Γ_4 represent the vortex shed from M/N. Once shed its strength, now a known quantity, diminishes according to the decay function (Sect. 2.3). The terms pertaining to Γ_3/Γ_4 are included in $C(\varsigma)$ only when a vortex is shed. Milne-Thomson theorem is used to transform $C(\varsigma)$ to $C_t(\varsigma)$. For brevity, we forgo the expression for $C_t(\varsigma)$ and leave it for the reader to derive. From $C_t(\varsigma)$, the velocity field in the z plane can be computed as

$$\mathbf{u}^* = dC_t(\varsigma)/dz = (dC_t(\varsigma)/d\varsigma)(dz/d\varsigma)^{-1} = (dC_t(\varsigma)/d\varsigma)\left(\varsigma^2/(\varsigma^2 - a^2)\right) \quad (5)$$

At $\varsigma = \pm a$, the denominator on the right hand side tends to infinity. For regularization therefore, two Kutta conditions, $dC_t(\varsigma)/d\varsigma|_{\pm a} = 0$ can be employed to determine the circulations Γ and Γ_1 or Γ_2.

Consider first the case where $\Gamma_{1/2}$ is attached to the leading or trailing edge. Smith [9] treats the spiral vortex sheet as a line vortex. He computes the center of circulation such that the overall force on the surface and the line vortex is zero. The modified Smith's condition derived for the *delta* planform of the wing is written as

$$\lim_{z \to z_v} \left(dC_t(z)/dz - \Gamma_{1/2}/2\pi(z - z_V)\right) = U_a(\bar{z}_V - \bar{z}_E)/\sigma \quad (6)$$

where z_V represents the center of the vortex, z_E is the edge to which the vortex is attached, U_a is the velocity incident to the wing and σ is the distance of the wing chord from the hinge. U_a is determined using the Maxworthy's model [5] as

$$U_a = 1.414\sigma V_{R_o}/S \quad (7)$$

where V_{R_o} is the wing tip velocity and S is the wing span. As the vortex detaches, it is assumed that the axial velocity U_a becomes zero since the free circulation destabilizes rapidly. Thus the velocity field in the core region around the shed vortex is give by the Saffman's model as,

$$U_{z_1^s} = \lim_{z \to z_1^s} \left(dC_t(z)/dz - \Gamma_{3/4}/2\pi i(z - z_1^s)\right) \quad (8)$$

where z_s^1 represents the center of the shed vortex.

2.3 Vortex Decay and Wake Capture

We model the decay of circulation into the wake. The axial feed is assumed absent ($U_a = 0$). $\Gamma_{3/4}$ will decay, either rapidly or gradually from the center of the core, outward, due to viscosity and other factors. Based on computational and experimental data, semi-empirical models exist that employ single-phase decay formulation for predictions. The model of Sarpkaya [8] and others predict decay rate, initially large but diminishing with time. Linear best-fit models of decay also exist that are developed based on the data from a range of weather conditions.

In this paper we model the vortex decay post detachment using a predefined decay function characteristic of a free vortex decay. Due to the reciprocating motion of insect wings, shed vortices from the previous stroke lingering in the wake might interact with the wings during the next stroke [7]. This phenomenon called as wake capture was observed to cause an increase in lift [1].

3 Computation of the Mean Lift: Blasius Theorem

Derivation of the analytical expression for the hydrodynamic force follows from Minotti [6]. The force **F** per unit length, on a wing chord MN in some crossflow plane is given by

$$F_x - iF_y = i\rho/2 \oint_{\mathbb{P}} [dC_T(z)/dz]^2 dz + i\rho\partial\left[\oint_{\mathbb{P}} C_T(z)dz\right]^* /\partial t \tag{9}$$

where F_x is the drag force acting along the horizontal of the attached reference frame and F_y is the lift force normal to F_x. $\mathbb{P}$ is the perimeter around the wing chord. The integrals can be computed using the Cauchy's residue theorem.

4 Mechanism Synthesis

Dragonflies and damselflies use contraction of synchronous flight muscles, namely the medial elevator and lateral depressor, for wing actuation. These are arranged vertically, with one end attached to the wings and the other to the floor of the thorax. Together, the contractions help produce wing pronation, downstroke, supination and upstroke. Wings also rotate and twist during these stages.

A mechanism such as Fig. 1 is proposed, where the slider A replicates the muscle contractions. Wing base is attached to AB and the wing projects out in the plane perpendicular to the paper. Two reference frames termed as global frame (XOY, capital letters) and a local frame (xOy, letters with subscript l) are defined for ease of kinematic analysis as presented below

$$\begin{aligned} &x_P = r_1 \cos A_1, \quad y_P = r_1 \sin A_1, \quad x_{P_l} = x_P \cos\gamma + y_P \sin\gamma, \\ &y_{P_l} = -x_P \sin\gamma + y_P \cos\gamma \end{aligned} \tag{10}$$

$$x_{A_l} = e, \quad y_{A_l} = y_{P_l} + l_1 \sin\beta_{1_l}, \quad \beta_{1_l} = \cos^{-1}(x_{A_l} - x_{P_l}/l_1) \tag{11}$$

Moving back to the global frame,

$$x_A = e\cos\gamma - y_{A_l}\sin\gamma, \quad y_A = e\sin\gamma + y_{A_l}\cos\gamma \tag{12}$$

$$x_Q = r_2\cos A_2, \quad y_Q = r_2\sin A_2, \quad x_B = x_Q + l_2\cos\beta_2, \quad y_B = y_Q + l_2\sin\beta_2 \tag{13}$$

where,

$$\beta_2 = \theta + \eta; \quad \theta = \tan^{-1}(y_A - y_Q)/(x_A - x_Q) \quad \eta = \cos^{-1}(AQ^2 + l_2^2 - l_C^2)/2l_2AQ \tag{14}$$

Temporal derivatives provide velocities and accelerations. These can be used to compute the lift forces using the aforementioned potential flow theory. Coupler link l_c is assumed to be of a constant length of 10 mm. The mechanism is synthesized using the following design parameters described by vector D_1. Thus,

$$D_1 = [r_1, r_2, A_1, A_2, e, l_1, l_2, \gamma]. \tag{15}$$

Mechanism locking and singularities are avoided by using the following constraints.

$$x_{A_l} - x_{P_l} < l_1; \quad AQ^2 + l_2^2 - l_C^2 < 2l_2AQ; \quad -0.95 \le \frac{l_c^2 + l_1^2 - BP^2}{2l_cl_1} \le 0.95;$$
$$-0.95 \le \frac{l_2^2 + QP^2 - BP^2}{2l_2QP} \le 0.95$$

5 Wing Profile

The shape of a flapping wing plays a vital role in governing the fluid dynamics in the length scales of insect flight. In this work, the wing profile is modelled as a single/multiple Rational Bezier/B-spline curves with C^0 continuity at the junction points (if any). Constraints are used to ensure that no loop is formed. A general point on a rational Bezier curve is defined by

$$\mathbf{P}(t) = \sum_{i=0}^{n}\left(\mathbf{P}_i^c w_i B_i^n(t) / \sum_{i=0}^{n} w_i B_i^n(t)\right) \tag{16}$$

where t is a parameter such that $0 \leq t \leq 1$. B_i^n is a Bernstein polynomial defined as $B_i^n = \binom{n}{i}(1-t)^{n-i}t^i$. $\mathbf{P}_i^c$ denotes a set of Bezier points for $0 \leq i \leq n$ and w_i is a corresponding weight assigned to the data point. Ideally a delta wing profile is sought implicitly by maximizing the lift. Bezier points and the corresponding weights constitute the design vector D_2 such that

$$D_2 = [\mathbf{P}_i^c, w_i], \quad 0 \leq i \leq n \tag{17}$$

6 Coupling

After developing the mechanism kinematics and the wing profile, we couple the flapping assembly with the 2D potential theory to calculate the lift on the wing. Frame $x''Ry''$ is the proposed Z plane which is mapped to the corresponding ς plane by the conformal transformation. The wing attached to the link AB in Fig. 1a is divided into chords in their respective crossflow planes (Fig. 1b). Total lift is computed as summation of individual lifts on the chords. V_M and V_N are tip

Table 1 Optimal design parameters for the flapping mechanism

	r_1 (mm)	r_2 (mm)	A_1 (deg)	A_2 (deg)	e (mm)	l_1 (mm)	l_2 (mm)	γ (deg)
D_1	6.04	5.95	48.04	71.62	7.24	15.05	9.06	70

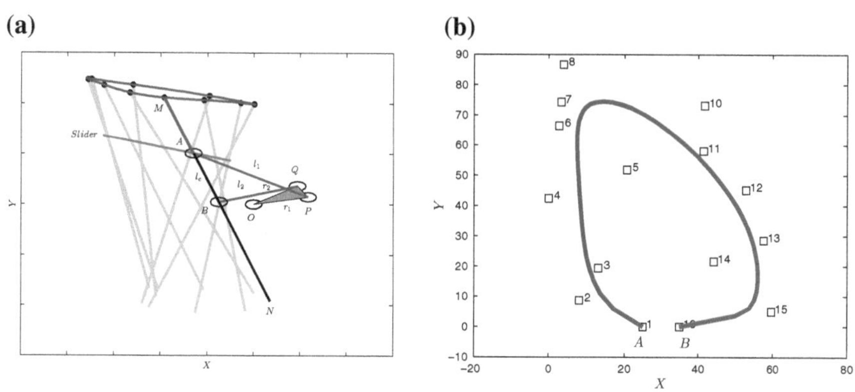

Fig. 2 **a** A planar view of the optimal mechanism. A wing is attached with the coupler AB which protrudes in a plane out of the paper on the other side. A wing chord MN can be seen as a projection on the mechanism plane. The slider attached at A (not shown here) slides along the *pink line*. Different poses of the wing chord MN are depicted in *gray* in the background. **b** Shows Bezier (control) points marked (*small squares*) around the same wing profile

Table 2 Optimal design parameters D_2 for the wing profile (Fig. 2b)

D_2	Bezier points ($\mathbf{P}_i^c$)															
i	1	2	3	4	5	6	7	8	9	10	11	12	13	14	15	16
X_i (mm)	25	8.1	13.1	0	20.7	2.8	3.3	3.9	42.8	41.7	41.4	52.7	57.6	44.2	59.6	35
Y_i (mm)	0	8.7	19.4	42.4	51.8	66.5	74.3	86.6	98.5	73.1	58.1	45.1	28.6	21.6	5	0
w_i	1	2.3	5.0	4.9	3.6	2.9	9.9	5.0	1.9	2.1	3.3	1.0	8.3	1	5.0	1

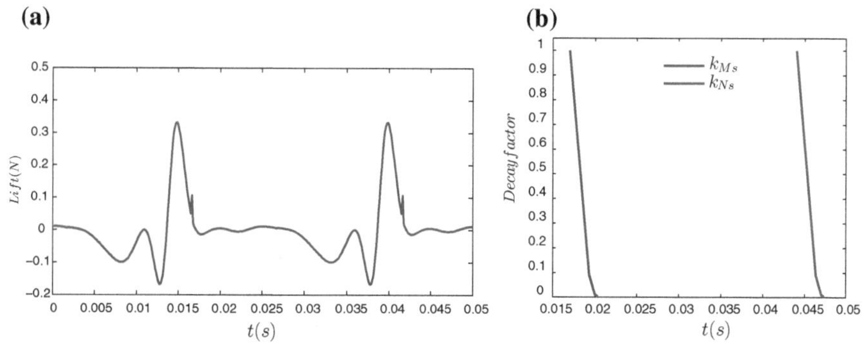

Fig. 3 **a** Denotes the lift generated on the whole wing of the mechanism, **b** is a plot of the decay model employed. The same model is employed for vortices shed from edges *M* and *N* and during supination and pronation

velocities, deduced from V_A and V_B, which are used in calculating U_a. The coupling between the mechanism kinematics, wing profile and the potential flow is then set up as an optimization problem

$$\underset{\mathbf{D}=[D_1,D_2]}{\text{Maximize}} \text{ Mean Lift per unit length}$$

A stochastic hill climbing genetic algorithm is used with 10,000 iterations and crossover and mutation rates as 1 and 0.02 respectively.

7 Results

Optimal mechanism design parameters are provided in Table 1 with mechanism in Fig. 2. Table 2 gives optimal Bezier points and weights. A delta shape of the wing, Fig. 2b, can be observed starting from the connection points *A* and *B* (attached to the coupler l_c). Figure 3a shows lift generated on the wing during two cycles of mechanism rotation. An overall positive lift can be observed during the cycle. Decay functions in Fig. 3b model the vortex decay of the vortices shed from edge *M* (blue) and edge *N* (red).

8 Conclusion

The synthesized mechanism mimics the role of elevator and depressor muscles. The synthesized wing profile, as expected, resembles a delta shape which is necessary to create and sustain the leading edge vortex. A practical model the mechanism should be motorized with a high frequency to obtain the wingbeat of an insect, e.g. a dragonfly. Since the mechanism is composed of pin joined rigid links vibrational instabilities may set in at such high speeds. A better design could involve using flexural components with their inherent advantages such as low friction and wear, easy manufacturability and vibrational stability.

References

1. Dickinson, M.H., Lehmann, F.O., Sane, S.P.: Wing rotation and the aerodynamic basis of insect flight. Science **284**(5422), 1954–1960 (1999)
2. Dudley, R., Ellington, C.: Mechanics of forward flight in bumblebees: Ii. quasi-steady lift and power requirements. J. Exp. Biol. **148**(1), 53–88 (1990)
3. Ellington, C.: The aerodynamics of hovering insect flight. i. the quasi-steady analysis. Philos. Trans. R. Soc. London. B, Biol. Sci. **305**(1122), 1–15 (1984)
4. Ellington, C.P., Van Den Berg, C., Willmott, A.P., Thomas, A.L.: Leading-edge vortices in insect flight. Nature **384**, 626–630 (1996)
5. Maxworthy, T.: The formation and maintenance of a leading-edge vortex during the forward motion of an animal wing. J. Fluid Mech. **587**, 471–475 (2007)
6. Minotti, F.: Unsteady two-dimensional theory of a flapping wing. Phys. Rev. Ser. E **66** (5; PART 1), 051–907 (2002)
7. Sane, S.P.: The aerodynamics of insect flight. J. Exp. Biol **206**(23), 4191–4208 (2003)
8. Sarpkaya, T., Robins, R., Delisi, D.: Wake-vortex eddy-dissipation model predictions compared with observations. J. Aircr. **38**(4), 687–692 (2001)
9. Smith, J.: Improved calculations of leading-edge separation from slender, thin, delta wings. Proc. R. Soc. Lond. A **306**(1484), 67–90 (1968)
10. Wang, Z.J.: Dragonfly flight. Phys. Today **61**(10), 74–75 (2008)

Conceptual Design of a New Neurosurgical Brain Retractor

Hsuan-Ping Kuan and Chin-Hsing Kuo

Abstract Brain retractor is an instrument that delivers a steady retraction for retaining a working channel within the brain during brain surgery. Yet, current brain retractors still look forward to some augmentations, e.g., the adjustable retraction distance and spatula angle, the fine and coarse adjustments and the changeable spatulas. Though there are already existing designs that possess one to three functions mentioned above, but none of them integrates all the aforementioned four design variables together. In this paper, we propose a new mechanism concept that possesses all of the above functions. To achieve adjustable retraction distance, a threaded slider is meshed with a screw. Two pairs of bevel gears and a rod make up the mechanism for adjustable angle of retraction spatula. The fine and coarse adjustments for the retraction distance are achieved by using planetary gear trains. We also designed a spatula holder with multiple sizes of spatulas to accomplish changeable spatulas. The CAD models of the new design are presented.

Keywords Brain retractor · Medical device · Surgical tool · Neurosurgery · Mechanism design · Screw mechanism · Gear train

1 Introduction

During intracranial surgery, brain retractor is an indispensable instrument that retracts the brain steadily and retains a working channel for surgical tools. Figure 1a–d show the operation procedure of a kind of brain retractors (i.e., the concentrically expansible needle retractor [5]). Figure 1a depicts a so-called concentrically expansible needle retractor [5]. At the beginning, the retractor is

H.-P. Kuan · C.-H. Kuo (✉)
National Taiwan University of Science and Technology, Taipei City, Taiwan
e-mail: chkuo717@mail.ntust.edu.tw

H.-P. Kuan
e-mail: B10031038@mail.ntust.edu.tw

B. Corves et al. (eds.), *Mechanisms, Transmissions and Applications*,
Mechanisms and Machine Science 31, DOI 10.1007/978-3-319-17067-1_27

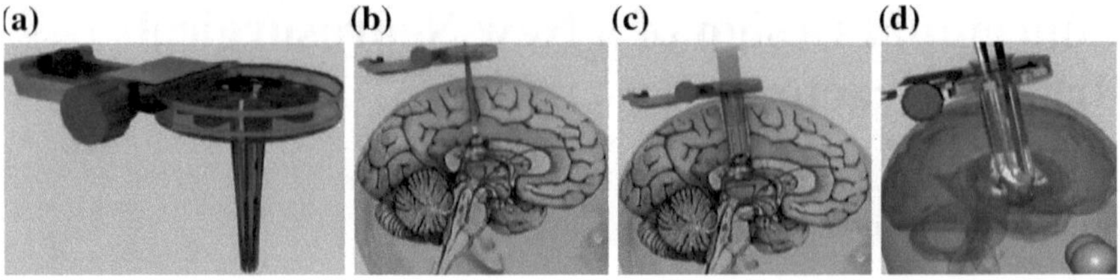

Fig. 1 Illustration of a brain retractor and its use [5]. **a** The retractor, **b** inserting stage, **c** retracting stage, **d** operating stage

inserted into the brain as shown in Fig. 1b. Then, it expands (as illustrated in Fig. 1c) and creates a channel for surgical instruments to reach the lesions inside the brain.

There are already some brain retractors in the market, such as LEYLA retractor system [1], minimally invasive tubular retractor [2], gelatin sponge retractor [4], concentrically expansible needle retractor [5], Miyake-Ohta brain retractor [6], cylindrical channel retractor [7], stitch retractor [8], microroll retractor [9], etc. These devices can all retain a working channel and each of them possesses some specific functions. For example, the LEYLA [1] and Miyake-Ohta systems [6] can change spatulas for accommodating different depth and channel sizes. For example, the LEYLA system [1] and the concentrically expansible needle retractor [5] can change the retraction distance, which is convenient for surgeons to expand the operating field. The LEYLA system [1] can adjust the angle of the spatulas. In that way, surgeons can retract less brain tissue to gain the same amount of space for operation. Although these retractors can fulfill the basic needs of brain surgery, their abilities can still be further improved to offer better assistance. For instance, if the retraction speed can be adjusted with fine and coarse tunes, it would suggest a more delicate operation.

This paper presents a new mechanism concept that delivers an alternative solution for adjustable retraction distance and angle of retraction spatula, fine and coarse adjustments, and changeable retraction spatulas. In the following, we propose the design concept first. Then, we describe how the proposed brain retractor works and explain the mechanism theories behind. Next, we compare the advantages and disadvantages of the new design with the existing retractors. Last, we talk about future improvements of our new design.

2 New Design

The proposed new brain retractor is shown in Fig. 2. It provides adjustable retraction distance and angle of retraction spatula, fine and coarse adjustments in adjusting the retraction distance and changeable spatulas.

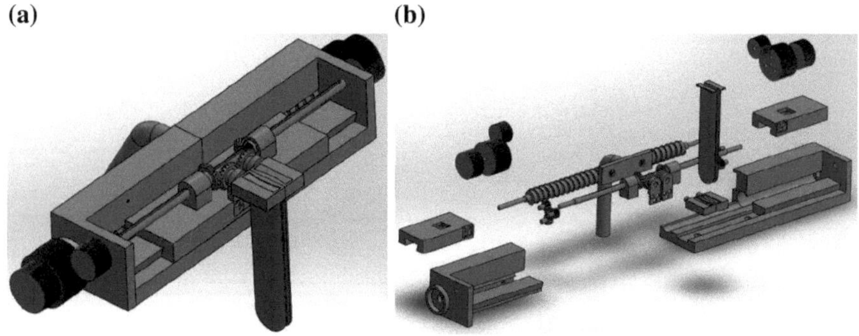

Fig. 2 Schematics of the proposed brain retractor. **a** Assembly view, **b** explode view

For achieving adjustable retraction distance, the mechanism is made of two sliders and an integrated right- and left-handed screw. The sliders are both threaded on one side. One slider has a left-handed thread, while the other has a right-handed one. The threads on the screw mesh with the ones on the sliders. When the screw rotates, the two sliders move oppositely as illustrated in Fig. 3a, b.

In order to make the angle of retraction spatulas adjustable, we attached two bevel gears to both spatula holders, respectively, as shown in Fig. 4a, b. Next, we mount two bevel gears on the rod and insert a pin through the hole on the bevel gear and the slot on the rod each as shown in Fig. 5. In this way, the bevel gear can move translationally with respect to the rod as depicted in Fig. 5a, b. Finally, we join the assembly of the rod and the bevel gears to the slider and mesh both gears to the gears on the spatula as shown in Fig. 4b. As a result, the spatula will rotate as the rod rotates. Figure 6 shows the schematic drawing of the overall mechanism.

Figure 6 is the mechanism sketch of the proposed brain retractor. I_1 and I_2 are the two inputs, which respectively control the translational and rotational motions of the spatulas. When I_1 is actuated, the integrated right- and left-handed screw is driven to push the bevel gear set, S_L and S_R, on either side so as to adjust the

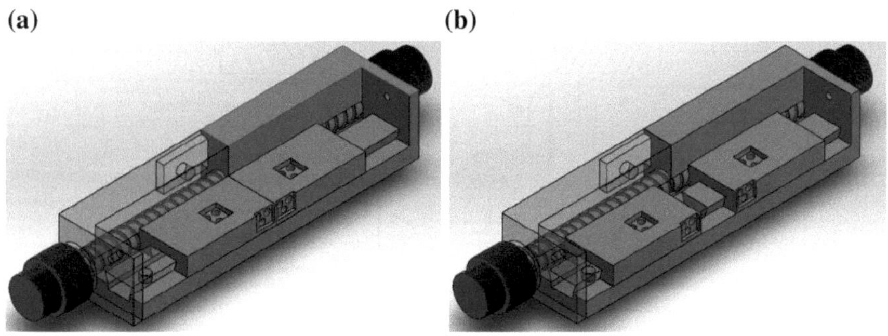

Fig. 3 Schematics of the mechanism for adjustable retraction distance. **a** Before rotating the screw, **b** after rotating the screw

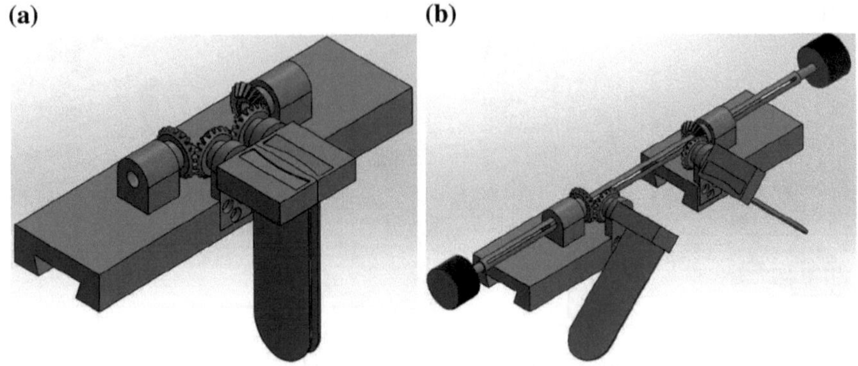

Fig. 4 Schematics of the mechanism for adjustable angle of retraction spatulas. **a** Spatulas and its holder assembled on sliders, **b** full assembly of the mechanism for adjustable angle of retraction spatulas

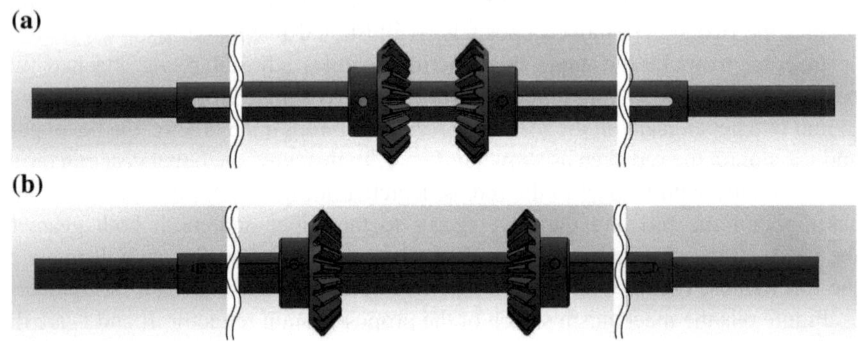

Fig. 5 Schematic of the assembly of the slotted rod and two bevel gears. **a** Bevel gears before translational and rotational moving, **b** bevel gears after translational and rotational moving

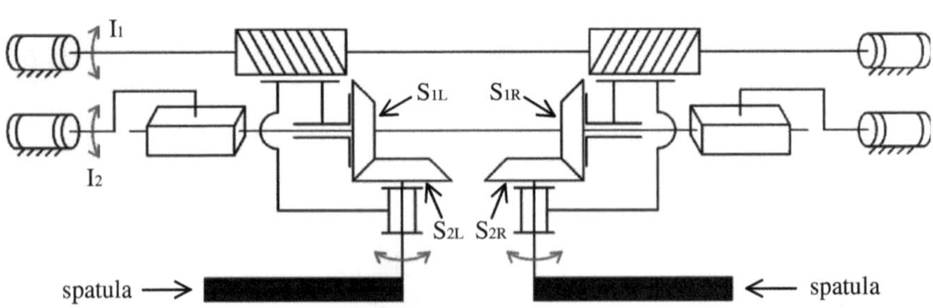

Fig. 6 Schematic drawing of the proposed brain retractor

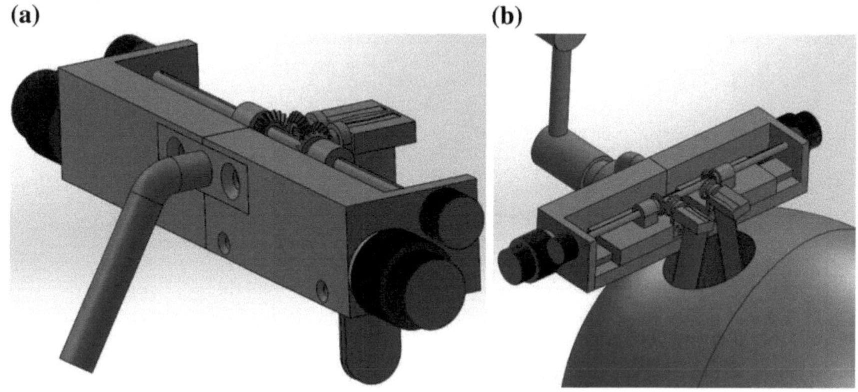

Fig. 7 Attachment module and its use. **a** Endoscopic holder module, **b** usage of the holder

distance between the two spatulas, i.e., the retraction width. On the other hand, when I_2 is actuated, the two bevel gears S_{1L} and S_{1R} will rotate accordingly, which drives the other two bevel gears and consequently regulates the orientation angles of the two spatulas, i.e., the opening angle.

For mounting the retractor on surgical tables, a fixation module is attached to the retractor. Different modules can be put on the retractor for different purposes. For instance, the one we have now is an endoscopic holder module, which is just a bent tube that can be attached to the back of the retractor as shown in Fig. 7a. This module allows an endoscopic holder to hold the retractor as shown in Fig. 7b.

Different sizes of retractor spatulas, as shown in Fig. 8, are provided. The surgeons can choose suitable sizes for their needs. We also used planetary gear trains, which are hidden in the red circles in Fig. 9, to accomplish fine and coarse adjustment. The structure of the hidden gear trains is elaborated in Fig. 10a, b. The theories behind it will be discussed in the next section.

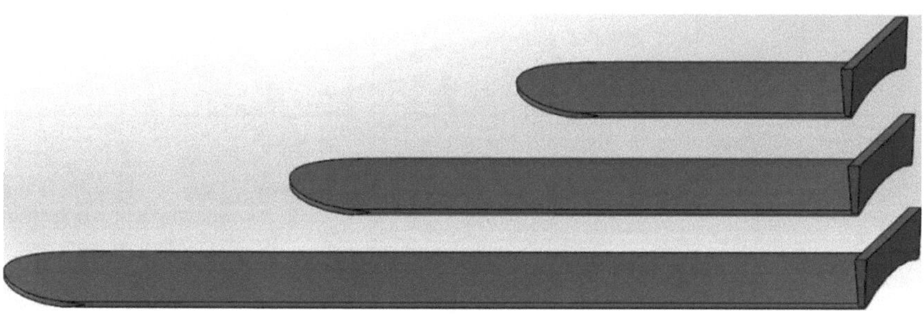

Fig. 8 Different sizes of retractor spatula

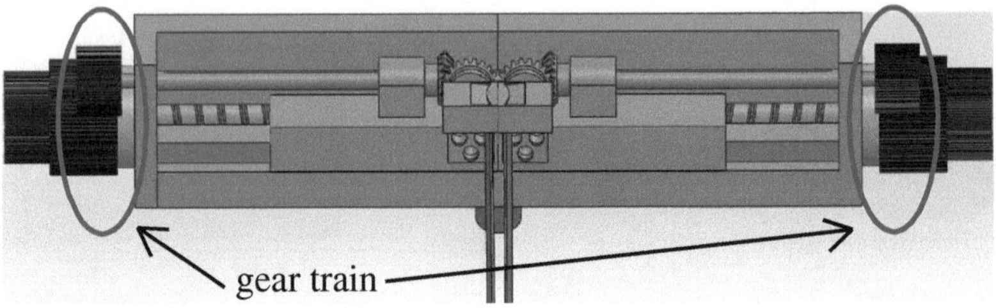

Fig. 9 Installation of the planetary gear trains

3 Speed Analysis

As shown in Fig. 10, the planetary gear train used in our brain retractor consists of a sun gear (yellow), three planet gears (blue) supported by the carrier (purple) and a ring gear (green). It is widely known that epicyclic gearing must obey the following two equations:

$$\frac{\omega_s - \omega_c}{\omega_a - \omega_c} = -\frac{N_a}{N_s} = -\frac{R_a}{R_s} \tag{1}$$

$$R_a = 2R_b + R_s \tag{2}$$

where ω_a, ω_s, ω_p and ω_c are the angular velocities of the ring gear, sun gear, planet gear, and planet carrier, respectively; N_a and N_s are the numbers of teeth of the ring gear and sun gear, respectively; R_a, R_s and R_p are the pitch radii of the ring gear, sun gear, and planet gear, respectively. Here, because the ring gear is fixed to the base of the retractor, $\omega_a = 0$. By setting $\omega_a = 0$ and the angle speed ratio $\omega_s : \omega_c = 4 : 1$, we can get the result of $R_p = R_s = 1/(3R_a)$ by solving Eqs. (1) and

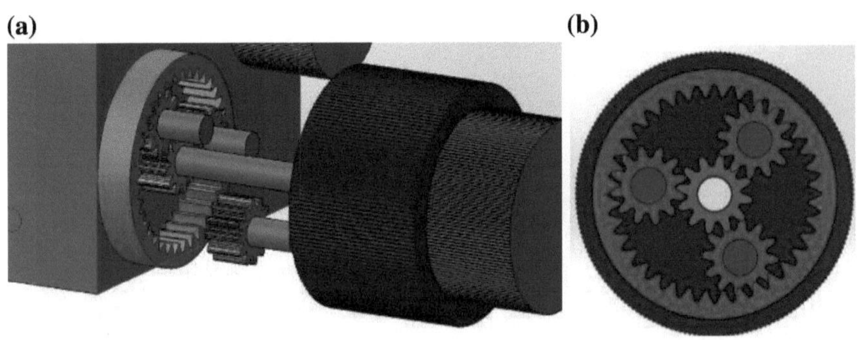

Fig. 10 The planetary gear train for fine and coarse adjustment. **a** Explode view, **b** front view

(2) together. So, when the radii of the planet gears are equal to that of the sun gear and one third of the ring gear, the fine adjustment distance will be one-fourth of the coarse adjustment distance.

4 Discussion

We have presented a new retractor design in the above. The unique part of this new design is that it gives the surgeon the same amount of working space and yet retracts less brain. This is done through the advantage that the new retractor can retract the brain in two different modes, which are achieved by moving the spatulas translationally and rotating them.

A comparison of our proposed retractor and other eight types are presented in Table 1. From the table, we observe that none of the other retractors can adjust retraction distance with different velocities; one can change the angle of retraction spatula; two can change spatulas and one can change retraction distance; and only our design has all four abilities as a whole.

Table 1 Comparison of the brain retractors

System	Adjustable retraction distance	Adjustable angle of retraction spatula	Fine and coarse adjustment	Changeable spatulas
LEYLA retractor system [1]	✓	✓	✗	✓
Minimally invasive tubular retractor [2]	✗	✗	✗	✗
Gelatin sponge retractor [4]	✗	✗	✗	✗
Concentrically expansible needle retractor [5]	✓	✗	✗	✗
Miyake-ohta retractor [6]	✗	✗	✗	✓
Cylindrical channel retractor [7]	✗	✗	✗	✗
Stitch retractor [8]	✗	✗	✗	✗
Microroll retractor [9]	✗	✗	✗	✗
Our proposed brain retractor	✓	✓	✓	✓

✓ Yes; ✗ No

5 Conclusions and Future Work

We have presented the conceptual design of a new brain retractor that provides adjustable retraction distance and angle of retraction spatula, fine and coarse adjustments, and changeable retraction spatulas. A threaded slider is meshed with a screw for achieving adjustable retraction distance. Two pairs of bevel gears and a rod comprise the mechanism for adjustable angle of retraction spatula. Fine and coarse adjustments are achieved by using planetary gear trains. We also designed a spatula holder and multiple sizes of spatulas to accomplish changeable spatulas. Then, we analyzed the velocity ratio of the gear train for understanding the displacement ratio between the fine and coarse adjustments.

In addition to the aforementioned abilities, there are still several things the retractor can further go with. For example, if the brain retractor is made of anti-glare materials, the reflected light that may distract surgeons would be minimized. Furthermore, if the retractor can be fixed on patients, unexpected damage caused by accidental movement of the patient's head can be reduced. Moreover, because brain is delicate tissue, the retractor is highly possible to harm the brain when its retraction width is being regulated. On the other hand, if the brain retractor has the ability of monitoring the brain pressure, we can prevent the damage from it [3, 10]. Accordingly, the next step we may take is to make the retractor smaller and lighter in order to reduce the stress when it is mounted on the patient's skull. Also, we need to design an attachment module for attaching the retractor to the patient. What is more, we will attach pressure sensors to the outer surface of the spatula in the future for providing the pressure monitoring ability. For minimizing the glare created by the retractor, we will use plastic material for spatulas instead of using metals in our future prototype.

Acknowledgments This work is financially supported by Ministry of Science and Technology, Taiwan under the MOST Undergraduate Research Scheme with project number MOST-103-2815-C-011-014-E.

References

1. Dujovny, M., Ibe, O., Perlin, A., Ryder, T.: Brain retractor systems. Neurol. Res. **32**(7), 675–683 (2010)
2. Greenfield, J.P., Cobb, W.S., Tsouris, A.J., Schwartz, T.H: Stereotactic minimally invasive tubular retractor system for deep brain lesions. Neurosurgery **63**(4 Suppl 2, 334–339 (2008)
3. Jadhav, V., Solaroglu, I., Obenaus, A., Zhang, H.: Neuroprotection against surgically induced brain injury. Surg. Neurol. **67**(1), 15–20 (2007)
4. Kashimura, H., Ogasawara, K., Kubo, Y., Kakino, S., Sasoh, M., Takahashi, H., Suzuki, K., Ogawa, A.: Brain retraction Technique Using Gelatin sponge in the subtemporal ap-proach. Neurol. Med. Chir. **48**(3), 143–146 (2008)
5. Michaeli, D.: Concentrically expansible needle retractor. WO/ 2001/003586 (2001)
6. Miyake, H., Ohta, T.: Revised new brain retractor: characteristics and new indications. Neurol. Med. Chir. **40**, 128–130 (2000)

7. Singh, L., Agrawal, N.: Cylindrical channel retractor for intraventricular tumour surgery—a simple and inexpensive device. Acta Neurochir. (Wien) **151**(11), 1493–1497 (2009)
8. Singh, L., Agrawal, N.: Stitch retractor—simple and easy technique to retract brain. World Neurosurg. **73**(2), 123–127 (2010)
9. Tsutomu, I., Takeo, G., Hiroki, M., Toshihiro, T., Kenji, O.: Microroll retractor for surgical resection of brainstem cavernomas. Acta Neurochir. (Wien) **151**(11), 1493–1497 (2009)
10. Zhong, J., Dujovny, M., Alfred, R.P., Eimir, P.A., Hun, K.P., Fernando, G.D.: Brain retraction injury. Neurol. Res. **25**(8), 831–838 (2003)

Mechanism Design for Haptic Handwriting Assistance Device

Mehmet İsmet Can Dede and Gökhan Kiper

Abstract One of the applications of haptic technology is in education and training. Handwriting for first year-elementary students has been included in the curriculum for some years in Turkey as the first and only writing skill to be taught. Providing these students with a haptic assistance device during the handwriting learning process is the global aim of this work. Among the other components of the design such as electronics, controls and communication, mechanism design is a critical component to be considered for optimization of the device at different levels. This paper aims to address a solution to meet the design criteria through ergonomic design for user along with optimized force exertion capabilities.

Keywords Haptics · Planar 5R mechanism · Mechanism design · Assistive device

1 Introduction

Haptics literally means sense of touch and haptics technology aims at transmitting this sense from one location to another location. Haptics technology finds application in assisting the blind [1], education [2], training [3], computer-aided design [4], medical field [5], entertainment [6], automotive industry [7], mobile phones [8] and even art [9]. Depending on the application specifications, a variety of haptic devices can be produced. If the application aims at stimulating cutaneous sensory system, in which the receptors under the skin are targeted, a group of devices called tactile haptic/cutaneous haptic interfaces are considered. In this case, the stimulated sensation is temperature, texture, slip, vibration, force or pain.

M.İ.C. Dede (✉) · G. Kiper
İzmir Institute of Technology, İzmir, Turkey
e-mail: candede@iyte.edu.tr

G. Kiper
e-mail: gokhankiper@iyte.edu.tr

B. Corves et al. (eds.), *Mechanisms, Transmissions and Applications*,
Mechanisms and Machine Science 31, DOI 10.1007/978-3-319-17067-1_28

Kinesthesia is the sense that detects bodily position, weight, or movement of the muscles, tendons, and joints. The receptors of this sensation are usually located in the joints and muscles of the human body. Therefore, if a haptic interface is developed to stimulate kinesthesia senses, it aims to stimulate sense of location/configuration, motion, force or compliance. In training or education type of applications, motion and rendering of slave side induced forces plays a vital role. Hence, most of the education or training type of applications calls for kinesthetic haptic devices. Haptic devices coupled with other human-computer interfaces, such as three-dimensional (3D) virtual reality (VR) visualization systems, are referred to as haptic systems. Many researchers have developed education or training purpose haptic systems based on the existing general purpose haptic devices [10–12].

One commercially available example of these type haptic systems is the dentistry trainer by Moog [13]. Main difference of this system from the previously mentioned ones is that a new haptic interface is specifically developed for the system. Workspace, motion and force rendering capabilities are selected according to the application and the device is constructed accordingly.

Application considered in this study is assistive system for handwriting education/rehabilitation. The need in education arises from recent changes in the education system of Turkey. Since couple of years, first graders in elementary school start to learn to write by handwriting. In a classroom of 30 or more students, it becomes impossible for the teacher to work with every student one on one. Hence, a possible teleoperated or automated system to assist the student at early stages can fill this gap in education. In the later stages, the same system can be used as a source of evaluation of the learnt handwriting skills.

In rehabilitation there are two potential needs at different levels. In a more severe case scenario, it can be used for post-stroke patients to re-gain motor neurons for performing finer motor skills through BCI-based initiation of the task. This requires a joint research with neuroscientists in order to correctly locate the neurons to work with. For course motor neuron training, various studies exist in the literature [14]. Another case for rehabilitation is the motor skill rehabilitation of post-injury patients.

Since the application is specifically focused on writing skills, modifying existing haptic devices is not the choice since they are manufactured usually for 3D motion. There have been studies on developing handwriting assistance systems with such general-purpose haptic interfaces [15]. As a result of this, a planar haptic device for handwriting assistance is designed and presented in this paper. Next section describes design criteria followed by the description of the initial design and optimization of the design with respect to force exertion capability throughout the workspace.

2 The Design Criteria

Complying with the aim of the study, the type of grasping is selected to be precision grasp. This type of grasp is used for handwriting and holding tools for precise operations. The amount of forces that a person can exert in this type of a grasp is limited to 1 N in average, hence the devices built for this type of grasp and precise operations (Phantom Omni® [16]) are developed around this range of force exertion. The maximum amount of force to be reflected to the user is selected to be at 1 N throughout the workspace of the device. Although an equal amount of maximum force exertion throughout the workspace cannot be achieved (unless the manipulator is not a PP type of mechanism), an optimization criteria is described as to have equal force ranges throughout the workspace.

Another design criterion is set for the workspace dimensions to cover a writing range on an A4 (297 mm × 210 mm) paper. The location of the workspace is to be optimized for the ergonomics of the user being right-handed. The performance criterion in this optimization is to enable a visual feedback of the written letters at all times.

Writing requires the positioning of the tip of the pen on a planar surface, therefore the task space is two dimensional and it is required to design a two degrees-of-freedom (dof) planar mechanism. For the structure of the device, a parallel mechanism is preferred over a serial one due to its higher precision capability [17]. We confined ourselves with single loop planar mechanisms with revolute (R) and prismatic (P) joints only. The following considerations are taken into account for the selection of the kinematic structure of the mechanism: (1) the actuated joints should be grounded, (2) If exists, prismatic joints should be actuated, (3) the topological structure of the mechanism should be symmetric. Of all possible alternatives only $\underline{R}RRR\underline{R}$ (5R) and $\underline{P}RRR\underline{P}$ comply with these conditions. For the planar writing application 5R mechanism is chosen due to compactness and force transmission characteristics.

3 Mechanism Analysis and Design Optimization

The mechanism structure to be designed and an A4 paper are illustrated in Fig. 1.

3.1 Static Force Analysis

The direct and inverse kinematic analysis formulation is performed, however for brevity, we do not present the kinematic formulation here. In this application, inertial forces are not comparable with the external forces (due to selecting high strength to weight ratio link structures), therefore dynamic effects are ignored in design stage. We shall use virtual work principle in order to obtain relation between the actuator torques, $\mathbf{T}_1$ and $\mathbf{T}_2$, and external force, $\mathbf{F}_{04} = F_{04} \measuredangle \phi$. The total virtual work done on the system:

Fig. 1 5R mechanism

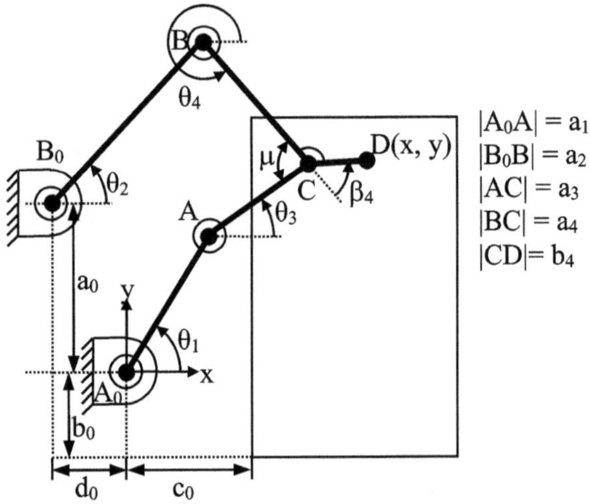

$$F_{04}c\phi\delta x + F_{04}s\phi\delta y + T_1\delta\theta_1 + T_2\delta\theta_2 = 0 \tag{1}$$

The velocity level kinematic analysis results in

$$\begin{aligned}
\delta x &= \frac{-a_1 s(\theta_3-\theta_1)[a_4 s\theta_4 + b_4 s(\theta_4+\beta_4)]\delta\theta_1 - a_2\begin{bmatrix} a_4 s(\theta_2-\theta_4)s\theta_3 \\ + b_4 s(\theta_4+\beta_4)s(\theta_3-\theta_2) \end{bmatrix}\delta\theta_2}{a_4 s(\theta_3-\theta_4)} \\
\delta y &= \frac{a_1 s(\theta_3-\theta_1)[a_4 c\theta_4 + b_4 c(\theta_4+\beta_4)]\delta\theta_1 + a_2\begin{bmatrix} a_4 s(\theta_2-\theta_4)c\theta_3 \\ - b_4 c(\theta_4+\beta_4)s(\theta_3-\theta_2) \end{bmatrix}\delta\theta_2}{a_4 s(\theta_3-\theta_4)}
\end{aligned} \tag{2}$$

where s and c stand for sine and cosine. Substituting Eq. (2) in Eq. (1) we get

$$\begin{aligned}
T_1 &= \frac{a_1 s(\theta_3-\theta_1)[a_4 s(\theta_4-\phi) + b_4 s(\theta_4+\beta_4-\phi)]}{a_4 s(\theta_3-\theta_4)} F_{04} \\
T_2 &= \frac{a_2[a_4 s(\theta_2-\theta_4)s(\theta_3-\phi) - b_4 s(\theta_3-\theta_2)s(\theta_4+\beta_4-\phi)]}{a_4 s(\theta_3-\theta_4)} F_{04}
\end{aligned}$$

3.2 Design Optimization

The 5R mechanism is simulated using Microsoft Excel® using the kinematic and force equations given in Sect. 3. When the end effector is grasped by a user, we aim to achieve uniform feeling throughout the workspace. Assuming a constant force along the x-direction (direction of major motion of hand while writing) acting on

Table 1 Designed link lengths

a_0	b_0	c_0	d_0	a_1	a_2	a_3	a_4	b_4	β_4
250 mm	50 mm	117.5 mm	50 mm	200 mm	250 mm	250 mm	250 mm	60 mm	60°

the end-effector, the motor torque requirements are computed and drawn as a surface plot on the workspace. The torque surfaces are desired to be as flat as possible. Taking avoidance of link collisions into account, a parametric design is performed by changing the link lengths in order to obtain flat torque surfaces for both of the motors. The optimization criterion is to minimize

$$r = \frac{\max(T_i) - \text{average}(T_i)}{\text{average}(T_i) - \min(T_i)}$$

where T_i for $i = 1, 2$ stands for the torque of the motors. An optimal solution is obtained by manually changing the link lengths in Excel®. Designed link lengths are given in Table 1. Torque requirements for a constant force of 1 N at the end-effector are given in Fig. 2. Accordingly, one of the motors demands 0–0.18 N m, while the other one demands –0.21 to 0.05 N m. Ratio r is ~0.86 for both motors.

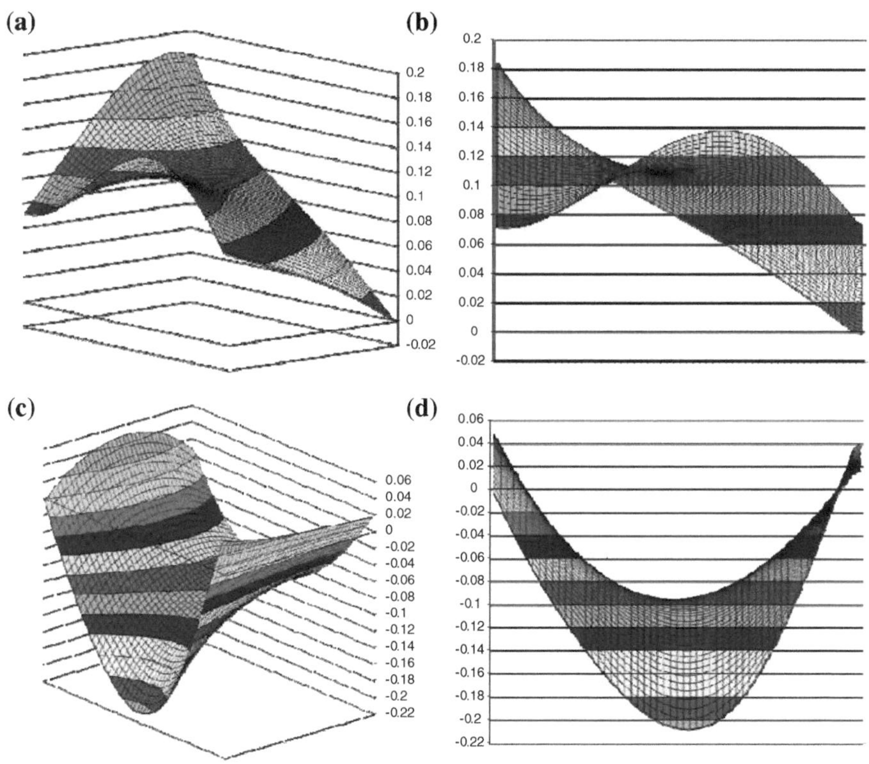

Fig. 2 Torque requirement of the motor at joint A_0: **a** auxiliary, **b** side views; torque requirement of the motor at joint B_0: **c** auxiliary, **d** side views

Minimizing the ratio r corresponds to optimizing the mechanical advantage throughout the workspace of the mechanism. It is well known that the mechanical advantage is lost when the mechanism is positioned near singularities. So, optimizing r also guaranties that the mechanism is far from the singular configurations as well.

It is also possible to apply numerical optimization techniques to obtain optimal dimensions for the mechanism. We did not formulate the problem as a constrained optimization problem for several reasons. First of all, there are 10 design parameters (see Table 1), which means that the optimization should be applied on a 10 dimensional space. The dimension of the optimization space can be reduced by assuming some of the parameters, but running the optimization algorithm for different assumed values is very time consuming. Also, the objective function should include all design criteria with well-chosen weights of the criteria, whereas with parametric design in Excel®, we monitor the ratio r, the variation of both of the motor torques over the workspace, possible link collisions and ergonomics of the design at the same time. Since everything is evaluated and printed on screen in the blink of an eye, Excel® enables a quite fast and intuitive design environment.

4 Constructional Design

The constructional design of the device is performed in CAD environment. The main concern in choosing link material is to obtain higher strength to weight ratio at links. In this way, the minimum impedance of mechanism will be as low as possible by having the smallest inertia properties. The maximum impedance that can be simulated by the mechanism should be as high as possible as well. In that case, the link rigidity receives importance.

Joints structure is also important since for a high precision application minimum joint clearance is required. In addition, in designing impedance type of haptic devices, another important issue is minimization of joint frictions to increase the transparency of the device, which is the desired to displayed impedance ratio. In order to have lower joint frictions, two ball bearings per joint are used for bedding and they are housed by Aluminum parts to minimize the joint clearance by having acceptable rigidity while having relatively low weight.

The assembly of the device is done in such a way that the control card and the power supply do not collide with the links at any instant of manipulation. Manufactured and assembled handwriting mechanism is shown in Fig. 3.

Fig. 3 Constructed and assembled first prototype

5 Conclusions and Discussions

Haptics technology has found application in various areas to provide a solution for different needs. This study aims to make use of this technology for providing assistance in learning handwriting skills. In this paper, work carried out for designing the mechanism of the handwriting assistance device is described. First the kinematic and force equilibrium equations are derived. Then the system is simulated and link lengths are determined in order to optimize force exertion characteristics.

There still remains some more work to do in this research. The design for the pen to be attached at the end-effector is to be carried out. The vertical motion of the pen for writing and traveling modes is to be achieved by means of a solenoid actuator. The electronic hardware for communication between the device and a computer and driving the motors is designed and manufactured and the software is still being developed. The experimental tests will be performed when the overall design is completed.

Acknowledgments This study has been funded by the Ministry of Science, Industry and Technology of Turkey (Project code: 0391.TGSD.2013) provided to CAROB Mühendislik Ltd. Şti., Turkey.

References

1. Yu, W., Ramloll, R., Brewster, S.: Haptic graphs for blind computer users. Lect. Notes Comput. Sci. **2058**, 41–51 (2001)
2. Minogue, J., Jones, M.G.: Haptics in education: exploring an untapped sensory modality. Rev. Educ. Res. **76**(3), 317–348 (2006)
3. Coles, T.R., Meglan, D., John, N.: The role of haptics in medical training simulators: a survey of the state of the art. IEEE Trans. Haptics **4**(1), 51–66 (2010)
4. Liua, X., Dodds, G., McCartney, J., Hinds, B.K.: Virtual design works—designing 3D CAD models via haptic interaction. Comput. Aided Des. **36**(12), 1129–1140 (2004)
5. Tarwala, R., Dorr, L.D.: Robotic assisted total hip arthroplasty using the MAKO platform. Curr. Rev. Musculoskelet. Med. **4**(3), 151–156 (2011)
6. Chang, D.: Haptics: Gaming's new sensation. Computer **35**(8), 57, 26–32 (2002)
7. Rydstrom, A., Brostrom, R., Bengtsson, P.: Can haptics facilitate interaction with an in-vehicle multifunctional interface? IEEE Trans. Haptics **2**(3), 141–147 (2009)
8. Brown, L.M., Kaaresoja, T.: Feel who's talking: using tactons for mobile phone alerts. In: CHI '06 Extended Abstracts on Human Factors in Computing Systems, pp 604–609 (2006)
9. Bergamasco, M., Frisoli, A., Barbagli, F.: Haptics technologies and cultural heritage applications. In: Proceedings of Computer Animation, pp 25–32 (2002)
10. Mohamad, A.E., Mohamed, M., Addulmotaleb, H.E.S., Iglesias, R.: A haptic multimedia handwriting learning system. In: Proceedings of the International Workshop on Educational Multimedia and Multimedia Education, pp 103–108 (2007)
11. Palluel-Germain, R., Bara, F., de Boisferon, A.H., Hennion, B., Gouagour, P., Gentaz, E.: A visuo-haptic device—telemaque—increases kindergarten children's handwriting acquisition. In Proceedings of EuroHaptics Conference, pp 72–77 (2007)
12. Teo, C.L., Burdet, E., Lim, H.P.: A robotic teacher of Chinese handwriting. In: Proceedings of 10th Symposium on Haptic Interfaces for Virtual Environment and Teleoperator Systems, pp 335–341 (2002)
13. Moog Simodont Dental Trainer. http://www.moog.com/markets/medical-dental-simulation/haptic-technology-in-the-moog-simodont-dental-trainer/newsensation. Accessed on 27 Oct 2014
14. Kim, Y.K., Yang, X.: Hand-writing rehabilitation in the haptic virtual environment. In: IEEE International Workshop on Haptic Audio Visual Environments and their Applications, pp 161–164 (2006)
15. Ramos-Murguialday, A., et al.: Brain–machine interface in chronic stroke rehabilitation: a controlled study. Annals of Neurology **74**(1), 100–108 (2013)
16. Silva, A.J., Ramirez, O.A.D., Vega, V.P. Oliver, J.P.O.: PHANToM OMNI haptic device: kinematic and manipulability. In: Proceedings of Electronics, Robotics and Automotive Mechanics Conference, pp 193–198 (2009)
17. Briot, S., Bonev, I.A.: Are parallel robots more accurate than serial robots? CSME Transactions **31**(4), 445–456 (2007)

3 DoF Haptic Exoskeleton for Space Telerobotic

Dan Teodor Margineanu, Erwin-Christian Lovasz, Valentin Ciupe, Marius Mateaş and Eugen Sever Zăbavă

Abstract For extravehicular activities in space, as mounting and dismounting devices and maintenance, a teleoperated robotic arm is frequently used to spare the human operator from the dangers in orbital space. For precision handling operations such as clamping, inserting and screwing bolts, etc., visual feedback is not sufficient. In this paper, a lightweight 3DoF haptic exoskeleton is developed to control a robotic arm in a natural way by copying operator movements and assisting the operator in feeling the loads in the robotic hand.

Keywords Space · Robotics · Haptic · Exoskeleton · Lightweight

1 Introduction

Exoskeletons are mainly used in rehabilitation of physically disabled persons [1]. A great number of patents [2–5] propose various structures and control solutions. Joutras et al. [3] patented a complex equipment covering almost all human body joints (shoulder, elbow, wrist, knee, ankle and neck) for rehabilitation of patients

D.T. Margineanu (✉) · E.-C. Lovasz · V. Ciupe · M. Mateaş · E.S. Zăbavă
Politechnica University of Timisoara, Timisoara, Romania
e-mail: dan.margineanu@upt.ro

E.-C. Lovasz
e-mail: erwin.lovasz@upt.ro

V. Ciupe
e-mail: valentin.ciupe@upt.ro

M. Mateaş
e-mail: marius.mateas@upt.ro

E.S. Zăbavă
e-mail: eugen.zabava@upt.ro

B. Corves et al. (eds.), *Mechanisms, Transmissions and Applications*,
Mechanisms and Machine Science 31, DOI 10.1007/978-3-319-17067-1_29

sitting in wheelchairs through body segments movement controlled in amplitude and reacting load. Dariush [2] found a control method by actuators of an exoskeleton.

Bionic body HAL-5 developed at Tsukuba University, Japan, is a robotic body suit helping the elderly and disabled persons to walk and move objects. Researchers at Korea Institute of Science and Technology in Seoul developed a master arm exoskeleton controlling the arms of a humanoid robot. An arm exoskeleton developed at Salford University, Manchester, England helps users at rehab exercises. A pneumatic exoskeleton developed at Kanagawa Institute of Technology in Atsugi, Japan allows a fitness trainer to hold a 20 kbar without effort. L-EXOS developed at Scuola Superiore Sant'Anna, Pisa, Italy, is an arm and hand exoskeleton allowing human interaction with virtual environments. Carlos Owens from Wasilla, Alaska, built Mecha, a 5.5 m and 1360 kilos exoskeleton. Profesor Jacob Rosen from University of Washington, Seattle, and his team designed a full arm exoskeleton to help neurologically disabled persons. A team from Saga University, Japan, developed an upper limb exoskeleton transforming neuromuscular signals in robotic movement.

At the Free University in Bruxelles, Belgium, in the Active Structures Lab ASL Sensoric Arm Master (SAM) [6] exoskeleton was designed as haptic interface with a serial structure isomorphic to human arm. SAM has 7 degrees-of-freedom (DoF) actuated according to the human arm joints (shoulder, elbow, shoulder and wrist flexion and extension, shoulder and wrist abduction and abduction, arm and forearm pronation and supination) and 6 prismatic joints allowing morphologic adaptation between the active and human joints. This allows a good compromise between using capabilities (maximized workspace and no singularities) and mechanical complexity. Each exoskeleton joint has a similar concept with a local actuator, a position and torque sensor, allowing some control strategies (impedance, admittance control).

In previous works [7, 8], the authors studied some variants of lightweight mechanical structures with servomotor or electromagnetic brake on the revolute joint, planetary mechanism with linear actuator or pneumatic cylinder, belt or cable drives. Potentiometres, encoders, CCD cameras, digital magnetic compases and 3-axes gyroscopes were compared as angular sensors or measuring devices. Actuation with DC servo- and vibration motors, brakes, linear stepper motors, pneumatic cylinders and miostimulators were discussed. The solution with DC electromagnetic brake in the revolute joint and potentiometer was chosen as most suitable for lightweigt space telerobotic applications.

2 Lightweight 3 DoF Haptic Exoskeleton Design

For design, the considered maximal haptic reaction felt in the palm of the operator is similar to manipulating a load of 2.5 kg in earth gravitational field. For the three modules of the exoskeleton, this desired reaction needs the maximal reactive

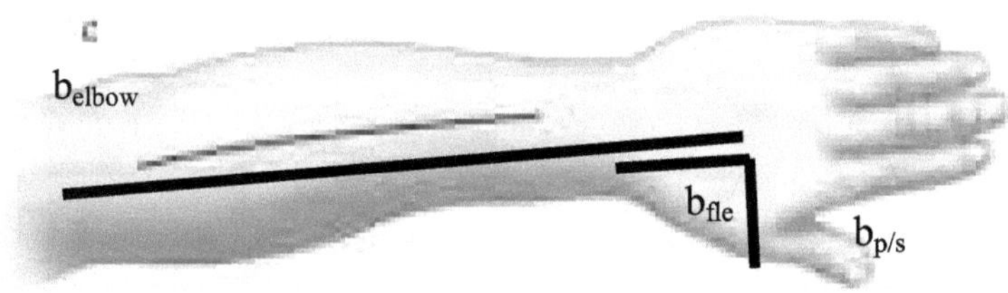

Fig. 1 Reference dimensions for the human forearm and hand

Table 1 Lever arms and maximal reactive torque in the exoskeleton's modules

Module	Wrist (flexion/extension)	Forearm (pronation/supination)	Elbow (flexion/extension)
Lever arm (mm)	100	100	400
Maximal reactive torque (Nm)	2.5	2.5	10

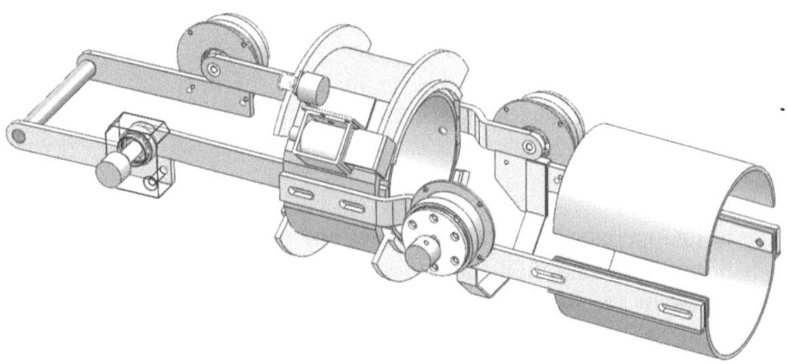

Fig. 2 CAD model of the 3 DoF lightweight exoskeleton

moments for the lever arms b_{elbow}, b_{flex} and $b_{p/s}$ (Fig. 1) given in Table 1, respectively.

The exoskeleton is composed of three modules (see Fig. 2): elbow flexion and extension, forearm pronation/supination and wrist flexion and extension. Wrist adduction/abduction was considered redundant for the tasks and to complicated to be included in a lightweight structure.

2.1 Elbow Flexion/Extension Module

The elbow module (Fig. 3) is clamped on operator's arm by the Velcro adjustable sleeve 1. The position of the elbow joint of the exoskeleton can be also adjusted

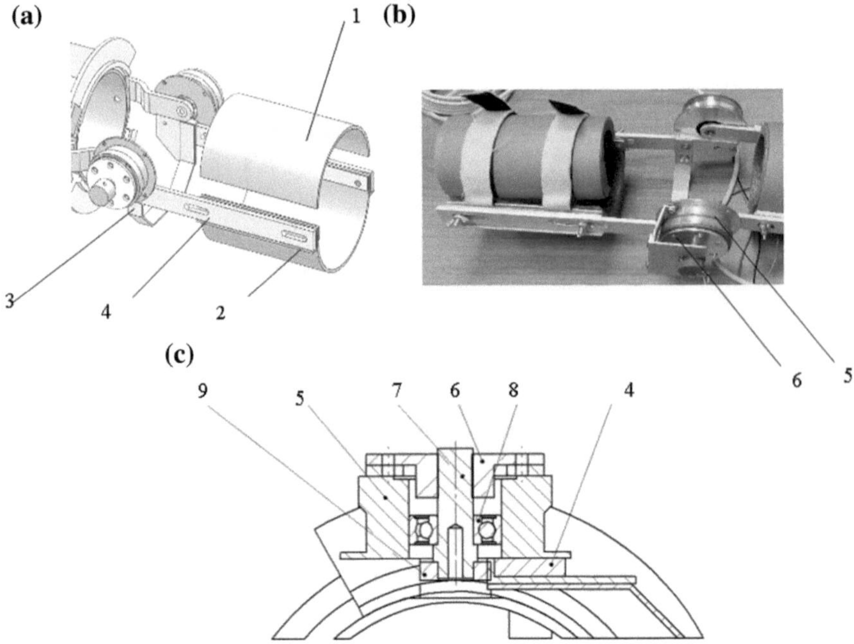

Fig. 3 The elbow flexion/extension module. **a** CAD model, **b** construction **c** cross-section through the elbow joint

according to the operator by sliding the inner and outer arm-to-elbow rods 4 along the prismatic guides 2 and then clamping into position. The structure is reinforced under the elbow by the ribs 3 connecting the two rods. At the rods' ends, the two electromagnetic breaks 5 are mounted. The breaks' rotors 6 mounted on axles 7 and ball bearings 8 are fixed on the elbow-to-forearm rods 9. On one side, the axle is connected also with the potentiometer 6 giving the rotation angle in the joint.

As the operator flexes the elbow, the signal from the potentiometer is sent to the slave robotic arm. As the torque in the robot's joint increases, the feedback signal is transmitted to the brake that creates the haptic response.

2.2 *Forearm Pronation/Supination Module*

The forearm pronation/supination module (Fig. 4) is built around operator's forearm. It is composed of two coaxial sleeves: the outer sleeve 1 is clamped on the elbow-to-forearm rods of the elbow module, the inner sleeve 2 on bushings 3 turns around forearm module axis and is clamped on the forearm-to-wrist rods 4.

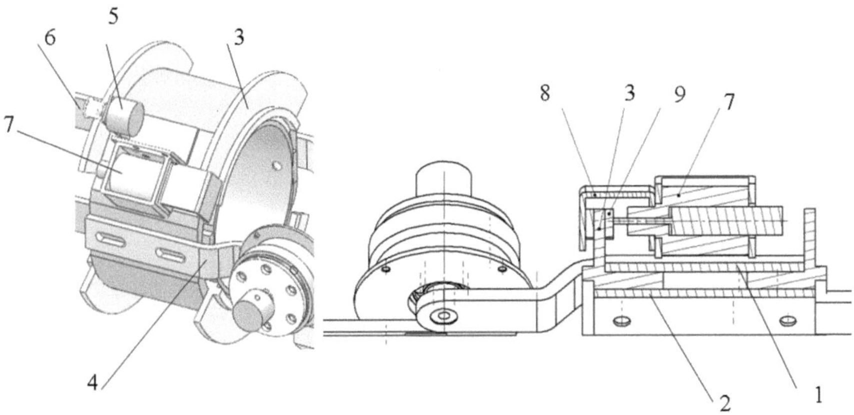

Fig. 4 Pronation/supination module

The rotation angle of the forearm module is measured by the potentiometer 5 mounted on the outer sleeve through the synchronous belt drive 6. The signal is then sent to the slave robotic arm.

The haptic response is produced by two electromagnets 7 mounted on the frames 8 on the outer sleeve. The electromagnetic force clamps the collars on the bushings between two friction plates 9 at 80 mm distance d_{ps} to the module axis.

Two electromagnets symmetrically positioned on the outer sleeve are used, as to balance the friction forces and to give only a haptic reactive torque T_{ps} in the forearm module.

For a friction coefficient μ_{ps} of 1.25 corresponding for a hard break pad friction material/steel friction pair, the necessary clamping force:

$$\left(F_{clamp}\right)_{nec} = \frac{\left(T_{ps}\right)_{nec}}{2 \cdot \mu_{ps} \cdot d_{ps}}$$

is 15.5 N, and the clamping pressure on the brake plates for a electromagnet diameter rod d_t of 2 mm is 4 MPa.

$$p_{ap\,ps} = \frac{4 \cdot \left(F_{clamp}\right)_{nec}}{\pi \cdot d_t^2}$$

2.3 Wrist Flexion/Extension Module

On the forearm-to-wrist rods' ends, the axles 1 of wrist flexion/extension module (Fig. 5) are mounted. On the outer side, the electromagnetic brake 2 fixed on a wrist-to-hand rod 4 gives the haptic response. The potentiometer 3, connected to the

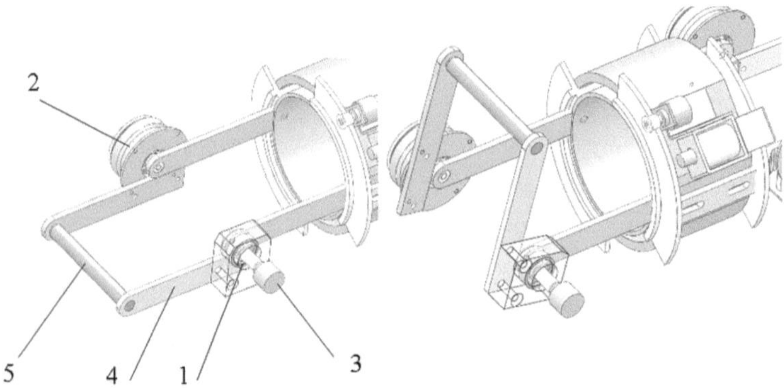

Fig. 5 The wrist flexion/extension module

inner axle reads the rotation angle in the wrist joint. The two rods are joined by the handle 5. Acting on the handle, the operator commands the slave robotic arm and feels the haptic response given by all three modules, as the exoskeleton structure is isomorphic to the human arm.

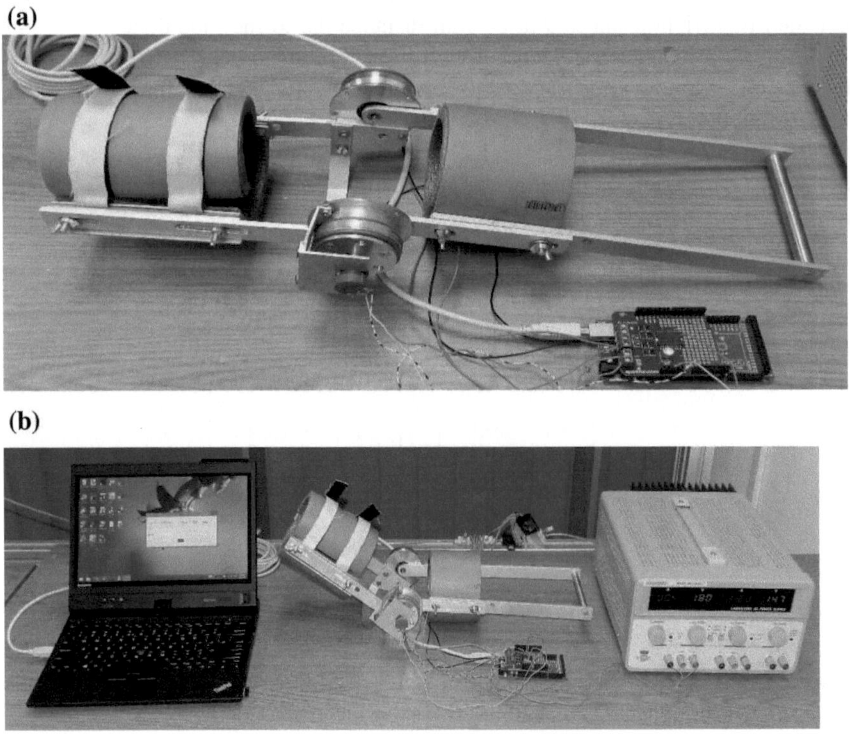

Fig. 6 The exoskeleton prototype with elbow flexion/extension module (**a**) and the test rig (**b**)

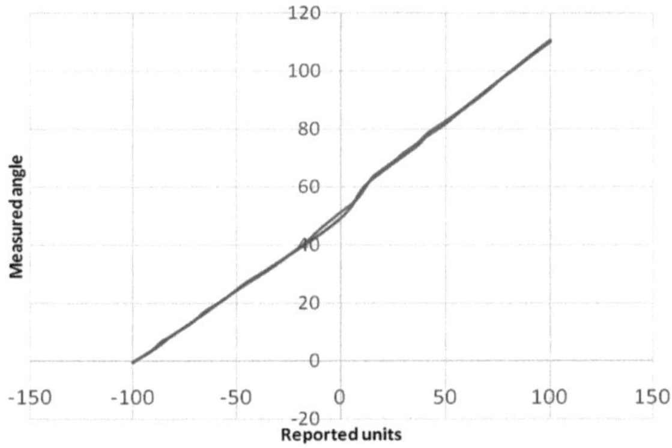

Fig. 7 Calibration diagram for rotation angle given by the potentiometer [8]

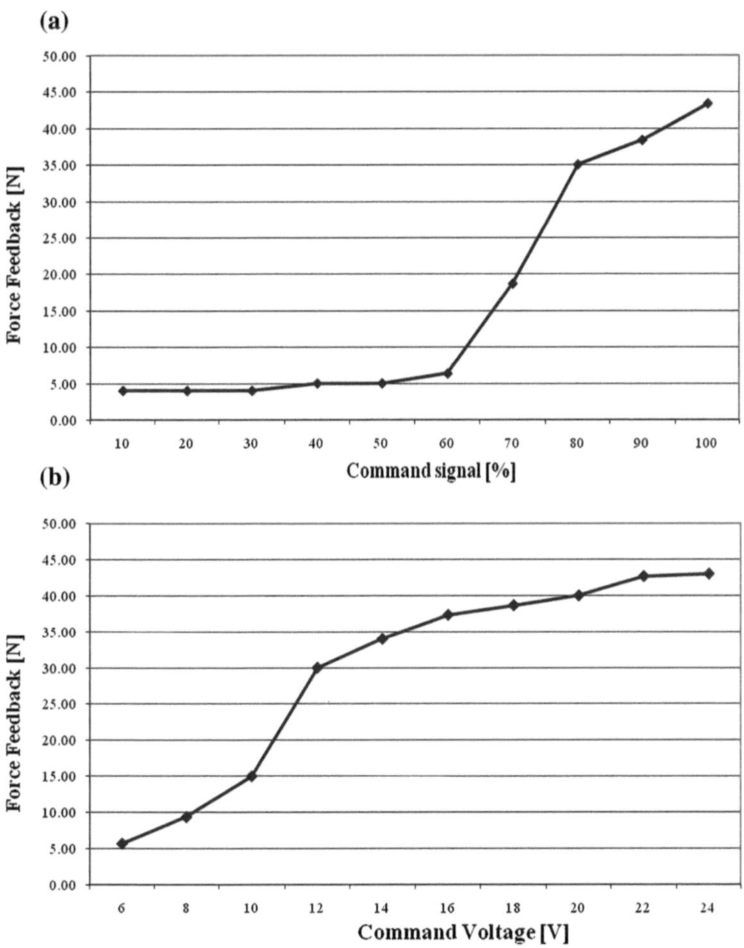

Fig. 8 Force feedback measured on the handle [8]

3 Exoskeleton Prototype and Testing

In the Mechatronics Research Laboratory at Politehnica University Timişoara, prototypes for haptic exoskeletons are built. In Fig. 6, a lightweight exoskeleton with one controlled DoF in the elbow joint [8] is presented.

In order to obtain an accurate command signal for the robotic arm, the module was calibrated by measuring the rotation angle for different values of the units reported by the potentiometer. The results are presented in Fig. 7.

For generating a haptic response according to the load in the joint, the saturation effect in the electromagnetic brake should be taken into account.

In Fig. 8a, b, the force feedback on the handle measured for given command signal and command Voltage is presented. Saturation effect is obvious on both diagrams, so an interface software for adapting the haptic response is needed. The force feedback repeatability was in the range of 5 % only for higher values (over 50 % command voltage), but it is considered sufficient for a proper haptic response.

4 Conclusions

Development of master lightweight exoskeletons with useful haptic response requires a compromise between structural complexity and precision. Electromagnetic brakes use reduces mass and dimensions, and potentiometers give accurate enough readings for rotation angles. Careful calibration for all joints is necessary in order to obtain the required performances. Further exoskeleton developments include the design of a 2 DoF shoulder module mounted on a fixed structure, to obtain a full arm exoskeleton.

Acknowledgments The authors would like to express their gratitude towards the Romanian Space Agency (ROSA) for the support through the project New Haptic Arm Exoskeletons for Robotics and Automation in Space (EXORAS).

References

1. Caldwell, G.D., Tsagarakis, G.N.: Development and control of a 'Soft-Actuated' exoskeleton for use in physiotherapy and training. Auton. Robots **15**(1), 21–33 (2003)
2. Dariush, B.: Exoskeleton controller for a human-exoskeleton system. US 7,774,177 B2. G06F 17/10 (2010)
3. Joutras, F.E., Hruska Jr., R.J.,: Exercise apparatus and technique. US005954621A. 5,954,621. A63B 21/021. (1999)
4. Patoglu, V.: Exoskeleton. US 20120330198A1. A61H 1/02 (2012)
5. Perry, J., Rosen, J.: Exoskeleton. US 2008/0009771 A1. A61B 5/103 (2008)

6. Letier, P., Motard, E., Ilzkovitz, M., Preumont, A., Verschueren, J.P., SAM: Portable haptic arm exoskeleton upgrade. Technologies and new application fields. Proceedings of ASTRA 2011, Noordwijk, Netherlands (2011)
7. Lovasz, E.-Ch., Mărgineanu, D.T., Ciupe, V., Maniu, I., Gruescu, C.M., Stan, S.D., Zăbavă, E. S.: Design and control solutions for haptic elbow exoskeleton module used in space telerobotics. In: Proceedings of 2014 IFToMM Asian Conference on Mechanism and Machine Science Tianjin, China (2014)
8. Mateaş, M.C., Lovasz, E.-Ch., Mărgineanu, D.T., Ciupe, V., Maniu, I., Mărgineanu, E.Z.: Control characteristics of haptic exoskeleton elbow module used in space robotised applications. Appl. Mech. Mater. **658**, 654–659 (2014)

[illegible]

Part VII
Industrial Applications

The Handling and Spreading Algorithms of a Multi-arm Robot System for Automated Cloth Sheet Ironing Machine

Hidetsugu Terada, Kazuyoshi Ishida, Koji Makino and Yasunori Atsumi

Abstract To realize the soft work handling application for automated cloth sheet ironing machine, a multi-arm robot system which can operate a cloth sheet handling and spreading operations is developed. This robot system consists of two rectangular coordinates type robot arms and one rotational joint type robot arm. And this system has the detection system using infra-red detection sensors to detect the cloth sheet edge and the sheet existence. And to avoid the collision, the motion plane of these robots is parallel to the particular plane on the Cartesian frame. In this report, considering the motion of a cloth sheet handling by human hands and the co-operative motion for the multi-arm robot system, the quantification method of the cloth sheet handling spreading operations by the geometrical calculation are proposed. Especially, the quantification method of the carton box assembling operations is applied to this motion. And considering the conventional finding approach of the cloth sheet edge by human hands, the cloth sheet existence and overlapping detection method without image-processing is proposed. Also, based on the motion test results, it is clear that the robot system can pick up and spread the cloth sheet edges.

Keywords Multi-arm robot · Quantification method · Handling · Spreading · Cloth sheet

H. Terada (✉) · K. Ishida · K. Makino
University of Yamanashi, Takeda 4-3-11, Kofu, Yamanashi 400-8511, Japan
e-mail: terada@yamanashi.ac.jp

K. Ishida
e-mail: isawa@yamanashi.ac.jp

K. Makino
e-mail: kojim@yamanashi.ac.jp

Y. Atsumi
Tosen-Machinary Corporation, Meguro-Ku, Tokyo, Japan
e-mail: y.atsumi@tosen.com

B. Corves et al. (eds.), *Mechanisms, Transmissions and Applications*,
Mechanisms and Machine Science 31, DOI 10.1007/978-3-319-17067-1_30

1 Introduction

In the end of 2012, a quarter of industrial robots of the world were used in Japan [1]. Especially, the Japanese particular application field is in assembling process for mechanical or electrical parts. Moreover since the labor costs are rising extremely, the labor saving is most important in Japan. It is one of the solutions that the operation fields of industrial robots are spread to the soft works handling process fields. Therefore, soft work handling methods by industrial robots have been studied [2, 3]. And the soft works handling process which is the operation of attaching the washed linen to the automated ironing machine, is one of the important problems in Japanese laundry. In this case, it is required that the workers don't touch these linens for hospital use to prevent an infectious disease, as possible.

To realize the soft works handling process, the various handling robot fingers and the handling algorithms have been developed, in which a lot of robot systems use image processing [3, 4]. However, the influence of overlap and the shadow of the object cannot be eliminated. So, the handling error has often occurred at conventional robot systems. On the other hand, we have developed motion planning approaches using multi-arm robot which doesn't use an image processing for carton box assembling and a cloth sheet wrapping operation [5, 6].

In this report, a multi-arm robot system of cloth sheet handling and spreading operations for the automated ironing machine is developed. Then, applying proposed motion planning approaches, the handling and spreading algorithms of the cloth sheet which is piled up at random is proposed. Especially, to realize the simplified algorithms, special robot end effectors which are attached the infra-red (IR) lighting and detection system are proposed. And the prototype robot system is tested to verify the usefulness of the proposed approach.

2 Structure of a Multi-arm Robot System

A structure of a multi-arm robot system is proposed as shown in Fig. 1. This robot system consists of two rectangular coordinates type robot arms (Robot arm 1 and Robot arm 2) and one rotational joint type robot arm (Robot arm 3). Especially, to avoid the collision at the handling shift motion, the motion plane is on the "*j*-*k*" plane of the Cartesian frame. It is perpendicular to the "*i*-*k*" plane, which is the motion plane of the Robot arm 1 and 2. And this conveyer feeds cloth sheets. And this conveyer feeds cloth sheets to the initial picking point. This robot system picks up a cloth sheet which is piled up at random. And that moves up along the diagonal line. And then, that spreads a cloth sheet for the next process of the automated ironing machine. Figure 2 shows the cloth sheet initial state and the final state at this procedure.

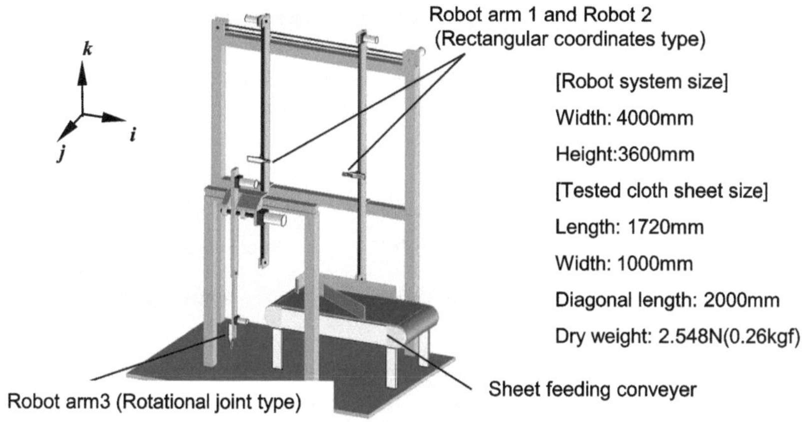

Fig. 1 Structure of a multi-arm robot system for cloth sheet handling and spreading operations

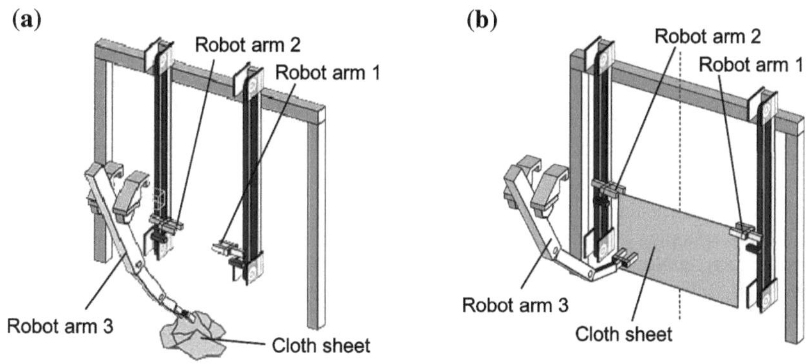

Fig. 2 Cloth sheet state of for the automated ironing machine. **a** Initial state. **b** Final state

3 Structure of the End Effectors and the Single Layer Cloth Sheet Detection Algorithm

In general, at operation of attaching the overlapped linens which are piled up at random to the automated ironing machine by human hand, workers always find the cloth sheet edges without visual information. In other words, it is not necessary to find cloth sheet using an image processing.

Therefore, to realize the simple cloth handling approach, considering the detection motion for the cloth sheet edge by workers, the end effectors of a multi-arm robot system are proposed as shown in Fig. 3. The Robot arm 1 and 2 are

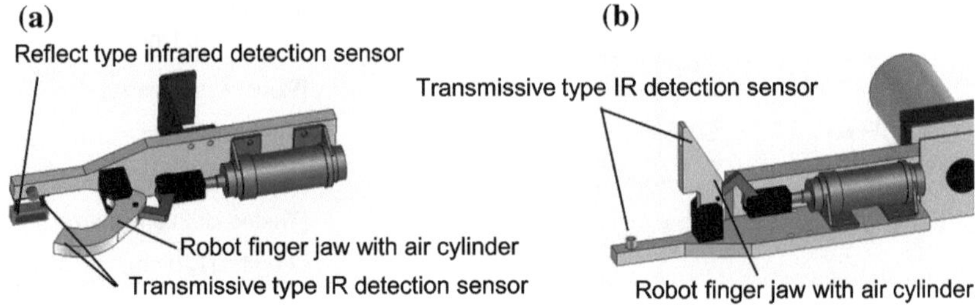

Fig. 3 End effectors. **a** For Robot arm 1 and Robot arm 2. **b** For Robot arm 3

operated as the main motions of a cloth handling. Also, the Robot arm 3 is operated as the support motion and the handling shift motion. So that, these end effectors have a different jaw shape. And clamping directions are orthogonal each other. And fingertip has a transmissive type IR detection sensor to detect a cloth sheet existence and to avoid the influences of the cloth sheet colors. Especially, non-uniformity of the moisture content after washing causes large dispersion of thickness detection using the conventional supersonic thickness sensor. This sensor can detect the overlap of the cloth as the difference of the signal level. As shown in Fig. 4, when the end effector clamps a single layer cloth sheet, the intensity of a transmissive signal level decreases from the "Not-existence" condition. These signal levels

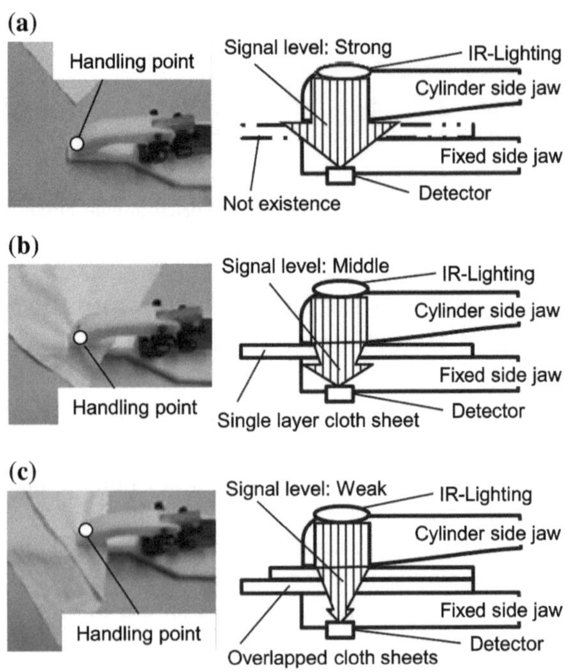

Fig. 4 Handling conditions of a sheet overlapping and signal intensity. **a** Not existence. **b** Single layer cloth sheet. **c** Overlapped cloth sheets

depend on the thickness and the overlap number of a cloth sheet. And in case that, the thickness of cloth sheet to use is fixed, the overlap number can estimate from the signal intensity level. And to find the appropriate single layer handling position, a search algorithm flowchart of Robot arm 1 and 2 is defined as shown in Fig. 5. And a search algorithm flowchart of Robot arm 3 is similar to this algorithm. That is only different point which Robot arm 3 moves on "*j*-*k*" plane. Therefore, a similar parameter is defined to the coordinate of the "*j*" axis.

Then, instead of using image processing systems, the reflect-type IR cloth sheet edge detection sensors are attached on Robot arm 1 and 2. When these sensors detect the existence of a cloth sheet which is hanged from the end effector, the height of the handling point is calculated geometrically. Also to find the appropriate cloth sheet edge position, a search algorithm is defined as the similar algorithm of a single cloth sheet handling position as shown in Fig. 6.

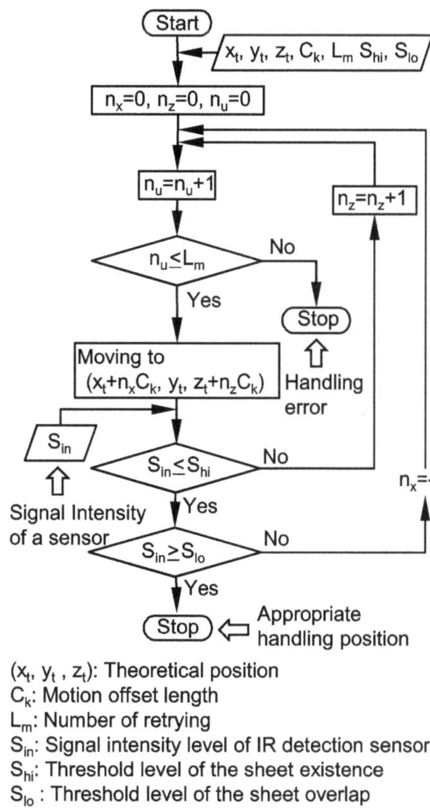

Fig. 5 Algorithm flowchart of finding the appropriate single layer handling position for Robot arm 1 and 2

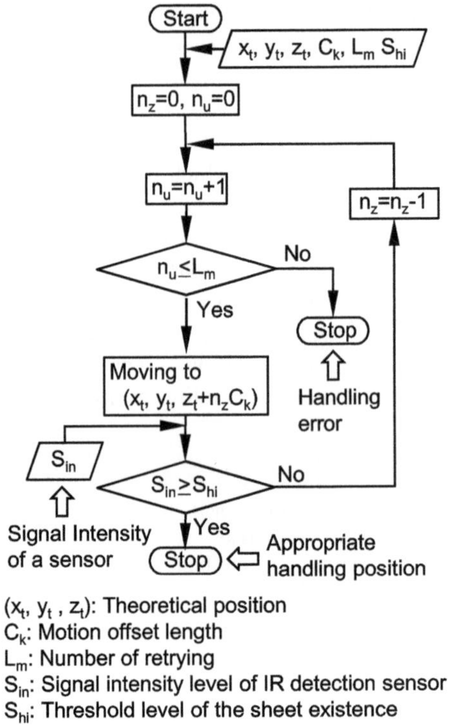

Fig. 6 Algorithm flowchart of finding the cloth sheet edge for Robot arm 1 and 2

4 Quantification of Handling and Spreading Operations Using a Multi-arm Robot System

To realize the cloth sheet attaching operation which is suitable for the multi-arm robot system, the handling and spreading operations have to be quantified using the geometrical vector equations. The geometrical vector model of each robot arm is defined as shown in Fig. 7. Each linkage length is defined as the constant vector $\mathbf{C}_{13}^{i}, \mathbf{C}_{23}^{i}, \mathbf{C}_{31}^{k}, \mathbf{C}_{32}^{k}$ and $\mathbf{C}_{33}^{k}$. And each liner motion vector is defined as the liner vector $\mathbf{L}_{11}^{i}, \mathbf{L}_{12}^{k}, \mathbf{L}_{21}^{i}, \mathbf{L}_{22}^{k}$ and $\mathbf{L}_{34}^{i}$. Also each joint rotation matrix of Robot arm 3 is defined as $\mathbf{E}^{i\theta_1}, \mathbf{E}^{i\theta_2}$ and $\mathbf{E}^{i\theta_3}$. At this robot system, it is assumed that tool center point (TCP) coincides with the detection point of a transmissive type IR detection sensor. So, it can replace to the positioning problem without the pose calculation. And the motion of each robot arm can be quantified as $\mathbf{P}_1$, $\mathbf{P}_2$ and $\mathbf{P}_3$ in Eqs. (1)–(3).

$$\mathbf{P}_1 = \mathbf{L}_{11}^{i} + \mathbf{L}_{12}^{k} + \mathbf{C}_{13}^{i} \tag{1}$$

$$\mathbf{P}_2 = \mathbf{L}_{21}^{i} + \mathbf{L}_{22}^{k} + \mathbf{C}_{23}^{i} \tag{2}$$

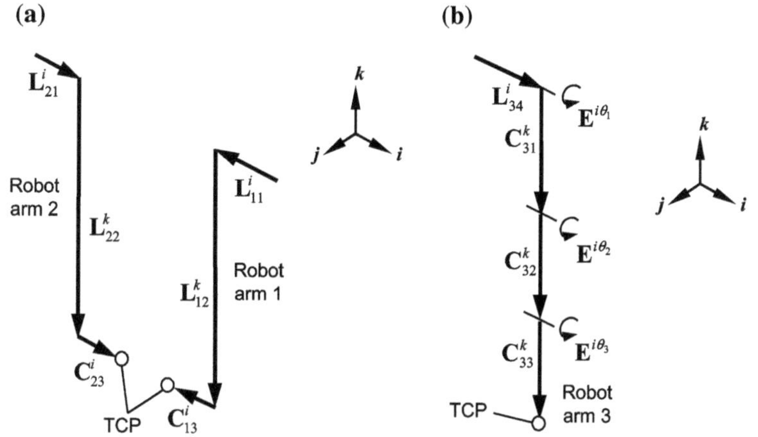

Fig. 7 Vector geometries of each robot arm. **a** Rectangular coordinates type. **b** Rotational joint arm type

$$\mathbf{P}_3 = \mathbf{E}^{i\left(-\frac{\pi}{2}\right)} \cdot \mathbf{E}^{i\theta_1}\left\{\mathbf{C}_{31}^{k} + \mathbf{E}^{i\theta_2}\left(\mathbf{C}_{32}^{k} + \mathbf{E}^{i\theta_3} \cdot \mathbf{C}_{33}^{k}\right)\right\} + \mathbf{L}_{34}^{i} \quad (3)$$

Then, the simple procedure is needed to the robot operation. Therefore, a cloth sheet handling and spreading operations from the initial to final state have to be divided in five main procedures, as follows;

Procedure 1: Pick up a cloth sheet on the arbitrary position,
Procedure 2: Detection and handling on the first edge,
Procedure 3: Detection and handling on the opposite side sheet edge,
Procedure 4: Extension of a cloth sheet along the diagonal line,
Procedure 5: Handling of the other edge points on the opposite diagonal line.

For example, at the procedure 1, the Robot arm 3 picks up a cloth sheet on the arbitrary position to transfer to the end effector of Robot arm 1 as shown in Fig. 8, at first. When rotation angles are defined as θ_{1a}, θ_{2a} and θ_{3a}, this motion $\mathbf{P_{3a}}$ is quantified as Eq. (4).

Fig. 8 Motion locus of pick up a cloth sheet on the arbitrary position

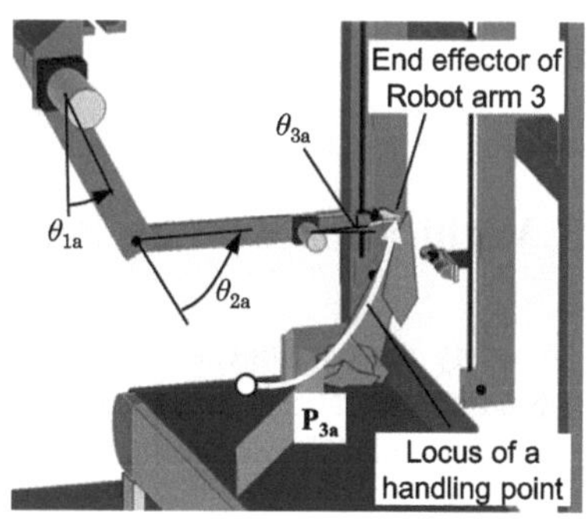

$$\mathbf{P}_{3a} = \mathbf{E}^{j\left(-\frac{\pi}{2}\right)} \cdot \mathbf{E}^{i\theta_{1a}}\left\{\mathbf{C}^k_{3a} + \mathbf{E}^{i\theta_{2a}}\left(\mathbf{C}^k_{32} + \mathbf{E}^{i\theta_{3a}} \cdot \mathbf{C}^k_{33}\right)\right\} + \mathbf{L}^i_{34} \tag{4}$$

And a cloth sheet is shifted to the Robot arm 1 at the picking point with a sheet handling algorithm which has been shown in previous section.

Then, at the procedure 2, the Robot arm 1 and 2 approach each other and move up to the start position as Eqs. (5), (6). In these equations, $\mathbf{L}^i_{11b}, \mathbf{L}^k_{12b}, \mathbf{L}^i_{21b}$ and $\mathbf{L}^k_{22b}$ are variable vectors.

$$\mathbf{P}_{1b} = \mathbf{L}^i_{11b} + \mathbf{L}^k_{12b} + \mathbf{C}^i_{13} \tag{5}$$

$$\mathbf{P}_{2b} = \mathbf{L}^i_{21b} + \mathbf{L}^k_{22b} + \mathbf{C}^i_{23} \tag{6}$$

And the Robot arm 2 moves down to detect a cloth edge with the proposed detection algorithm as Eq. (7). In this equation, $\mathbf{L}^k_{22c}$ is only variable vector.

$$\mathbf{P}_{2c} = \mathbf{L}^i_{21b} + \mathbf{L}^k_{22c} + \mathbf{C}^i_{23} \tag{7}$$

Then, the end effector of Robot arm 2 moves up with holding a cloth sheet. This motion is similar to Eq. (7); $\mathbf{L}^k_{22c}$ is only variable vector, too. And the end effector of Robot arm 1 releases the cloth sheet simultaneously. The execution order of these equations coincides with the procedure of handling and spreading operations. And then, applying the motion planning approaches of a carton box assembling and a cloth sheet wrapping [7, 8], motion loci of each tool center point of robot arm can be calculated. Also, each procedure can be described in the same equation. Therefore, at the procedure 3, 4 and 5, variable vectors and joint rotation matrices which are substituted for these equations are shown.

At the procedure 3, to detect the opposite side sheet edge, the end effector on Robot arm 2 moves up as $\mathbf{L}^k_{22d}$. And the end effector of Robot arm 1 with the cloth sheet edge detection sensors moves down simultaneously as $\mathbf{L}^k_{12d}$, too. At this procedure, the motion of the Robot arm 2 is replaced to the motion of the Robot arm 1 at the Procedure 2. Then, at the procedure 4, to spread the cloth sheet, the Robot arm 1 and 2 move outside and downward simultaneously with a little tension as $\mathbf{L}^k_{11e}, \mathbf{L}^k_{21e}, \mathbf{L}^k_{12e}$ and $\mathbf{L}^k_{22e}$. At the procedure 5, a Robot arm 3 clamps the other edge points on the opposite diagonal line just like Eq. (4) as $\mathbf{E}^{i\theta_{1f}}, \mathbf{E}^{i\theta_{2f}}$ and $\mathbf{E}^{i\theta_{3f}}$. And then, to eliminate tangling or winding of the sheet itself, these robots shift the hands as $\mathbf{L}^k_{11g}, \mathbf{L}^k_{21g}, \mathbf{L}^k_{12g}, \mathbf{L}^k_{22g}, \mathbf{E}^{i\theta_{1g}}, \mathbf{E}^{i\theta_{2g}}$ and $\mathbf{E}^{i\theta_{3g}}$.

5 Verification Test by Prototype Robot System

To verify the usefulness of a quantification method for a cloth sheet handling and spreading operations, the prototype of a multi-arm robot system is tested. Table 1 shows the success probability of each handling procedure at the 30 times trial. So, it

Table 1 Success probability of each procedure at 30 times trial using prototype robot system

Procedure	Ratio (%)
Pick up a cloth sheet on the arbitrary position	93.3
Detection and handling on the first edge	75.0
Detection and handling on the opposite side sheet edge	81.1
Extension of a cloth sheet along the diagonal line	94.1
Handling of the other edge points on the opposite diagonal line	56.3

is clear that the multi-arm robot system can pick up and spread the cloth sheet. However, this robot system sometimes fails in the handling or spreading operations. Because, the un-uniformity of the moisture content after washing causes the dispersion of a cloth sheet pose. And a position of a cloth sheet is estimated using a geometrical calculation. In addition, the times of detection for sheet existence are restricted to 3 times. That is not enough, so the numbers of a adjust motion have to be improved. Also, at the final procedure, the cloth sheet edge might sometimes entangle itself near the diagonal line. So, the prevention method of the cloth sheet tangle must be investigated.

6 Conclusions

To realize soft works handling and spreading processes, a multi-arm robot system has been developed. Especially, the quantification method of the carton assembling and cloth sheet wrapping operation has been applied to the cloth sheet handling operation. And the handling and spreading algorithms of a cloth sheet attaching to the automated ironing machine are shown. And based on these algorithms, the prototype of robot system has been tested. So, it is clear that the robot system can pick up and spread the cloth sheet edges. In future work, to reduce the handling error, the robot finger shape has to be improved. Also, considering finding motion of the cloth sheet edge by human finger tips, the recovery motion methods of the handling error will be investigated.

References

1. Japan Robot Association: Industrial robot operation number in the world in 2012. http://www.jara.jp/data/dl/kado.pdf
2. Buckingham, R.: Ind. Rob. **23**(1), 16–20 (1996)
3. Hata, S., Hojo, H.: Development of handling system for randomly piled washing clothes. J. Robot. Soc. Jpn. **27**(10), 1093–1096 (2009)
4. Karakerezisa, A., Doulgeri, Z., Petridis, V.: A gripper for handling flat non-rigid materials. In: Automation and Robotics in Construction XI: Proceedings of the 11th International Symposium on Automation and Robotics in Construction, Brighton, U.K., pp. 593–601 (1994)

5. Japanese Patent: Robot system making the sheet knot, and that control method. In: No. 5256457 (2013)
6. Japanese Unexamined Patent Application Publication: Spreading method of cloth sheets. In: No. 2012-210357 (2012)
7. Terada, H., Kobayashi, T.: Motion planning approach of a multi-fingered robot for carton folding operations. In: Proceedings of the First International Conference on Informatics in Control, Automation and Robotics, Setubal, Portugal, pp. 353–360 (2004)
8. Terada, H., Yagata, K.: Motion planning approach of a multi-robot system for "Furoshiki" wrapping operation. J. Jpn Soc. Precis. Eng. **76**(5), 546–551 (2010)

AutoHD—Automated Handling and Draping of Reinforcing Textiles

Burkhard Corves, Jan Brinker, Isabel Prause, Mathias Hüsing, Bahoz Abbas, Helga Krieger and Philipp Kosse

Abstract In almost all industrial sectors handling processes are automated through the use of robotic systems. However, in the manufacture of fiber-reinforced structures with complex geometries, the handling of dry, pre-impregnated semi-finished textiles is still performed mainly manually resulting in long processing times, low reproducibility and high manufacturing costs. A previous AiF research project "AutoPreforms" aimed at the automation of the entire production process of components with uniaxial curvature. The scope of this AiF research project "AutoHD" is to fully automate the draping and handling process of complex, three-dimensional fiber composite structures with high degrees of deformation and multiaxial curvature (e.g. car wings). Based on a draping simulation wrinkles can

B. Corves · J. Brinker (✉) · I. Prause · M. Hüsing
Department of Mechanism Theory and Dynamics of Machines (IGM), RWTH Aachen University, Aachen, Germany
e-mail: brinker@igm.rwth-aachen.de

B. Corves
e-mail: corves@igm.rwth-aachen.de

I. Prause
e-mail: prause@igm.rwth-aachen.de

M. Hüsing
e-mail: huesing@igm.rwth-aachen.de

B. Abbas
Institute for Management Cybernetics E.V. (IfU), RWTH Aachen University, Aachen, Germany
e-mail: bahoz.abbas@ima-zlw-ifu.rwth-aachen.de

H. Krieger
Institute of Textile Technology (ITA) at RWTH Aachen University, Aachen, Germany
e-mail: Helga.Krieger@ita.rwth-aachen.de

P. Kosse
Laboratory for Machine Tools and Production Engineering (WZL) of RWTH Aachen University, Aachen, Germany
e-mail: p.kosse@wzl.rwth-aachen.de

B. Corves et al. (eds.), *Mechanisms, Transmissions and Applications*, Mechanisms and Machine Science 31, DOI 10.1007/978-3-319-17067-1_31

already be recognized during the draping process and counteracted by the developed mechanical structure. This is achieved by the utilization of the bending stiffness of textile semi-finished products, a flexible end-effector and a built-in optical quality assurance process. In this paper the main aspects of preforming processes are described revealing the challenges of the project. With examples of currently existing systems, the objective and innovative contribution of the project are described. The paper serves as initial presentation of the project and its solution approaches.

Keywords Automated handling · Draping · Preforming process · Mechanism design · Composite materials

1 Introduction

For medium-sized suppliers of automotive and engineering industries, the production of technically sophisticated fiber-reinforced plastics (FRP) is an important and economically rewarding field of activity. However, further development of FRP so far fails due to the low degree of automation of the production process; particularly the automated handling. Although in almost all industrial sectors handling processes are automated through the use of robot and gripper systems, the handling and draping of dry and pre-impregnated semi-finished textiles into complex geometries is still mostly performed manually. The reasons for this are high demands on technology and lack of knowledge about the behavior of flexible semi-finished products [1, 2]. This in turn results in long processing times, low reproducibility and high manufacturing costs.

First approaches for the efficient handling of textile reinforced structures (preforms) for FRP components are presented in a joint AiF project "AutoPreforms" [3]. The research project "AutoPreforms" aimed at the automation of the entire production process of components with low geometric complexity like uniaxial curvatures. Complex, three-dimensional parts with high degrees of deformation, such as car wings, impose high requirements particular to the handling and quality assurance (Fig. 1).

The aim of the research project "AutoHD" is to fully automate the draping and handling process during preforming of composite structures with complex multiaxial geometries for the first time. This is achieved by the utilization of the bending stiffness of textile semi-finished products, a flexible end-effector and a built-in optical quality assurance process. On the basis of draping simulation an optimal draping strategy and process control is determined in order to, for example, minimize the occurrence of wrinkling and other drape defects. Additional rules are derived for wrinkle elimination in order to correctively adjust the draping process. The project solely focuses on the manufacturing steps of handling and draping, since automated solutions for the remaining process steps (e.g. cutting, consolidation) have already been developed [4–6].

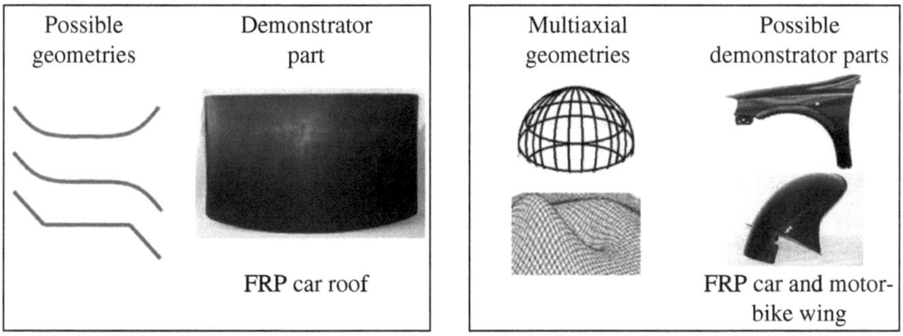

Fig. 1 Geometries realized within the project "AutoPreforms" (*left*) and examples of complex multiaxial geometries within "AutoHD" (*right*), (image ref. upon request)

2 Preform Processes

In the following paragraph, aspects of preform processes are described revealing the main challenges of the project. First, the general manufacturing process of a preform is introduced. Next, the draping and handling in textile preforming processes are presented. Finally, the need of sensor systems in order to avoid wrinkling is described.

A preform is a non-impregnated, textile, multi-layer reinforcement structure. In the preforming process, complex, three-dimensional, near net shaped textile reinforcement structures can be realized. These structures have a high potential for the automated mass production of high performance fiber-reinforced plastics [4]. Initial approaches for a semi-automated production have already been developed [7–9]. However, these approaches consider single steps of a process chain such as the investigations of sewing parameters for the automated sewing of fiber composites. The economical use of preforming technologies, e.g. within the medium-sized automotive industry, is only possible if the entire process chain of multiaxial handling is automated for mass production. The automation of the three-dimensional placing of the reinforcing textiles thereby represents a key role.

2.1 Draping of Fiber-Reinforced Plastics

Forming or draping of textile structures is an essential component in the manufacturing process of textile preforms. The deformation behavior of flexible textile semi-finished products is very complex. Therefore, the material selection of the semi-finished textile is typically performed for a given component geometry using a trial-and-error process. Draping is performed manually by experienced professionals correcting the resulting draping defects (e.g. wrinkles, gaps and loops). This procedure is technically and economically deficient with respect to a large-scale production of FRP components. The time and cost-intensive process of textile selection by trial-and-error and manual production can be improved by material modeling and process simulation. The objective of the simulation is to predict the deformation behavior of the reinforcing textile during preforming.

There are several computer-based methods for the so-called draping simulation. The available methods differ in their computational complexity and level of detail. Purely kinematic or on the finite element method (FEM) based models can be applied [10]. If external process forces (e.g. the motion and gripping of the robot) that act on the textile should be considered, FEM-based models must be used. As input for the simulation, the mechanical material properties of the semi-finished textile (e.g. Young's modulus, bending stiffness, friction coefficients between the reinforcing fibers as well as between the reinforcing fibers and the tool) are required.

2.2 *Handling in Textile Preform Processes*

Previously developed semi-automated systems provide solutions for specific handling tasks (cf. among others [10–14]). The forming of the reinforcing textiles by these systems is partly three-dimensional, but the individual transformations are realized only uniaxial [i.e. there are no spherical deformations (e.g. hemisphere)]. The placing of multiaxial reinforcing textiles in curved geometries with existing systems is, if at all, only possible at very low degrees of deformation.

When selecting a suitable handling system for textile semi-finished products, different criteria must be considered (e.g. the geometry of the gripping and placing position, the dimensions of the workpiece as well as its material properties [15]). An important criterion is the reliability of the gripping process. According to [16] the reliability depends on a variety of factors, such as the material flexibility, operating speed, gripping technology and the environment.

2.3 *Process Monitoring by Means of Sensor Systems*

For a continuous process-relevant protection of significant quality features only non-destructive, manufacturing-integrated audit procedures (referred to as inline inspection systems [17]) are of importance. For testing of multiaxially curved preforms—and in particular for the detection of wrinkles—it is essential to capture the 3D geometry. The recognition of wrinkling must take place early during the draping process in order to be able to counteract without loss of time. The strict requirements under industrial conditions (e.g. short cycle times, robustness, and reliability) can solely be met by integrating the sensors directly into the end-effector. A simultaneous quality control and feedback of the test results to the corrective intervention process (process control) is not achieved by currently existing methods.

3 Handling Devices in Textile Preform Processes

An adaptive multi-functional end-effector (Fig. 2), which was developed as part of a previous joint research project called "AutoPreforms", is able to handle different geometries [such as flat, bent, concave, convex or a combination thereof (see also Fig. 1)]. The component geometries treated by this end-effector can be divided into three groups (plane components with curved sides, curved components with constant radius of curvature and bent components). The gripping elements of the end-effector can be moved relatively to each other, where the distances between the gripping elements remain constant, so that the reinforcing textile is not damaged. The movement of the gripping elements is active and controlled by the robot controller. In addition, the gripping elements are passively pivotable and displaceably mounted in height. So they can orient themselves when placed towards the surface of the mold and in addition to the actively shaped curvature [3, 18–20].

A prototype for the automated handling of clothing textiles was developed within the scope of the joint research project "Integrated 3D sewing system" funded by the Federal Ministry of Education and Research (BMBF) [14]. Monitoring the gripping operation was carried out with the help of a camera system attached to a stationary system of linear axes [21]. For the conformal gripping of reinforcing textiles, a large-area vacuum gripping system whose gripping surface can be changed by sliding chamber walls was realized at the Bremer Werk for assembly systems [22]. At the University of Louisiana at Lafayette, USA a gripping system with suction surfaces attached to the movable kinematic units was developed. The system serves both the separation and handling of limp materials [23]. The Brötje-Automation GmbH, Wiefelstede developed a handling system for similar applications that is based on selectively controllable suckers. These suckers are arranged in a matrix form. Thus, an adaptation of the gripping surface is possible [11]. Within the Fraunhofer Innovation Cluster "KITe hyLITE" a handling unit was designed and developed with which unidirectional fiber semi-finished products can be formed and positioned in a mold [13]. Another example of a rigidly constructed end-effector is the "CFKTex" system with a selectively activatable gripping surface [12]. The Institute for Machine Tools and Industrial Management of the Technical

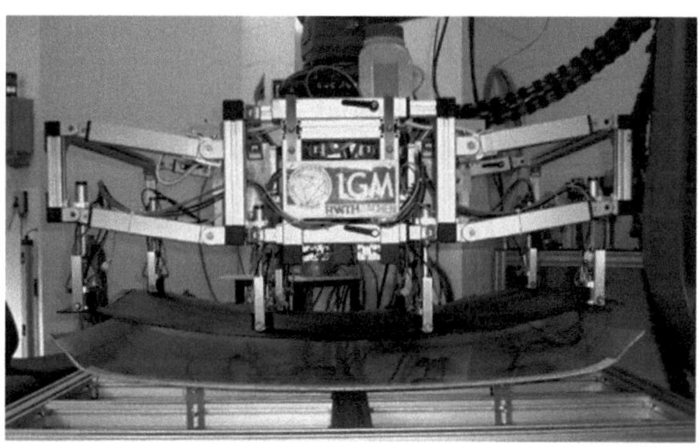

Fig. 2 Adaptive multi-functional end-effector [20]

University of Munich realized a system that enables the automated production of three-dimensional preforms, wherein also fixing of the reinforcing textiles is possible by binder activation. In a current research project of the German Aerospace Center called “EVo”, a continuous process chain for the automated manufacture of FRP components for automotive and aerospace applications is realized. The developed preform unit also includes an end-effector with 90 separately switchable vacuum grippers and is quality assured to handle and stack blanks. The transformation of the resulting layer stacks is done in a consolidation process, whereby an on-line correction of the deformation is not possible [24]. The mentioned systems, if at all, allow draping only with very low degrees of deformation. The quality of the conversion can be detected comprehensively, but is not used as a basis for on-line correction of the process. Regarding the flexibility of the production of automotive components, this represents a limitation that need to be addressed. Examples of such complex, multiaxial geometries are shown in Fig. 1 (right). The aforementioned sub-goal of an inline-wrinkle detection and correction has been integrated into any forming unit of the projects mentioned. Extensive reviews of literature about mechanisms for automated handling and draping of multiaxial reinforcing textiles are presented by [12, 25].

4 Objective and Innovative Contribution

The aim of the recently started joint research project “AutoHD” is to develop an automated handling and draping technology for fiber-reinforced semi-finished products for the series production of multiaxially curved structural components. The main task of the approach lies in the specific combination of draping, material-appropriate utilization of bending stiffness of reinforcing textiles, an end-effector for multiaxial draping and process-integrated quality control. The complete sequence of this process is shown in Fig. 3. First, a planar semi-finished textile is grasped by means of a suitable gripper. The gripper is located on an end-effector which in turn is mounted on an industrial robot for transport and coarse positioning of the textile. Using external cameras, the process is monitored, so that the actual position is directly adjusted in accordance to variations from the nominal position. After successful coarse positioning, fine positioning is carried out through the end-effector by reshaping and draping the semi-finished textile so that it can be placed in the multiaxially curved mold—preferably without wrinkling.

An optical sensor system monitoring the draping is also integrated within this process step. A second control loop, the draping control loop, can intervene directly in the handling process once deviations between the data of the sensor system and the data of draping simulation and especially wrinkles occur. The described process flow ensures a highly automated draping and handling process. The on-line quality assurance provides a zero defect production. This results in increased output rates and the saving of expensive reinforcing textiles.

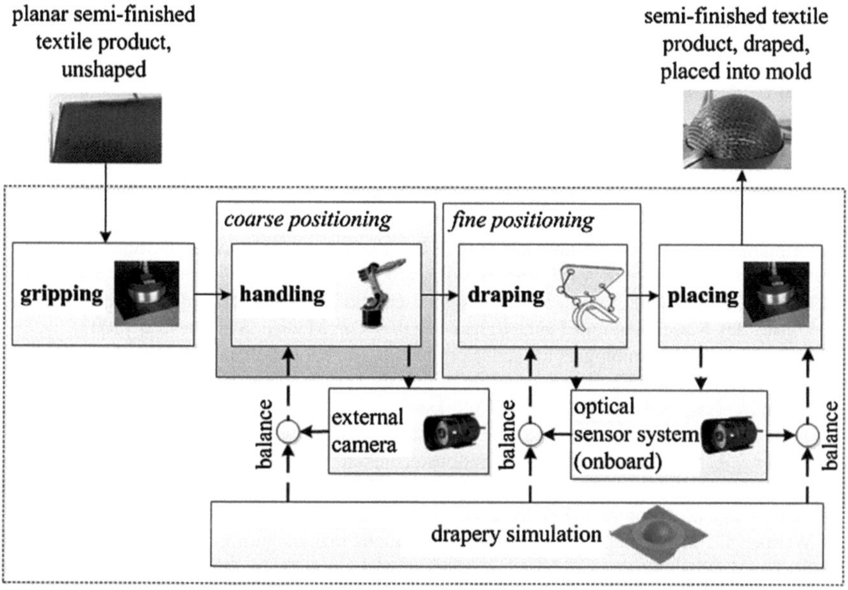

Fig. 3 Automated drapery and handling process (image ref. upon request)

The automated handling of textile preforms aims at the economical production of FRP parts of different geometries. The innovative aspect of the approach is to automate the current manual handling and multiaxial draping by a suitable robot end-effector system with integrated gripping elements. By integrating a sensor system, changes of the position of the textiles or the emergence of wrinkles when placed into the mold are registered. The connection of the sensor system to the control loop of draping allows for a direct intervention of the end-effector on these errors. A tool for process monitoring, without the need of separate sensors within additional process steps, ensures integrated optical inspection of wrinkles, correct positioning as well as correct fiber orientation of the placed semi-finished products. Thereby, potential errors can already be detected in the running process resulting in increased process quality and reduction of discard. For the first time, the measured 3D fiber orientation allows the balance between the real and the simulated draped preforms. The degree of automation in the production of FRP parts is thus increased significantly and represents an important contribution to the establishment of FRP as a material for high volume applications.

Acknowledgments The authors thank the German Federation of Industrial Research Associations "Otto von Guericke" eV (AiF) for the financial support of the research project "Automated handling and draping of reinforcing textiles for multiaxially curved composite structures—AutoHD" Project No 18264 N /4 (cf. www.auto-hd.de).

References

1. Berger, R.: VDMA: serienproduktion hochfester Faserverbundbauteile. Perspektiven für den deutschen Maschinen-und Anlagenbau (2012)
2. Schnabel, A.: Entwicklung von lokal angepassten textilen Halbzeugen für die Großserienfertigung von faserverstärkten Kunststoffen. RWTH Aachen, Diss (2013)
3. Henning, K., Gries, T., Flachskampf, P.: AutoPreforms: gemeinsamer abschlussbericht. Aachen (2008)
4. Grundmann, T.C.: Automatisiertes Preforming für schalenförmige komplexe Faserverbundbauteile. RWTH Aachen, Diss (2009)
5. Hufenbach, W. (ed.): Textile Verbundbauweisen und Fertigungstechnologien für Leicht-bau-konstr. des Maschinen-und Fahrzeugbaus. SDV—Die Medien AG, Dresden (2007)
6. Gries, T. et al.: Technologies for series production of near-net-shape textile preforms. Hui, D., Loboda, P., Chen, X., Wang, Y. (eds.): Use of Textile Composites Technology for Safer Vehicles; Nato Advanced Research Workshop, Kiev, Ukraine May 18–21, 2009. National Technical University of Ukraine, Kiev Polytechnic Institute, Kiev (2009)
7. Geßler, A. et al.: Textile Integrationstechniken zur Herstellung vorkonfektionierter Verstärkungsstrukturen für FVK "INTEX". BMBF-Projekt, Förderkennzeichen 03N3060. Schlussbericht, Ottobrunn: EADS (2002)
8. Weimer, C., Mitschang, P., Neitzel, M.: Continuous manufacturing of tailored reinforcements for liquid infusion processes based on stitching technologies. In: 6th International Conference on Flow Processes in Composite Materials, Auckland, New Zealand, 15–16 June 2002, Paper No. FPCM6-DE-2 (2002)
9. Gries, T. et al.: Technologies for series production of near-net-shape textile preforms. World J. Eng. **7**(1), 67–74 (2010)
10. Girdauskaite, L., Haasemann, G., KrzywinskiIn, S.: Modellierung und Simulation. In: Cherif, C. (ed.), Textile Werkstoffe für den Leichtbau, pp. 573–636. Springer, Berlin (2011)
11. Brötje-Automation GmbH: Automated handling system for fibre composites. JEC Composites Magazine **47**(57), 38 (2010)
12. Ehinger, C.A.: Automatisierte Montage von Faserverbund-Vorformlingen. Herbert Utz Verlag, Diss (2012)
13. Köhler, G., Ochs, A., Schneider, M.: Automatisierung in der Leichtbauproduktion. wt Werkstattstechnik online **99**(9), 614–617 (2009)
14. Zöll, K.: Nähtechnik zur Fertigung textiler Hüllen. RWTH Aachen, Diss (2002)
15. Biermann, D., Hufenbach, W., Seliger, G.: Serientaugliche Bearbeitung und Handhabung faserverstärkter Hochleistungswerkstoffe. Untersuchungsbericht zum Forschungs-und Handlungsbedarf, TU Dresden (2008)
16. Stephan, J.: Beitrag zum Greifen von Textilien. IPK Technical University, Diss, Berlin (2001)
17. Schmitt, R., Orth, A. Damm, B.: A knowledge based method to evaluate image processing algorithms—a case sample with texture analysis. EOS Conference on Industrial Imaging and Machine Vision, European Optic Society (EOS), World of Photonics Congress 2005, Munich, June (2005)
18. Corves, B., Kordi, M.T., Hüsing, M., Flachskampf, P.: Economic manufacturing of fibre reinforced composite structures using robot-supported automation. In: Proceedings of MUSME 2008, the International Symposium on Multibody Systems and Mechatronics, 8–12 April 2008, Paper no 03 (2008)
19. Kordi, M.T., Hüsing, M., Corves, B.: Development of a multifunctional robot end-effector system for automated manufacture of textile preforms. IEEE/ASME International Conference on Advanced Intelligent Mechatronics (2008)
20. Kordi, M.T.: Entwicklung von Roboter-Endeffektoren zur automatisierten Herstellung textiler Preforms für Faserverbundbauteile. RWTH Aachen, Diss (2009)

21. Feldhoff, J.: Texturbasierte Bildsegmentierung am Beispiel der Steuerung von Handhabungsprozessen für das automatisierte Nähen. Berichte aus der Produktionstechnik; Bd. 2001, 24, RWTH Aachen, Diss (2001)
22. Gebauer, I., Dörsch, C., Eckner, F.: Bremer Werk für Montagesysteme GmbH. Vorrichtung zum Aufnehmen und Ablegen flächiger Teile. Patent DE102005047645B4 (2010)
23. Kolluru, R., Valavanis, K.P., Smith, S.A., Tsourveloudis, N.: Design fundamentals of a reconfigurable robotic gripper system. IEEE Trans. Syst. Man Cybern. Part A: Syst. Humans **30**, 181 (2000)
24. Torstrick, S., Kruse, F., Wiedemann, M.: RTM Processing for Net-Shaped Parts in High Quantities. 7. CFK-Valley Stade Convention, Stade, 12 June 2013, URL: http://www.cfk-convention.com/uploads/media/130521_CFK-CONV_DLR_Torstrick.pdf
25. Straßer, G.: Greiftechnologie für die automatisierte Handhabung von technischen Textilien in der Faserverbundfertigung. H. Utz Verlag, Diss (2012)

Singularity and Workspace Analyses of a 3-DOF Parallel Mechanism for Vehicle Suspensions

Fernando Malvezzi and Tarcisio Antonio Hess Coelho

Abstract This work deals with the singularity and workspace analyses of a three-degree-of-freedom mechanism for rear independent suspensions, capable to adjust simultaneously the camber, rear steering and roll angles. This mechanism is able to improve handling performance when compared to the vehicle with a passive suspension mechanism. The inverse kinematic model was developed for the velocity kinematic analysis. The Jacobian matrices are applied to investigate the mechanism singularity configuration. Finally, the reachable workspace is obtained using the discretization method. The obtained results have shown there is not singular configuration inside the specified workspace. The available workspace can attend the camber and rear steering angles stroke. Moreover, this workspace can be obtained by using standard joints and actuators.

Keywords Singularity · Workspace · Jacobian analysis · Parallel mechanism · Automotive suspension

1 Introduction

Parallel mechanism is a mechanical system characterized by the presence of independent kinematic chains, actuating in-parallel on the end-effector. Among its mains advantages, one can mention low inertia, fast dynamic response, high load capacity, high rigidity and precision [1]. Consequently, it has been employed in flight simulators, robotic manipulators and machine tools. Despite the fact that the use of active parallel mechanisms in the automotive field is still not found, Malvezzi

F. Malvezzi (✉)
Maua Institute of Technology, São Paulo, Brazil
e-mail: fernando.malvezzi@maua.br; fermaldo@ig.com.br

T.A.H. Coelho
University of São Paulo, São Paulo, Brazil
e-mail: tarchess@usp.br

B. Corves et al. (eds.), *Mechanisms, Transmissions and Applications*,
Mechanisms and Machine Science 31, DOI 10.1007/978-3-319-17067-1_32

and Hess-Coelho [2–5] presented a novel type of independent rear suspensions, based on a 3-DOF parallel mechanism, which is able to adjust simultaneously the camber, rear steering and roll angles. The simulations results demonstrated that a vehicle equipped with such active mechanism really improves its handling performance in a comparison with the same vehicle but assembled with a passive suspension.

The design of a parallel mechanism for any application in general, and for automotive vehicles in particular, demands special cares. From this perspective, two important kinematic issues must be taken into account: (a) verification of occurrence of singular configurations; (b) evaluation of the available workspace. With respect to singularities, when a parallel mechanism reaches such configurations, it can either gain or lose degrees of freedom [1]. According to Gosselin and Angeles [6], Balchanowski [7], Azulay et al. [8], Choi and Ryu [9], the conditions for the occurrence of singularities can be identified by the Jacobian analysis. On the other hand, the workspace can be defined as the region reached by the end-effector. In the case of automotive suspensions, the end-effector corresponds to the tire-wheel set. The size and shape of the workspace is constrained by the link lengths, the joint and actuator strokes, and singular configurations, as well [10]. The available workspace can be determined either by the geometric [10] or discretization [11, 12] method. In the discretization method, the process begins with a solid, assumed larger than the feasible workspace, discretized by a regular mesh. Then, a procedure checks whether or not each mesh node violates the physical and kinematic constraints. Hence, the workspace boundaries are composed by a set of nodes that have at least one neighbour node that does not belong to the workspace [12, 13].

This work deals with the singularity and workspace analyses of a three-degree-of-freedom parallel mechanism for rear independent suspensions. Section 2 briefly describes the mechanism topology and how it actuates on the roll, camber and rear steering angles. Then, Sect. 3 deals with the kinematic model. The results are shown in Sect. 4, while Sect. 5 presents the conclusions.

2 Description of the Mechanism Topology

The parallel mechanism analyzed here was proposed in [2–4] and performs three independent motions. Topologically, the mechanism is parallel and asymmetric with three distinct active kinematic chains ($\underline{P}US + \underline{P}RS + U\underline{P}SRS$).

Figure 1b shows that the action on the body roll angle is achieved by the force exerted by the actuator 1 on the sprung mass (body). On the other hand, when the car does not perform evasive maneuvers or cornering, the actuator 1 does not exert force on the sprung mass and the wheel travel (passive movement) is influenced only by the action of the spring and the damper. In addition, the camber angle is changed by the simultaneous movement of the actuators 2 and 3, while the rear steering angle is modified by the actuator 3.

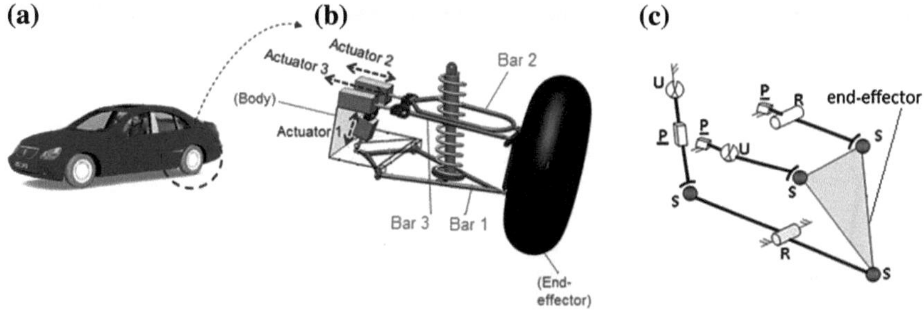

Fig. 1 **a** Mechanism in the vehicle; **b** Mechanism CAD model; **c** Kinematic scheme [2]

3 Kinematic Model

In order to analyze the singularity and the workspace of the mechanism, the inverse kinematic model is derived. The goal of the position kinematic model is to obtain the mathematical transformation between the end-effector location, defined by the vector $\vec{\chi} = [\Theta, \Gamma, \Psi]^T$ (Figs. 2a and 3b) and the displacements provided by the actuators, defined by the vector $\vec{S} = [s_1, s_2, s_3]^T$ (Fig. 2a, b). The actuators 1 and 2, and the bars 1 and 2 are in the Oy_1z_1 plane.

The null functions f_1, f_2 and f_3, dependent of the vector $\vec{S}$ and $\vec{\chi}$, developed in [2], are reproduced here. Equations (1)–(3) represent the mathematical transformation between the end-effector location and the actuators displacements. These equations are second-degree polynomial, independent and decoupled in the variables s_1, s_2 and s_3. Besides, there are relationships between the angles Γ and Ψ and the camber and the rear steering angles, respectively, which are presented in [2].

$$f_1(\vec{s}, \vec{x}) = s_1^2 + 2z_{E1}s_1 + 2L^2\cos\Theta - 2Lz_{E1}\sin\Theta - 2L^2 = 0 \tag{1}$$

$$\begin{aligned} f_2(\vec{s}, \vec{x}) = {} & s_2^2 - 2(\ell\cos\Theta + h\cos\Gamma)s_2 \\ & + 2[h^2 + \ell h\cos(\Theta - \Gamma) - \ell h\sin\Theta - h^2\sin\Gamma] = 0 \end{aligned} \tag{2}$$

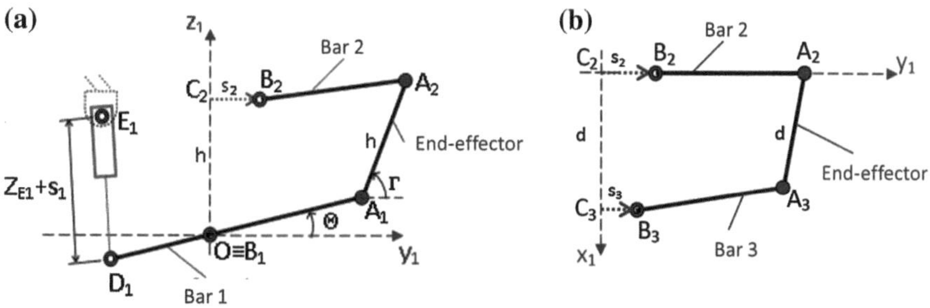

Fig. 2 Symbology for position analyses and location of the A_i, B_i, C_i points: **a** front view; **b** top view [2]

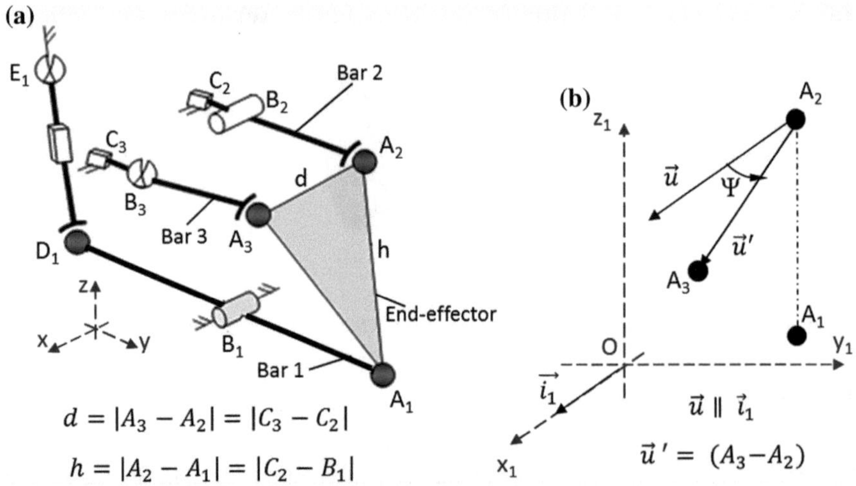

Fig. 3 **a** Symbology for position analyses-3D view; **b** Rotation angle (Ψ) of the vector $\vec{u}$ around the king pin (axle defined by the points A_1 and A_2)

$$\begin{aligned} f_3(\vec{s},\vec{x}) &= s_3^2 - 2(\ell\cos\Theta + h\cos\Gamma + d\sin\Gamma\sin\Psi)s_3 + d^2(1-\cos\Psi)^2 \\ &\quad + (\ell\sin\Theta + h\sin\Gamma - d\cos\Gamma\sin\Psi - h)^2 - \ell^2 + (\ell\cos\Theta + h\cos\Gamma)^2 \\ &\quad + (d\sin\Gamma\sin\Psi)^2 + 2d\sin\Gamma\sin\Psi(\ell\cos\Theta + h\cos\Gamma) = 0 \end{aligned} \tag{3}$$

Differentiating Eqs. (1)–(3) with respect to the time,

$$\frac{df_i}{dt} = \frac{\partial f_i}{\partial\Theta}\cdot\frac{d\Theta}{dt} + \frac{\partial f_i}{\partial\Gamma}\cdot\frac{d\Gamma}{dt} + \frac{\partial f_i}{\partial\Psi}\cdot\frac{d\Psi}{dt} + \frac{\partial f_i}{\partial s_i}\cdot\frac{ds_i}{dt} = 0\,(i = 1,2,3) \tag{4}$$

Equation (4) can be arranged in matrix form, which is the relationship between the end-effector and actuators velocities (Eq. 5).

$$J_x\,\dot{\vec{x}} + J_s\,\dot{\vec{s}} = \vec{0} \tag{5}$$

where $\dot{\vec{x}} = \left[\dot{\Theta},\dot{\Gamma},\dot{\Psi}\right]^T$ and $\dot{\vec{s}} = [\dot{s}_1,\dot{s}_2,\dot{s}_3]^T$ represent the end-effector and actuators velocities, respectively. The matrices J_s and J_x are Jacobians.

$$J_s = \begin{bmatrix} \frac{\partial f_1}{\partial s_1} & 0 & 0 \\ 0 & \frac{\partial f_2}{\partial s_2} & 0 \\ 0 & 0 & \frac{\partial f_3}{\partial s_3} \end{bmatrix} \quad J_x = \begin{bmatrix} \frac{\partial f_1}{\partial\Theta} & \frac{\partial f_1}{\partial\Gamma} & \frac{\partial f_1}{\partial\Psi} \\ \frac{\partial f_2}{\partial\Theta} & \frac{\partial f_2}{\partial\Gamma} & \frac{\partial f_2}{\partial\Psi} \\ \frac{\partial f_3}{\partial\Theta} & \frac{\partial f_3}{\partial\Gamma} & \frac{\partial f_3}{\partial\Psi} \end{bmatrix} \tag{6}$$

where

$$\frac{\partial f_1}{\partial s_1} = 2s_1 + 2z_{E1} \qquad \frac{\partial f_2}{\partial s_2} = 2s_2 - 2(\ell\cos\Theta + h\cos\Gamma)$$

$$\frac{\partial f_3}{\partial s_3} = 2s_3 - 2(\ell\cos\Theta + h\cos\Gamma + d\sin\Gamma\sin\Psi)$$

$$\frac{\partial f_1}{\partial \Theta} = -2L^2\sin\Theta - 2Lz_{E1}\cos\Theta \quad \frac{\partial f_1}{\partial \Gamma} = 0 \quad \frac{\partial f_1}{\partial \Psi} = 0$$

$$\frac{\partial f_2}{\partial \Theta} = 2\ell s_2\sin\Theta - 2\ell h\sin(\Theta - \Gamma) - 2\ell h\cos\Theta$$

$$\frac{\partial f_2}{\partial \Gamma} = 2hs_2\sin\Gamma + 2\ell h\sin(\Theta - \Gamma) - 2h^2\cos\Gamma \quad \frac{\partial f_2}{\partial \Psi} = 0$$

$$\begin{aligned}\frac{\partial f_3}{\partial \Theta} = {} & 2\ell s_3\sin\Theta + 2\ell\cos\Theta(\ell\sin\Theta + h\sin\Gamma - d\cos\Gamma\sin\Psi - h) \\ & - 2\ell\sin\Theta(\ell\cos\Theta + h\cos\Gamma) - 2d\ell\sin\Gamma\sin\Psi\sin\Theta\end{aligned}$$

$$\begin{aligned}\frac{\partial f_3}{\partial \Gamma} = {} & -2s_3(-h\sin\Gamma + d\cos\Gamma\sin\Psi) \\ & + 2(\ell\sin\Theta + h\sin\Gamma - d\cos\Gamma\sin\Psi - h)(h\cos\Gamma + d\sin\Gamma\sin\Psi) \\ & - 2h\sin\Gamma(\ell\cos\Theta + h\cos\Gamma) + 2d^2\sin\Gamma\cos\Gamma\sin^2\Psi \\ & + 2d\ell\cos\Theta\cos\Gamma\sin\Psi + 2dh\sin\Psi\cos(2\Gamma)\end{aligned}$$

$$\begin{aligned}\frac{\partial f_3}{\partial \Psi} = {} & -2s_3d\sin\Gamma\cos\Psi + 2d^2(1 - \cos\Psi)\sin\Psi \\ & - 2d\cos\Gamma\cos\Psi(\ell\sin\Theta + h\sin\Gamma - d\cos\Gamma\sin\Psi - h) \\ & + 2d^2\sin^2\Gamma\sin\Psi\cos\Psi + 2d\sin\Gamma\cos\Psi(\ell\cos\Theta + h\cos\Gamma)\end{aligned}$$

4 Singularity and Workspace Analyses

According to Tsai [1], an inverse kinematic singularity occurs when the determinant of J_s is null. On the other hand, the direct kinematic singularity occurs when the determinant of J_x is null. In Eq. (6), we can see that the determinant of J_s is null if one of the diagonal elements of J_s is equal to zero. Besides, regarding to the direct kinematic singularity, as the elements J_{x12}, J_{x13} and J_{x23} of J_x are nulls, the determinant of J_x is null if J_{x11} or J_{x22} or J_{x33} are equal to zero.

Table 1 shows the mechanism parameters employed in the singularity and workspace analyses.

Table 1 Mechanism parameters employed in the singularity and workspace analyses

L (mm)	ℓ (mm)	d (mm)	h (mm)
120	300	120	240

Figure 4 shows the inverse singular configurations (Fig. 4a, b) and the direct singular configurations (Fig. 4c, d) obtained in the singularity analysis. In Fig. 4a and b, the actuator 2 axis is orthogonal to the bar 2, which limits the workspace because the camber angle cannot be increased, even if the actuator 2 moves to any direction. Moreover, it is not possible increase the steering angle even if the actuator 3 moves to any direction. Figure 4c shows the singular configuration where the actuator 1 and the bar 1 are in the same line, so the vertical movement of the wheel becomes uncontrollable. Figure 4d shows that when all actuators are locked, the end-effector gains two degrees of freedom. It can perform infinitesimal rotation

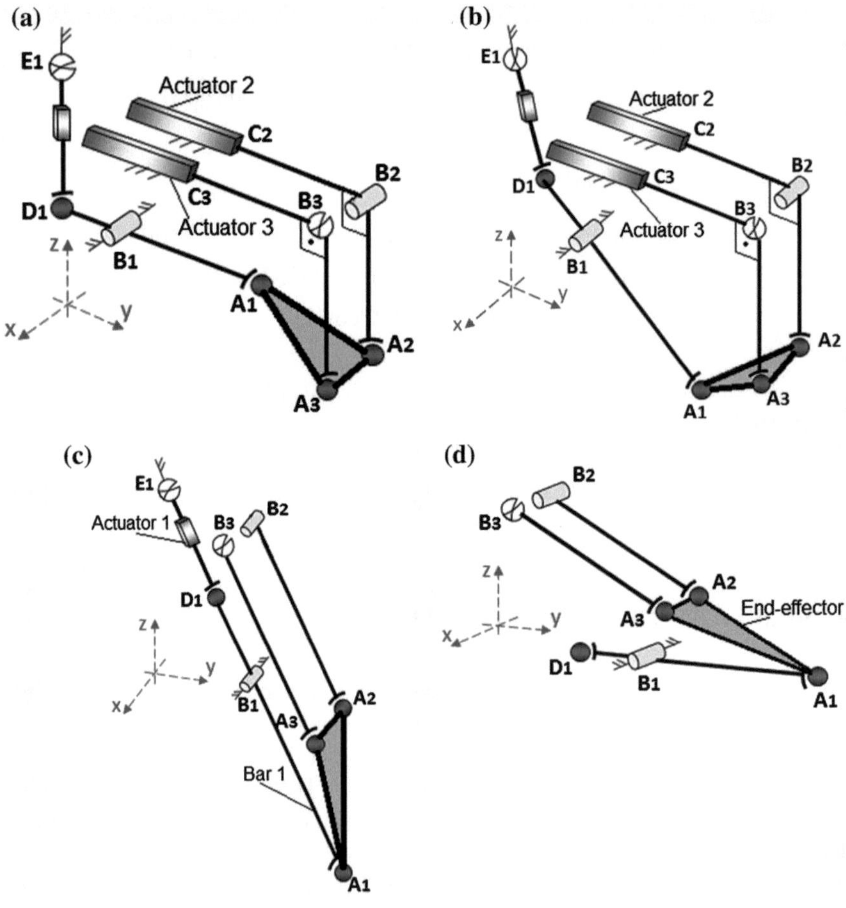

Fig. 4 Singular configurations

around the axis defined by the points A_1 and A_2. Similarly, the end-effector can also perform infinitesimal rotation around the line which is parallel to the x-axis as well passing through the point A_1.

Regarding to determine the mechanism workspace, the discretization method was employed. From a domain composed of a wide range of camber, rear steering and roll angles, a mapping is made in order to determine the loci of singularities and the available workspace. The camber and rear steering angles ranges are between −90° and +90°. So, an algorithm implemented in MATLAB software checks

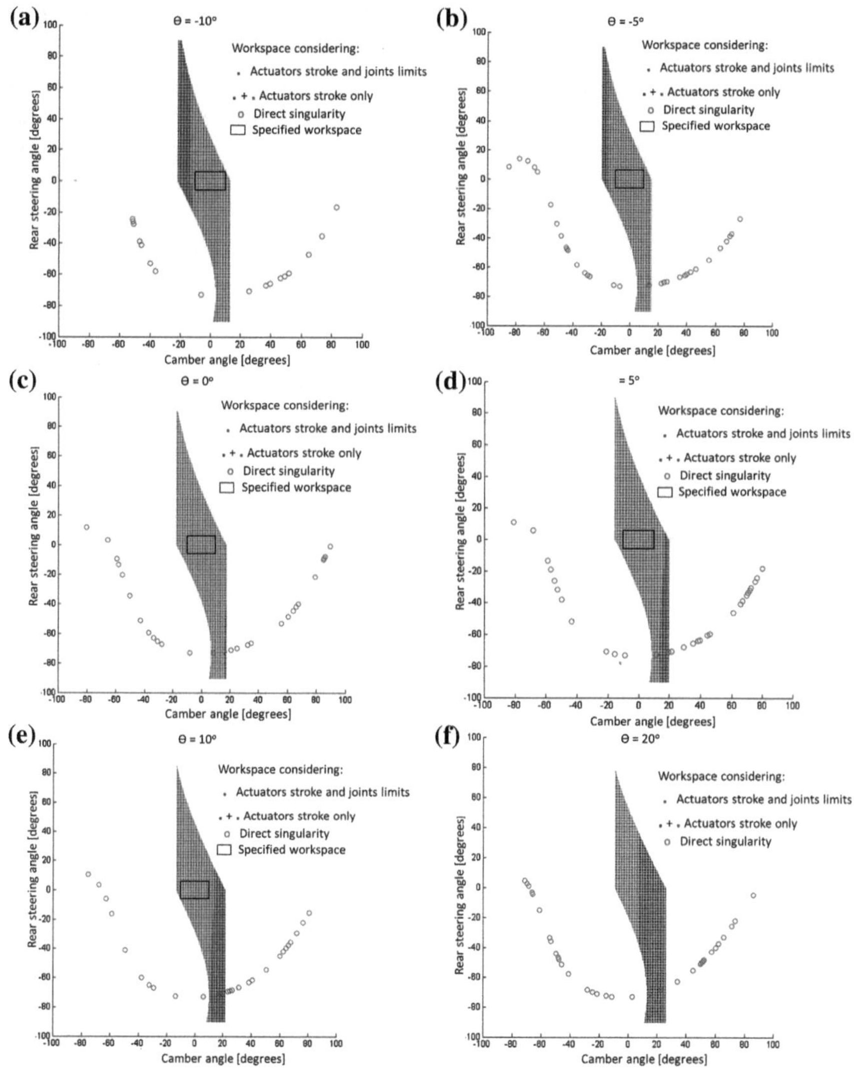

Fig. 5 Reachable and specified mechanism workspaces

whether or not each mesh node violates the physical and kinematic constraints. Consequently, workspace boundaries are composed by a set of nodes that has at least one neighbour node that does not belong to the workspace.

The physical constraints are represented by the bars lengths, actuators stroke and the passive joints limits. Moreover, the algorithm calculates the determinants of Jacobian matrices *Js* and *Jx* as well it verifies if their values are null. These null determinants correspond to singular configurations.

In order to attend the vehicle dynamic performance presented in [2], the specified workspace must allow both a camber variation between −10° and +10° and a rear steering variation between −6° and +6°. This workspace can be achieved if the strokes of the actuators 2 and 3 are equal to 145 mm and if the amplitude of the joints that connect the end-effector to the bars is equal to 20°.

Figure 5 shows both the reachable and the specified workspaces for several wheel travel positions. The red region represents the workspace limited by both the actuators stroke and the joint limits. On the other hand, the workspace limited only by the actuators stroke is defined by the addition of the red and blue regions. Moreover, there are no singular configurations inside the specified workspace.

It is noteworthy that when the car does not perform evasive manoeuvres or cornering, the camber and rear steering angles will be approximately null.

5 Conclusions

In this paper the singularity and workspace analyses of a three-degree-of-freedom mechanism for rear independent suspensions, capable to adjust simultaneously the camber, rear steering and roll angles, were developed. While the Jacobian analysis was applied to identify the conditions for the occurrence of singular configurations, the discretization method was employed to evaluate the available workspace. From a domain composed of a wide range of camber, rear steering and roll angles, a mapping was made in order to determine the loci of singularities and the available workspace, that influenced by constraints such as link lengths and joint/actuator strokes. The performed simulations have shown that possible singular configurations remain far from the specified workspace, which provides typical values of the wheel-tire travel and camber/rear steering ranges. Moreover, such workspace can be achieved by using standard joints and actuators.

References

1. Tsai, L.W.: Robot Analysis: The Mechanics of Serial and Parallel Manipulators. Wiley, Hoboken (1999)
2. Malvezzi, F., Coelho, T.A.H.: Modeling, feasibility and performance of a 3-DOF parallel mechanism employed in a rear vehicle suspension. Int. J. Veh. Syst. Model. Test. (2014) (in press)

3. Malvezzi, F., Coelho, T.A.H.: Topological synthesis of a novel parallel mechanism for vehicle rear suspensions. New Advances in Mechanism, Transmissions and Applications, Mechanism and Machine Science 17. Springer (2014). doi: 10.1007/978-94-007-7484-8_5
4. Malvezzi, F., Coelho, T.A.H.: A novel 3-dof parallel mechanism employed in a vehicle suspension for the improvement of handling performance. In: ECCOMAS Multibody Dynamics 2013, Thematic Conference. Zagreb (2013)
5. Malvezzi, F.: Novel approach to development of vehicle suspension systems: the employment of parallel mechanisms. Ph.D. Thesis, University of Sao Paulo, Brazil (2014) (In Portuguese)
6. Gosselin, C.M., Angeles, J.: Singularity analysis of closed-loop kinematic chains. IEEE Trans. Robot. Autom. **6**, 281–290 (1990)
7. Balchanowski, J.: Topology and analysis of the singularities of a parallel mechanism with three degrees of freedom. Arch. Civil Mech. Eng. **14**, 80–87 (2014)
8. Azulay, H. et al. Comparative analysis of a new 3_PPRS parallel kinematic mechanism. Robot. Comput.-Integr. Manuf. **30**, 369–378 (2014)
9. Choi, H.B., Ryu, J.: Singularity analysis of a four degree-of-freedom parallel manipulator based on an expanded 6 × 6 Jacobian matrix. Mech. Mach. Theor. 51–61 (2012)
10. Bonev, I., Ryu, J.: A Geometrical method for computing the constant-orientation workspace of 6-PRRS parallel manipulators. Mech. Mach. Theor. **36**, 1–13 (2001)
11. Fichter, E.F.: A stewart platform-based manipulator: general theory and practical construction. Int. J. Robot. Res. **5**(2), 157–182 (1986)
12. Rezaei, A., Akbarzadeh, A., Nia, P.M., Akbarzadeh-T,M.R.: Position, jacobian and workspace analysis of a 3-PSP spatial parallel manipulator. Robot. Comput.-Int. Manuf. **29**, 158–173 (2013)
13. HESS-Coelho, T.A, Malvezzi, F.: Workspace optimization of 3RRS+CP parallel mechanisms. In: 12th IFTOMM World Congress. Besançon, France (2007)

Part VIII
Mechatronics

The Mathematical Model of a Weaving Machine

Jiří Ondrášek

Abstract This paper deals with the mathematical model creation of a controlled mechanical system (weaving machine). The controlled mechanical system is composed of a constrained mechanical system and its drive and regulation. Motion equations of a constrained mechanical system are formed on the basis of the Langrangian equations of a mixed type—LEMT through an expert system, such as MSC.ADAMS. Drives of weaving machine mechanisms are implemented by synchronous servomotors. For mathematical description of this type of electric motor, a D,Q coil model was used. For the description of a permanent magnet excited synchronous machine the idealized equations of synchronous machines were used. The goal of a simulation model creation, was to determine the characteristic parameters of electric motors in the dependence on the heald shaft number and working speed.

Keywords Controlled mechanical system · Drive · Electric motor · Controller

1 Introduction

In order to describe the analyzed object unbiasedly, it is necessary that it is mathematically modelled and solved as a whole inclusive its subsystems: mechanical, electrical, hydraulic, pneumatic, thermal, etc., thus, the mechatronic system is created. In the practical implementation, partial systems work together and influence each other. To describe these systems well, it is necessary to create a good model of the given system.

J. Ondrášek (✉)
VÚTS, a.s., Liberec, Czech Republic
e-mail: jiri.ondrasek@vuts.cz

B. Corves et al. (eds.), *Mechanisms, Transmissions and Applications*,
Mechanisms and Machine Science 31, DOI 10.1007/978-3-319-17067-1_33

The complex solution of the joint systems will enable us to predict, to evaluate and to test all the possible running states of a developed machine or device in the models. In such a way, that it is possible to avoid unpredictable situations which can substantially slow down and increase expenses of the development of the machinery in practice. This will create time and financial savings.

2 Controlled Mechanical System

This section gives only the basic information on how to create mathematical models of a controlled mechanical system. Detailed knowledge of this issue may be found for example in [1–3].

One of the possible procedures for generating a simulation model of a controlled mechanical system is the procedure of composing abstract dynamical systems with causal orientation input–output, see Fig. 1. The particular abstract dynamical systems are most often described either with a state or transfer characterization. This link is simple because the outputs of one model are the inputs of another model. The main goal of creating a model is determining the time course of the dynamical behavior of a system which lies in the numerical solution of the generated system of differential equations, or the system of algebraic and differential equations. A Controlled mechanical system is mainly composed of abstract systems, which describe the mechanical system itself, its drive and regulation.

The next section is devoted to the definition of mathematical models of these systems.

2.1 Constrained Mechanical Multi-body System

The motion equations of a constrained mechanical system with flexible bodies are set on the basis of the Langrangian equations of a mixed type—LEMT. A summary record of motion equations in matrix form can be included [1]:

$$\begin{bmatrix} \mathbf{M} & -\mathbf{J}^T \\ \mathbf{J} & \mathbf{0} \end{bmatrix} \begin{bmatrix} \ddot{\mathbf{q}} \\ \boldsymbol{\lambda} \end{bmatrix} = \begin{bmatrix} \mathbf{p}_1 - \mathbf{D}\dot{\mathbf{q}} - \mathbf{K}\mathbf{q} \\ \mathbf{p}_2 \end{bmatrix}, \tag{1}$$

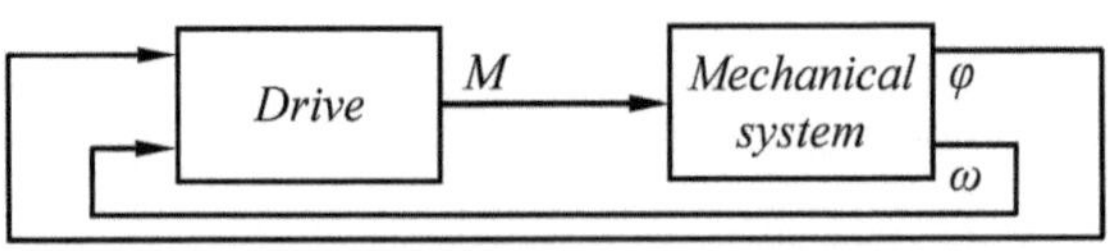

Fig. 1 Block diagram of a controlled mechanical system

in which:

$$\begin{aligned}\mathbf{p}_1 &= \mathbf{Q} - \mathbf{f}^g - \dot{\mathbf{M}}\dot{\mathbf{q}} + \frac{1}{2}\left[\frac{\partial \mathbf{M}}{\partial \mathbf{q}}\dot{\mathbf{q}}\right]^T \dot{\mathbf{q}}, \\ \mathbf{p}_2 &= -\frac{\partial}{\partial \mathbf{q}^T}(\mathbf{J}\dot{\mathbf{q}})\dot{\mathbf{q}} - 2\frac{\partial}{\partial \mathbf{q}^T}\left(\frac{\partial \mathbf{f}^V}{\partial t}\right)\dot{\mathbf{q}} - \frac{\partial^2 \mathbf{f}^V}{\partial t^2}.\end{aligned} \tag{2}$$

To the constrained mechanical system description, generally dependent physical coordinates $\mathbf{q}$ are used with dimension r, which are constrained by a set of scalar constraint conditions in the number of s:

$$\mathbf{f}^V(\mathbf{q}, t) = \mathbf{0}. \tag{3}$$

For i-flexible body, vector $\mathbf{q}_i$ may be described in the form as follows:

$$\mathbf{q}_i = \left[\mathbf{r}_i, \mathbf{p}_i, \mathbf{q}_{ei}\right], \tag{4}$$

in which $\mathbf{r}_i$ represents the vector of the coordinates which determine the position of the given body in the stationary coordinate system. The orientation of the body in this basic space is given by Euler parameters $\mathbf{p}_i$. The elastic deformations of the body are expressed by the vector of the elastic coordinates $\mathbf{q}_{ei}$.

In Eq. (1), matrices **M**, **K** and **D** represent the mass, stiffness and damping matrix of the mechanical system as a whole. Vector **Q** represents the vector of generalized forces, whose components associate with relevant generalized coordinates q_j and vector $\mathbf{f}^g$ is the gravity forces vector. Furthermore, in Eq. (1), $\boldsymbol{\lambda}$ vector of Lagrange multipliers exists in the number of s. They are related to the forces and moments required to maintain the constraints. The matrix $\mathbf{J} = \partial\mathbf{f}^V/\partial\mathbf{q}^T$ expresses the Jacobi matrix of constraint equations set.

In the case of Eq. (1), it is a set of $(r + s)$ differential/algebraic equations for the $(r + s)$ unknowns—the r generally dependent physical coordinates $\mathbf{q}$ and the s Lagrange multipliers $\boldsymbol{\lambda}$. The constrained mechanical multi-body system is characterized by $(i = r - s)$ degrees of freedom. At present it is possible to set and to solve the motion equations of a constrained mechanical systems through a commercially available expert systems, such as MSC.ADAMS.

2.2 Drive

In many cases, a machinery drive is implemented by a 3 phase synchronous electric motor with permanent magnets with which an exciting magnetic rotor flux is produced. This motor type is used namely in servo drives with which high dynamics and accuracy are required. For simulations, a DQ coil model was used, which is based on the mathematical description of a synchronous machine and uses the transformation of stator coordinates into rotor coordinates. The model of this motor including its control structure leads to a system of differential equations.

The voltage equations of the synchronous servomotor expressed in the component shape in rotor reference frame (*d, q*) have this form [2]:

$$U_d = R_s I_d + L_d \frac{dI_d}{dt} - \omega_{el} L_q I_q, \quad U_q = R_s I_q + L_q \frac{dI_q}{dt} + \omega_{el}(L_d I_d + \psi_m), \tag{5}$$

where ω_{el} is the electric angular velocity of the rotor and also of the rotating coordinate system (*d, q*), I_d and I_q are current components expressed in the rotor reference frame (*d, q*) and R_s denotes stator coil electrical resistance. L_d and L_q are inductances of stator winding in the direct and quadrature axis and $\psi_m = const$ represents the flux linkage due to permanent magnet. The electromagnetic torque is then given by the expression [2]:

$$M = \frac{3}{2} p_p \left[\psi_m I_q + (L_d - L_q) I_d I_q \right], \tag{6}$$

in which p_p is the number of pole pairs. The block diagram of the synchronous machine DQ-model is shown in Fig. 2. It was created on the basis of Eqs. (5) and (6).

The vector control of the synchronous motor is given by the condition $I_d = 0$ A, then the vector of stator current **i** is perpendicular to the flux linkage vector **ψ** and the electromagnetic torque *M* is maximum. During the vector control of this electric motor type, the cascade control loop is almost exclusively used with three hierarchically arranged feedbacks: current, speed, and position. Maintaining the required values of position φ^*, revolutions ω^* and current I_d^*, I_q^* is ensured by PID linear controllers. The control action is thus the sum of the three terms: proportional feedback, the integral term and derivative action. The relation between the output

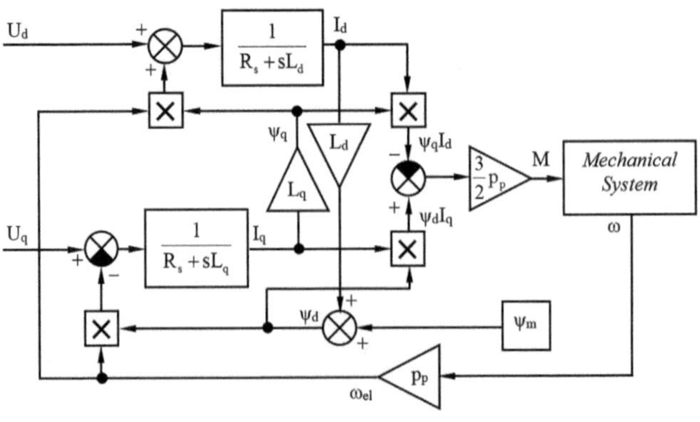

Fig. 2 Block diagram of a synchronous machine DQ-model

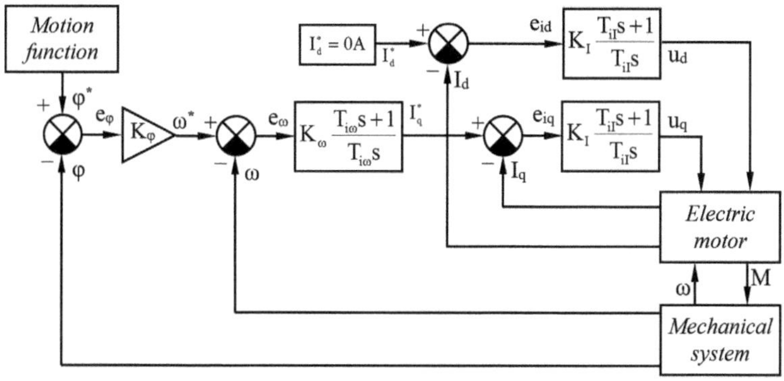

Fig. 3 Block diagram of cascade control loop with ideal P and PI controllers

value—the control value $u(t)$ and the input value—the tracking error $e(t)$ is given by differential equation [3]:

$$u(t) = K\left(e(t) + \frac{1}{T_i}\int_0^t e(\tau)\mathrm{d}\tau + T_d\frac{\mathrm{d}e(t)}{\mathrm{d}t}\right), \quad e(t) = w(t) - y(t). \tag{7}$$

The controller parameters are the proportional gain K, the integral gain K/T_i and the derivative gain $K \cdot T_d$. The time constants T_i and T_d are called integral time constant and derivative time constant. The tracking error $e(t)$ expresses the difference between the desired input value ($w(t) \equiv \varphi^*, \omega^*, I_d^*, I_q^*$) and the actual output ($y(t) \equiv \varphi, \omega, I_d, I_q$) of the controlled system, see Fig. 3.

3 Weaving Machine

The working cycle of a weaving machine consists of four basic phases: shed opening, weft insertion, shed closing, and weft beat-up. With the pneumatic weaving machine VERA [5], those phases are ensured by two fundamental mechanisms: a slay mechanism and a shedding mechanism, which are mutually synchronized with an electrical connection, see Figs. 4 and 5. The motion synchronization of both mechanical systems results from the requirement of the equality of rotations of the slay box cam shaft φ_R and shaft 2 of the shedding mechanism φ.

In the case of the slay mechanism, it is a cam mechanism with radial conjugate cams and an oscillating roller follower with a displacement law $\vartheta = \vartheta(\varphi_M)$. It is composed of a slay cam shaft and a batten, see Fig. 4. Its drive is implemented by a synchronous servomotor—*master*. The shedding device is a combined cam mechanism with a displacement law $\vartheta_{CM} = \vartheta_{CM}(\varphi_S)$ of a Stäubli cam motion [7],

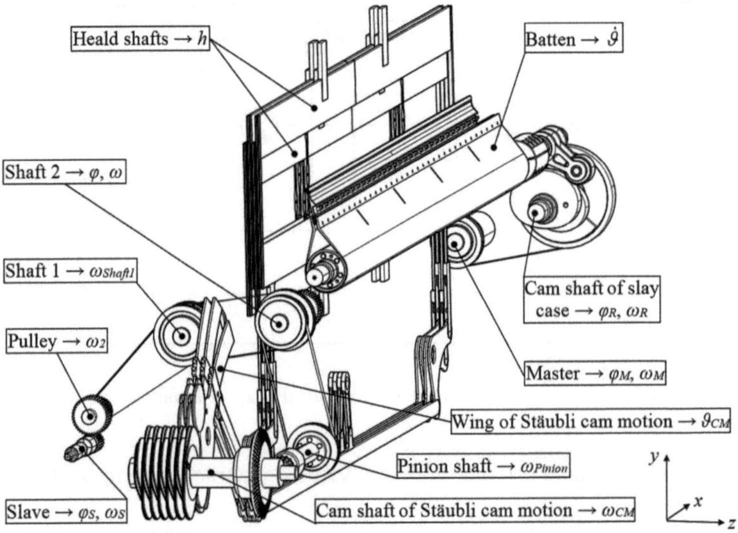

Fig. 4 Model of the slay and shedding mechanism with 4 heald shafts

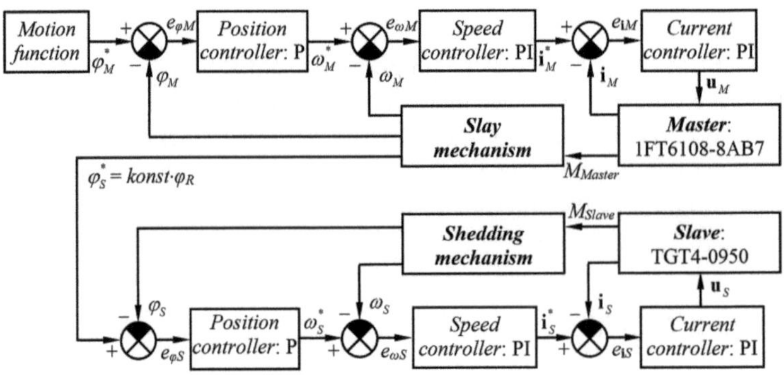

Fig. 5 Block diagram of weaving machine control

which consists of a cam mechanism and linkages. Its drive is performed by means of a servomotor—*slave*. The displacement law of the Stäubli cam motion was identified on the basis of the cam shape measurement. The cam profile coordinates for the requirements of the weaving machine mathematical model were determined by the kinematic synthesis of combined cam mechanisms [6]. The kinematic synthesis was implemented on the basis of the displacement law knowledge of the given cam mechanism and its dimensional parameters.

In the case of the mathematical model of the weaving machine, this is a dynamic analysis of a spatial multi-body system with flexible elements—timing belts. Between the bodies, there were defined kinematic pairs with friction in such a way

Table 1 Drives parameters

Parameter	Unit	Slay mechanism: 1FT6108-8AF7	Shedding mechanism: TGT4-0950
Number of pole pairs	p_p [−]	4	3
Stator coil electrical resistance	R_s [Ω]	0.065	1.95
D-axis stator inductance	L_d [mH]	1.5	11.4
Q-axis stator inductance	L_q [mH]	1.5	11.4
Rotor permanent magnetic flux linkage	ψ_m [NmA^{-1}]	0.5954	0.5183
Current controller proportional component	K_I [VA^{-1}]	5	2
Current controller time integration constant	T_I [s]	0.0005	0.0005
Speed controller proportional component	K_ω [As/rad]	5	0.2
Speed controller time integration constant	T_ω [s]	0.08	0.08
Position controller proportional component	K_φ [s^{-1}]	5	0.2

as the analyzed mechanical system is free of redundant constraints. The dynamic friction factor value of revolving joints was defined according to bearing type. Values are generally taken from $\mu = 1 \times 10^{-3}$ to $\mu = 3 \times 10^{-3}$ [4]. The beat-up forces were ignored in calculations because they were inconsiderable due to the inertial effects of the slay mechanism. The inertial effects and loads of warp threads were integrated into the inertial effects of the shedding mechanism. The intensity of these forces depends on the type of woven fabric.

With the use of knowledge included in Sect. 2.2, the mathematical models of the drives and regulations were created. The appropriate parameters of PID controllers were tuned on the basis of the Ziegler–Nichols tuning rules [3], see Table 1. The motion equations of the weaving machine controlled mechanical system were set and numerical computed by MSC.ADAMS. The main aim of the calculations was to determine the characteristic parameters of the electric motors dependent on the heald shaft number, working speed and the control accuracy.

3.1 Sample of Results

For the electric motor design, it is necessary to know the basic power parameters, which are one of the many outputs of the tasks of multi-body systems dynamics. They are values of driving torque M_h and power P. In the machine working cycle, these variables have a time-varying course. Therefore, they used to be quantitatively expressed as its root mean square values M_{hEff} and P_{Eff}. Root mean square (RMS) of

Table 2 Power parameters of electric motors

Weaving machine mechanism	Transient state		Steady state	
	$M_{h\text{Max}}$ [Nm]	P_{Max} [kW]	$M_n \equiv M_{hEff}$ [Nm]	$P_n \equiv P_{Eff}$ [kW]
Slay mechanism	110	28.5	31	9.0
Shedding mechanism	30	9.5	8	2.6

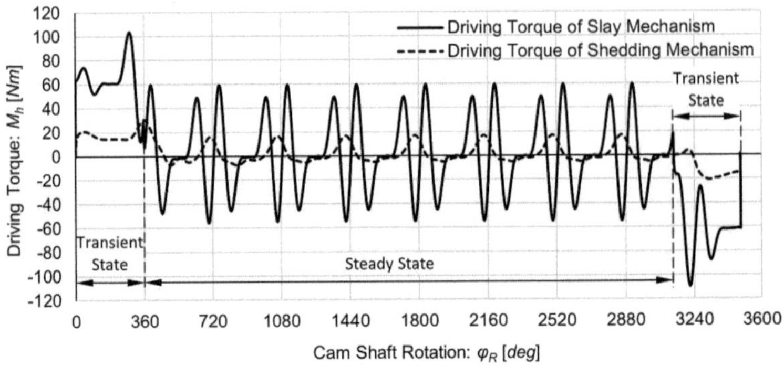

Fig. 6 Driving torque of slay and shedding mechanism

a quantity expresses an imaginary substitutive constant quantity that has the same average power as the time-varying quantity. In the case of electric motors, they can be assigned to the rated typical parameters such as rated driving torque *Mn* and the rated power P_n.

The power demands on the drive of the slay and shedding mechanism are numerically expressed with the RMS values M_{hEff} and P_{Eff} during the steady state and the maximum values $M_{h\text{Max}}$ and P_{Max} during the transient state, see Table 2. These values were computed on the basis of the weaving machine mathematical model with four heald shafts and the working speed reached the value n_n = 700 rpm. The hypothetical courses of required driving torque M_{Master} and M_{Slave} are shown in Fig. 6. The synchronous servomotors 1FT6108-8AF7 [8] and TGT4-0950 [9] were chosen on the basis of computational simulations.

4 Conclusions

A mathematical model of the weaving machine was created. A complex model is composed of the slay and shedding mechanism model and a drive model of both mechanical systems. The drive model includes the synchronous servomotor model and the model of the weaving machine control structure. The complex model is expressed with a set of differential/algebraic equations. The power parameters of the electric motors were determined on the basis of computational simulations. By means of these parameters, the drive electric motors of both mechanical systems were selected. A synchronous servomotor 1FT6108-8AF7 [8] was chosen for the drive of the slay mechanism and an electric motor TGT4-0950 [9] for the drive of the shedding mechanism. The motor size is dependent on the control accuracy.

References

1. Stejskal, V., Valášek, M.: Kinematics and Dynamics of Machinery. Marcel Dekker Inc., New York (1996). ISBN 0-8247-9731-0
2. Bimal, B.K.: Modern Power Electronics and AC Drives. Prentice-Hall Inc., New Jersey (2001). ISBN 0-13-016743-6
3. Åström, K.J., Murray, R.M.: Feedback Systems: An Introduction for Scientists and Engineers. http://www.cds.caltech.edu/~murray/FBSwiki
4. http://www.nskamericas.com/cps/rde/xbcr/na_en/Technical_Data.pdf
5. http://www.vuts.cz/novy-web/docs/en/VERA_EN.pdf
6. Koloc, Z., Václavík, M.: Cam Mechanisms. Elsevier, Amsterdam (1993). ISBN 0-444-98664-2
7. http://www.staubli.com/en/textile/textile-machinery-solutions/sms/cam-motions1600-1700/
8. http://w3app.siemens.com/mcms/infocenter/dokumentencenter/mc/Documentsu20Catalogs/nc61_2010_e2.pdf
9. http://www.tgdrives.cz/en/servomotors/tgt-series/tgt4/

Approximation of Periodic Displacement Law with Fourier Series in the Applications of Mechanisms with Electronic Cam

Petr Jirásko, Pavel Dostrašil and Miroslav Václavík

Abstract When investigating movements of working links of manufacturing and handling machine mechanisms, we often face a serious problem in complying with positional accuracy in relation to the production technology. The design of a working mechanism seeks to minimize inertial forces or to reduce the weights of mechanism links. Then, the consequence is a reduction of stiffness with significant deviations from the prescribed trajectory. Deviations are caused by natural oscillations of links of the kinematic chain of a working mechanism. The paper outlines the methodology of designing a displacement law for the periodic working movements of electronic cams. As an exemplary displacement law, it is used VDI_2143 Praktische Anwendung.

Keywords Electronic cam · Positional deviation · Harmonic analysis · Fourier transform

1 Introduction

When implementing *periodic* displacement laws with large compliance in the *driven* part of a cam mechanism, there often occurs an unacceptable positional deviation. The issues concern manufacturing and handling mechanisms, for example, a yarn traversing mechanism in OE—spinning machines. Tasks of similar type are characterized by requirements for the highest speed of rotation (periods of displacement law), which leads to considerable dynamic forces in the driven part of a mechanism.

P. Jirásko · P. Dostrašil (✉) · M. Václavík
VÚTS, a.s., Liberec, Czech Republic
e-mail: pavel.dostrasil@vuts.cz

P. Jirásko
e-mail: petr.jirasko@vuts.cz

M. Václavík
e-mail: miroslav.vaclavik@vuts.cz

B. Corves et al. (eds.), *Mechanisms, Transmissions and Applications*,
Mechanisms and Machine Science 31, DOI 10.1007/978-3-319-17067-1_34

The aim is to decrease those forces or to reduce the weights of moving elements and with regard to the required speed to control the kinematic quantities of a working link, which is the control of a positional error due to natural harmonic oscillations, excited by the motion function of a cam mechanism (conventional or electronic). At the same time, requirements of the positional accuracy of a working link *are not* extreme and inertia forces are dominant. Same problem, but with a different approach is resolved in [5].

2 The Methodology of Investigation

The task of this type is modeled on a dynamic stand with a mechanism with electronic cam, *Yaskawa* (Japan) *hw* configuration, and with mass parameters, as to Fig. 2. Displacement law is a function as to *VDI_2143* [1], which is shown graphically in Fig. 1 in the form of displacement (the zero derivative), the 1st and the 2nd derivatives. The required quantities obtained from three sources (*A, B, C groups*) are processed to functional dependencies and present a degree of conformity of virtual modeling (discrete dynamic models) with reality acquired by independent measurement and system monitoring in the *PLC* development environment (*MP2300* controller). Due to a limited extent of the paper, only some results according to the following groups are presented. We do not indicate the results of the measured kinematic quantities of working mass I_1, we only state conformity with the theoretical values. *HW* means of measuring equipment and measurement results themselves are in [3, 4]. There are mentioned interesting measurement results of *PERR* positional error and its comparison with virtual computing models. This comparison is a criterion of accuracy of computational models.

A Group. Evaluated quantities of measurement in the 5th cycle period are:

- The sensed position of working inertia mass I_1 in [deg], see Fig. 3
- Speed of working inertia mass in revolutions [min^{-1}]

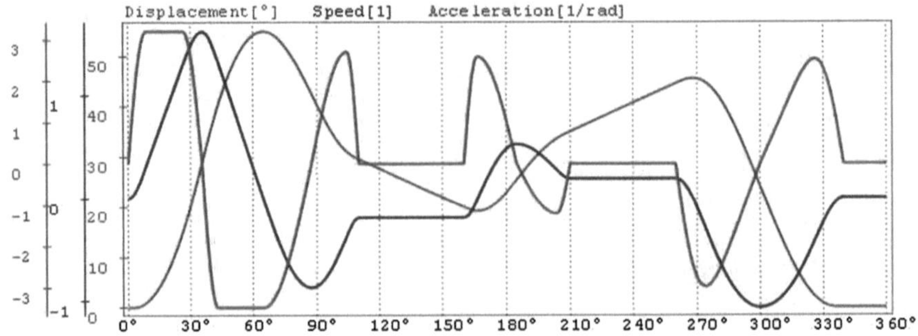

Fig. 1 Displacement (zero derivative), 1st and 2nd derivatives as to VDI_2143 [1]

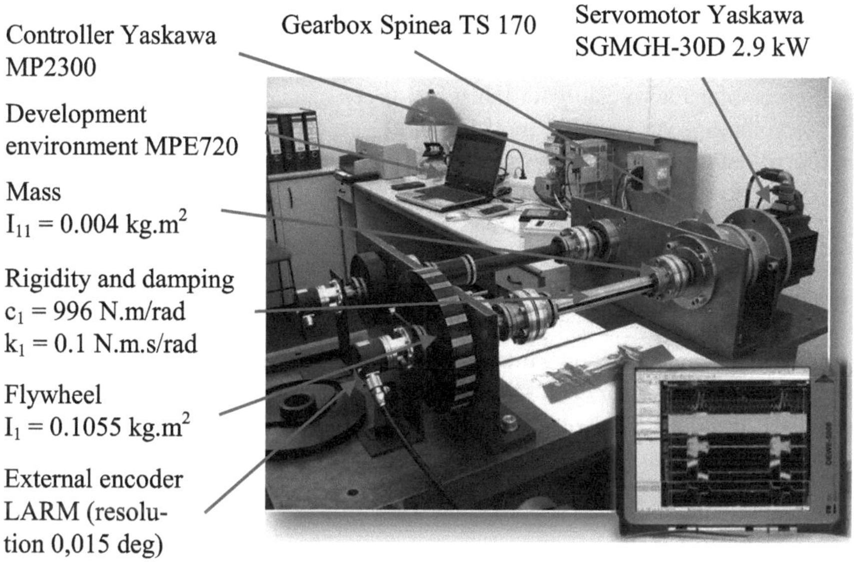

Fig. 2 Yaskawa electronic cam stand

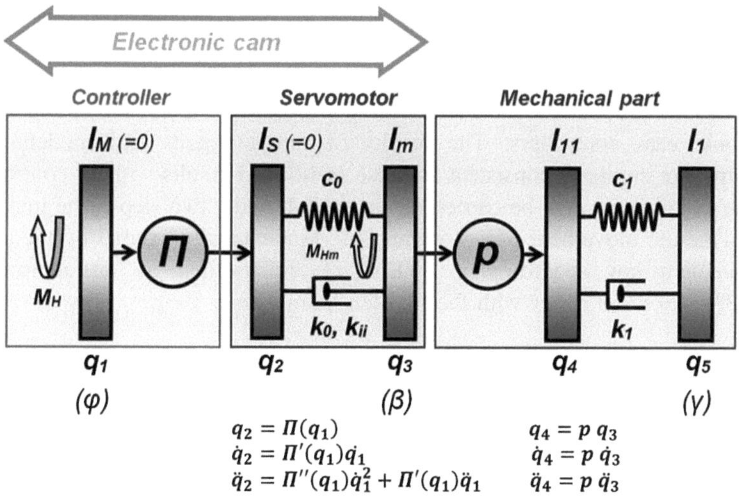

Fig. 3 Discrete model of electronic cam 0-Π-1.1

- Accelerating torque on the servomotor shaft (analog value from the drive inverter in [%] of *EKM* efficient torque, catalog value)
- *P-error* positional deviation of motion function on the servomotor shaft from the theoretical values in [deg] (analog value from the drive inverter)

B Group. Evaluated quantities by the *MPE720* system by means of *PLC* monitoring functions (digital values) are (Figs. 5 and 8):

- Servomotor rotor position in [pulse] … ***green***
- Servomotor rotor speed in [rad/s] … ***blue***
- Servomotor rotor acceleration in [rad/s^2] … ***red***
- *PERR* positional deviation in [pulse], equivalent to *p-error* … ***violet***

C Group. Virtually modeled quantities are:

- Kinematic quantities of servomotor rotor (position [deg], speed [rad/s], acceleration [rad/s^2])
- Kinematic quantities of working inertia mass (position [deg], speed [rad/s], acceleration [rad/s^2])
- PERR positional deviation in [deg] (equivalent to *PERR* from *PLC* and *p-error* from the measurement)
- Torsion [deg] of compliant shaft between working inertia mass and servomotor shaft

Furthermore, it will be indicated a comparison of *PERR* positional deviation of servomotor rotor with the theoretical data of the motion function from the measurement, from a computational model using *Lagrange equations of the 2nd kind* [2, 3] and from a *sw MSC.ADAMS/EASY5* modeling. Thus, *PERR* is the difference of the actual position on the servo shaft from the theoretical position, according to Fig. 3, it is $q_3 - q_2$ [deg].

Because of the limitations of the extent of this paper, there will be mentioned the results with *PI* modeled controller in speed constraint of a cascade arrangement of electronic cam controllers. The results of measurements and modeling with *P* control are similarly consistent as those mentioned results with *PI* control.

The examination will be carried out in the following two steps. The first step is to analyze the movement of a working link, kinematically excited by the original displacement law according to *VDI_2143* and excited by an approximated *VDI_2143 Fourier series* with the first ten harmonics.

3 Solution

3.1 SW Modeling and Measuring on a Dynamic Stand with the Original Displacement Law According to VDI_2143

Sw developed in VÚTS for investigating the kinematics and dynamics of electronic cam mechanisms solves ***0-П-1.1*** model of electronic cam described in [3] according to Fig. 3. Input parameters are the displacement law of the working link according to *VDI_2143* in terms of its derivatives, depending on the position of virtual axis,

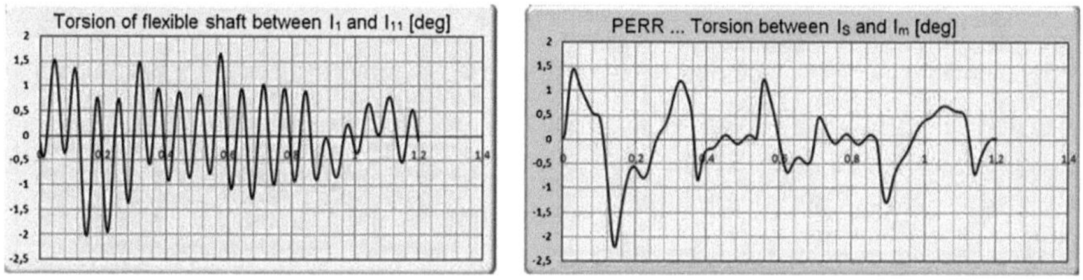

Fig. 4 SW model (VDI_2143), selected quantities as to **C group**

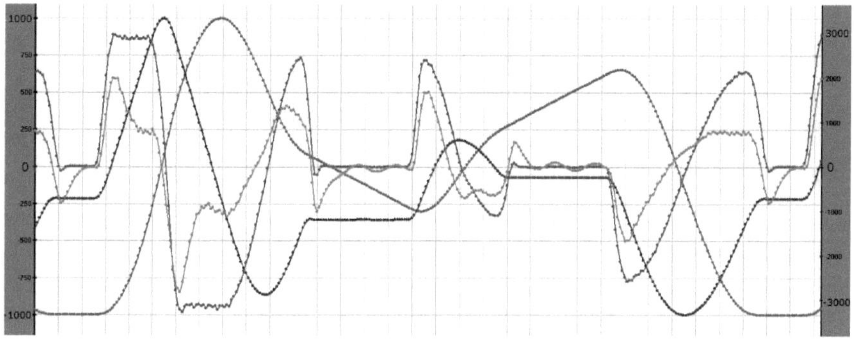

Fig. 5 MPE720, quantities as to **B group** (color figure online)

speed, mass, stiffness, damping, etc. The obtained results are functional dependences according to the above mentioned ***C*** group. In Fig. 4, there are only displays of torsion [deg] between masses $I_1 - I_{11}$ and *PERR* value. An independent variable is time [s] of an interval of one cycle.

Figure 5 shows the results according to ***B*** group. The description of the environment, monitoring functions and the features of the *MPE720 (Yaskawa)* development environment is given in [3]. The main axis *Y1* (left) refers to *PERR* positional deviation and it is in units *[pulse]*. The used *SGMGH-30D* servomotor *(Sigma II type)* has *Encoder*, which sends 2^{17} pulses per 1 revolution. The secondary axis *Y2* (right) refers to acceleration in units [rad/s^2]. Other quantities are in *auto* scale and are given for guidance and monitoring the nature of functional dependencies.

In Fig. 6, there are provided the measured and modeled *PERR* positional deviations from the servomotor rotor motion function in two ways of virtual simulation.

A more detailed description of the numerical solution of the discrete model, which is depicted by *Lagrange equations of the 2nd kind* as to Fig. 3, is not listed here due to the limited scope of this paper. The principles of controlling servo drives are not described as well, it is only stated that the majority of servo drives show a cascade control structure with torque, speed and positional feedback.

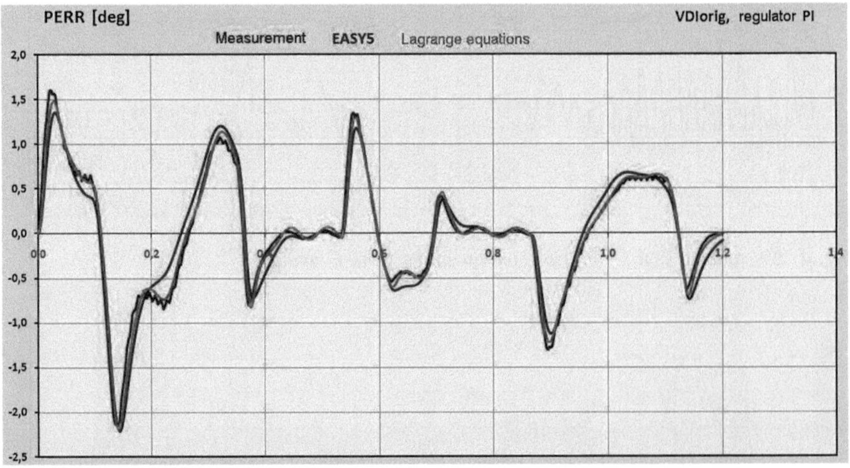

Fig. 6 PERR positional deviations (measurement, 2 methods of virtual simulation)

Controllers are generally proportional (*P*) and proportional-integral (*PI*). Detailed descriptions of the controlling and modeling of servo drive can be found in [6]. Therefore, we aim at such an intervention in the numerical solution of motion equations so that *PERR* characteristic quantity as a positional deviation of the servomotor itself should correspond to the *PI* actual mode utmost. *PERR* is a criterion for the accuracy of the given *sw* model and its waveform (course) is compared with two independent sources. One source is measurement and the other is the virtual model of a controlled mechanical system created in *MSC.ADAMS* and *MSC.EASY5* program (software) systems. Compliance is remarkable. Thus, the results provide a virtual study of the influence of the dynamic properties of displacement laws of kinematically excited compliant systems, as described below.

3.2 Harmonic Analysis of Displacement Law as to VDI_2143. Sw Modeling and Verifying the Dynamic Properties of the Approximated Displacement Law on the Stand

The aim of the *harmonic analysis* of displacement law is to replace its original data with approximated values of *Fourier series* with the least final number of first harmonic components. From this new approximated displacement law, we expect *smaller* values of oscillation amplitudes of working inertia mass. Thus, by choosing the number of harmonic components, the error of approximation is monitored with respect to the kinematic quantities of a working link. In this exemplary solution, it is used the first ten harmonics from the *original VDI_2143* harmonic analysis and a

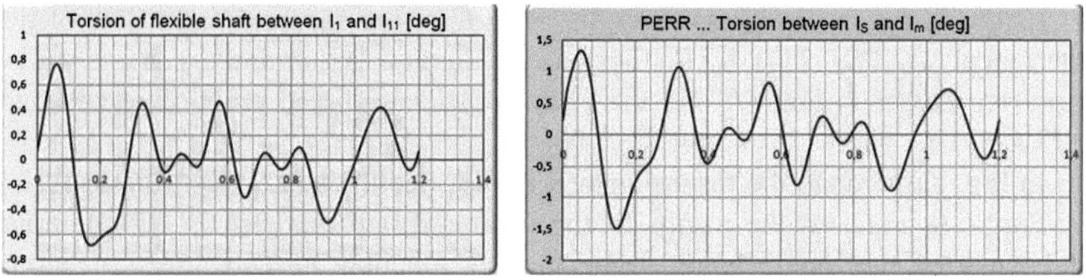

Fig. 7 SW model (10 harm. Components VDI_2143), selected quantities as to **C group**

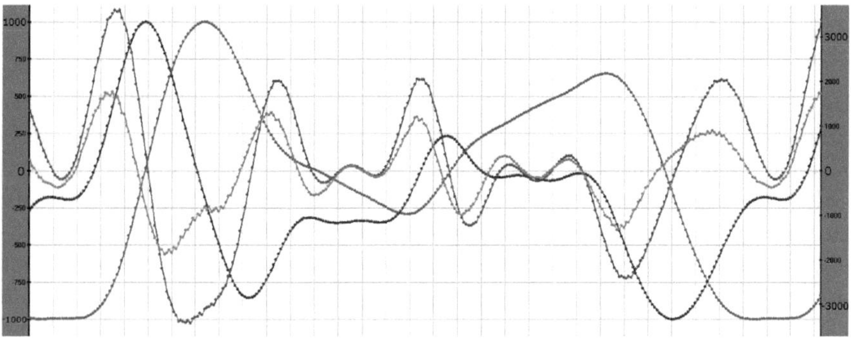

Fig. 8 MPE720, quantities as to **B group** (color figure online)

new displacement law is implemented on the stand as to Fig. 2. Measurement and virtual testing on *sw* models are the same as with the *original VDI_2143* by the above Sect. 3.1. Furthermore, there are mentioned only two basic results.

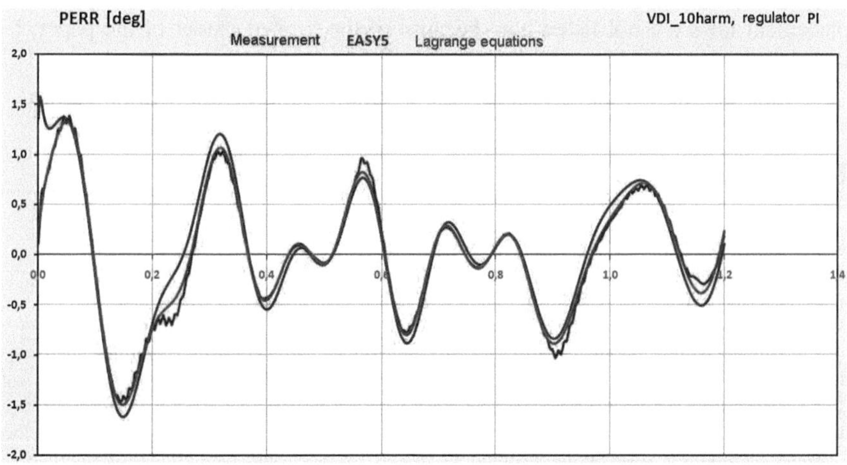

Fig. 9 PERR positional deviations (measurement, virtual simulations)

Figure 7 shows the solution results as to ***C group*** (approximated data of displacement law *VDI_2143* by Fourier series—the first 10 harmonics). It is a torsion between masses I_1 and I_{11} and the characteristic positional error of servo motor *PERR*.

In Fig. 8, there are the results as to ***B group***. See the text to Fig. 5 in Sect. 3.1.

Figure 9 indicates the results of the measured and modeled *PERR* positional deviation of the servo motor rotor motion function in two ways of virtual simulation.

4 Conclusions

By comparing Figs. 5 and 8, *an improvement is demonstrable.* The mentioned method of replacing the original displacement law with a new approximated function with an intentional "positional error" can achieve smaller positional deviations of a working link, it is effective, and is easy feasible and modifiable particularly in a mechanism with electronic cam on the basis of a different number of harmonic components. Based on the virtual simulation, whose conformity with the reality is excellent, it can be studied various characteristic influences of mechanisms with electronic cam (for example, the effect of speed ratios, control properties or compliance and damping of an electromagnetic constraint stator-rotor, etc.) with highly compliant driven output links of working mechanisms.

The methodology of replacing the original displacement law by a certain number of harmonics is gaining importance in the implementation of non-periodic motion functions, the so-called stepping ones with significant rest intervals. There are important mechanisms in the form of rotary tables and stepping mechanisms where the requirement for the minimum step time increases technological production time or productivity. Those issues associated with the residual spectra of stepping displacement laws are not listed here because of the limited extent of the paper, but they will be presented at the conference.

References

1. Bewegungsgesetze für Kurvengetriebe (Praktische Anwendung), VDI 2143, Jan 1987
2. Jirásko, P., Václavík, M.: Residual spectra of displacement of conventional and electronic cams. In: National Conference Engineering Mechanics 2009, Svratka (CZ), May 2009
3. Jirásko, P.: Methodology of electronic cam applications in drives of working links of mechanisms of processing machines. Dissertation. TUL, FM, Liberec (2010), 207 pp
4. Dostrašil, P., Jirásko, P., Bušek, M.: An analysis of planetary gearboxes for their use with electronic cam, IFToMM Liberec (TU) (2010)
5. Dostrašil, P.: Finding the parameters of the compensation pulse with the use of iterative Nelder-Mead method for suppressing residual vibrations of electronic cam. In: 18th International

Conference CSCC. Santorini, Greece. 2014. ISBN 978-1-61804-242-2. http://www.europment.org/library/2014/santorini/bypaper/ROBCIRC/ROBCIRC-24.pdf
6. Lindr, D.: Control of Servodrives in Dynamic-Intensive Application. Dissertation. TUL, FM, Liberec (2011)

High-Speed and High-Resolution Linear Microstepper Based on Toggle Mechanism Actuated by Electromagnet

Takaaki Oiwa, Yuichiro Toyoda and Junichi Asama

Abstract Novel linear microsteppers are designed for high-speed and high-resolution positioning actuators. In general, most microsteppers equipped with a multi-layered piezoelectric actuator for positioning have difficulty to improve their feed rate because of the small stroke of the actuator, despite their excellent positioning resolution. The microstepper in this study employs two electromagnets for clamping the rail, and one electromagnet for longitudinal actuation of the mechanism. Use of electromagnets with a larger stroke enables a high-speed feed rate and allows for the use of inexpensive low-voltage amplifiers. Moreover, the toggle mechanisms reduce the actuator's displacement and enlarge the actuator's force. Because the relation between the input and output displacements of the toggle mechanism is non-linear, the microstepper achieves both high-speed and high-resolution positioning as necessary. This paper deals with experimental mechanisms equipped with two types of actuators based on one or two gaps between the armature and the iron core of the electromagnet. First, the fundamental design of the microstepper with two toggle mechanisms actuated by an electromagnet is introduced. Next, an experiment using each of the two sets of microsteppers we manufactured is described. As a result, the microsteppers are capable of of 1 mm/s maximum speed and a positioning resolution better than 0.1 μm.

Keywords Displacement reducer · Inchworm mechanism · Microstepper · Positioning resolution · Toggle mechanism

1 Introduction

In recent years many types of microsteppers have been developed as miniature linear actuators for precision positioning [1–3]. The conventional microstepper shown in Fig. 1a can move over a long distance regardless of the positioning

T. Oiwa (✉) · Y. Toyoda · J. Asama
Shizuoka University, Shizuoka, Japan
e-mail: tmtooiw@ipc.shizuoka.ac.jp

B. Corves et al. (eds.), *Mechanisms, Transmissions and Applications*,
Mechanisms and Machine Science 31, DOI 10.1007/978-3-319-17067-1_35

actuator's stroke because they move like an inchworm by sequentially using two clamping elements and one positioning actuator. The positioning resolution of the microstepper usually depends on the resolution of the positioning actuator, so the maximum feed rate of the microstepper is restricted by the positioning actuator's stroke and maximum driving frequency. Because most microsteppers employ a multi-layered piezoelectric actuator as the positioning actuator, it is possible to achieve nanometer-order positioning resolution. However, positioning speed more than 1 mm/s requires a driving frequency higher than 100 Hz because the stroke of the piezoelectric actuator is around 10 μm at most. Moreover, the drive of the piezoelectric actuator requires high-voltage amplifiers. The microstepper in this study employs an electromagnet actuator with a larger stroke for high-speed positioning, driven by low-voltage amplifiers. Furthermore, a displacement reducer based on toggle mechanisms [4–6] enables high positioning resolution. Since the output displacement of the reducer exponentially increases with an increase in input displacement, this microstepper can achieve not only high-speed positioning but also high-resolution positioning as necessary.

This paper deals with experimental mechanisms equipped with two types of actuators based on one or two gaps between the armature and the iron core of the electromagnet. First, we introduce the fundamental design of the microstepper with two toggle mechanisms actuated by one electromagnet. Second, we describe the design of the electromagnets both for positioning and clamping. Third, we relate the results of evaluations of the positioning speed and resolution for the two types of the microstepper we manufactured.

2 Fundamentals

2.1 Outline

Figure 1b depicts schematic illustration of the proposed microstepper consisting of two magnetic clamps and one positioning actuator, as well as the conventional microstepper. Unlike it, however, ours uses two toggle mechanisms to reduce the displacement of the positioning actuator. The output displacement of the toggle mechanisms changes the distance between the clamps, and moves the microstepper along the rail. Since each clamping element has not only an electromagnet but also a permanent magnet, the microstepper can magnetically hold to the rail even in an electric power outage.

2.2 Displacement Reducer

Figure 2a depicts one toggle mechanism consisting of two flexure elements for simplicity although the microstepper is equipped with two such mechanisms. When

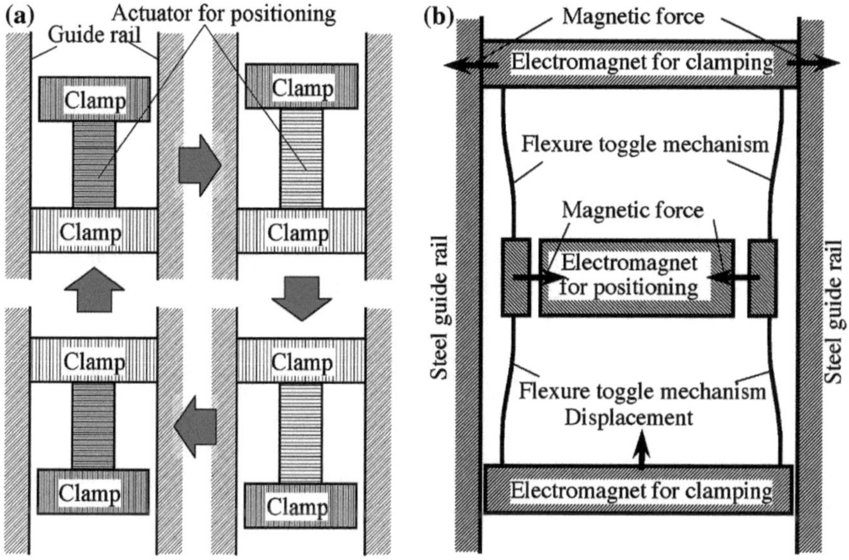

Fig. 1 Linear microsteppers: **a** conventional, and **b** proposed in this paper, consisting of flexure toggle mechanisms, positioning electromagnet and clamping electromagnet

the input displacement of the toggle x is small relative to the length of the springs l, the output displacement y is approximately expressed [5, 6] by equation,

$$y \approx \frac{x^2}{l}. \tag{1}$$

Figure 2b shows the output displacement and reduction ratio when the input displacement x varies from 0 to 1.5 mm. This figure also indicates that small input displacement less than 0.3 mm is extremely reduced to less than 5 μm for fine positioning. On the other hand, larger displacement of 1.5 mm produces a 110-μm-step displacement for fast feeding. When the input displacement x varies from 0.3 to 1.5 mm, the reduction ratio increases 5 times from 0.015 (≈1/67) to 0.075 (≈1/13). This means that positioning resolution depends on the input displacement of the toggle mechanism.

2.3 Electromagnet

In general, when increase of the gap enables larger displacement of the electromagnet actuator, a considerably large electric current is needed to close the wide gap because the attractive force drastically decreases inversely with the square of the gap. To avoid the above defect, in this study, two types of electromagnet

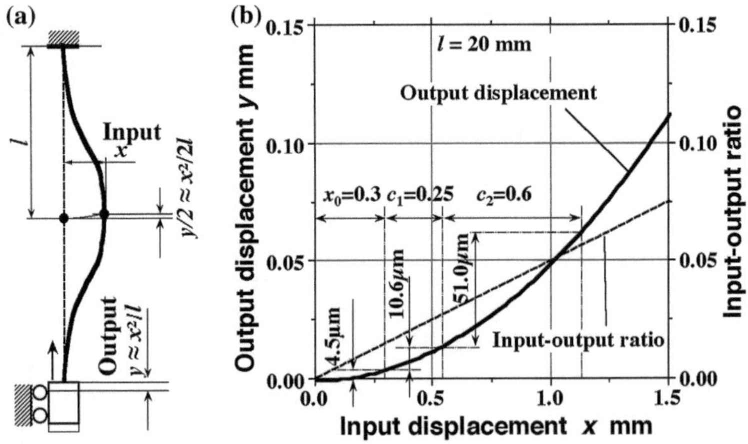

Fig. 2 **a** One flexure toggle mechanism, and **b** relationship between input and output displacements of toggle mechanism

actuators, with one and two gaps, respectively, were designed as the positioning actuators. The actuator shown in Fig. 3a has two gaps of 0.25 and 0.6 mm among one electromagnet and two pole pieces. Increasing the current of the electromagnet firstly closes the smaller gap, and thereafter decreases the larger gap. This means that the maximum displacement of the actuator is 0.85 mm. From these design values it is expected that input displacements of 0.25 and 0.85 mm would produce the reducers' output displacement of 10.6 and 61.6 μm, respectively, when the initial input displacement x_0 is 0.3 mm. Consequently, these actuator's strokes yielded microstepper steps of 5.3 and 31 μm, respectively, and a maximum positioning speed of 0.78 mm/s at a 25-Hz driving frequency.

Another actuator shown in Fig. 3b has one gap between an electromagnet and a pole piece. However, because the pole piece is made of a permanent magnet, bias flux is provided, and slightly decreases the gap without an electric current. The current flow in the coil to further increase the flux closes the gap, and results in a large output displacement of the actuator. On the other hand, an inverse current decreases the bias flux, and increases the gap until the point at which there is no permanent magnet. It is expected that this causes very small input displacements of the reducer, and extremely small steps of the microstepper. From these design values it is expected that an input displacement of 1.2 mm produces the reducers' output displacement of 67 μm when the initial input displacement x_0 is 0.3 mm. Consequently, this stroke yields microstepper steps of 34 μm and a maximum positioning speed of 0.84 mm/s at a 25-Hz driving frequency.

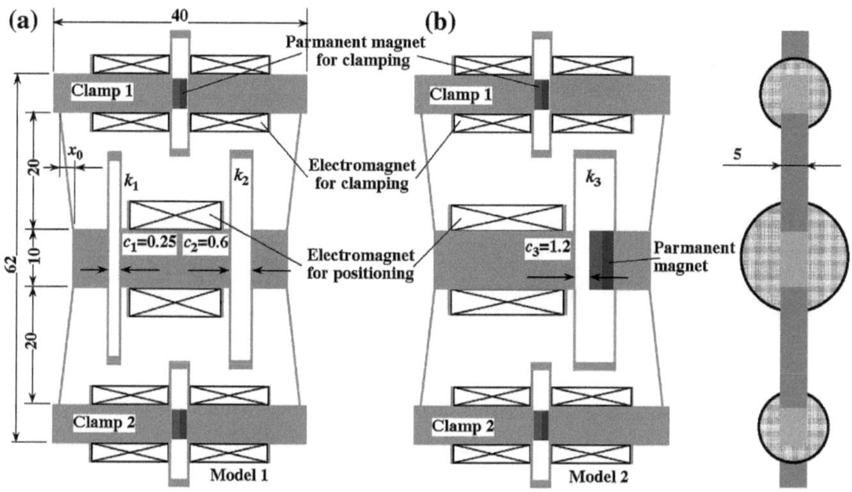

Fig. 3 Microsteppers with two types of positioning actuators

3 Experimental Setup

To avoid assembly error, monolithic structures consisting of the cores of the two clamps, the core of the positioning actuator, and the two toggle mechanisms were fabricated from a plate (annealed carbon steel, S50C, 5-mm thickness) by the wire electric discharge machine. Moreover, a neodymium permanent magnet was installed into the core of each clamp in both model 1 and model 2. Two types of microsteppers equipped with the electromagnetic coils described in the Sect. 2.3 are shown in Fig. 4a. Each microstepper was placed on a rail consisting of parallel blocks of ground carbon steel (S45C). Figure 4b depicts the driving and measuring setup for the microstepper. A DC servo-driver (I–V amplifier, Servo techno LA320) voltage-controlled by a PC with a 12-bit D/A converter regulated the coil current of the positioning actuator from –0.8 A to 2.0 A. Another servo-driver (Servo Techno LA220) drove the series-connected electromagnetic coils for the two clamps with a current of ±0.4 A. Each sequential motion of the positioning actuator and clamps, shown in Fig. 1, was repeated at a frequency of 25 Hz for a period of 2.0 s. Figure 4 also illustrates the timing chart for the two drivers, consisting of 90°-out-of-phase square waves. Meanwhile, a plunger-type linear encoder (Heidenhain MT1201 + IK121; measuring range, 12 mm; measuring resolution, 2 nm) with zero measuring force measured the displacement of the microstepper at a sampling interval of 1 ms. At the current stage, this PC for measuring is not synchronized with the PC for driving.

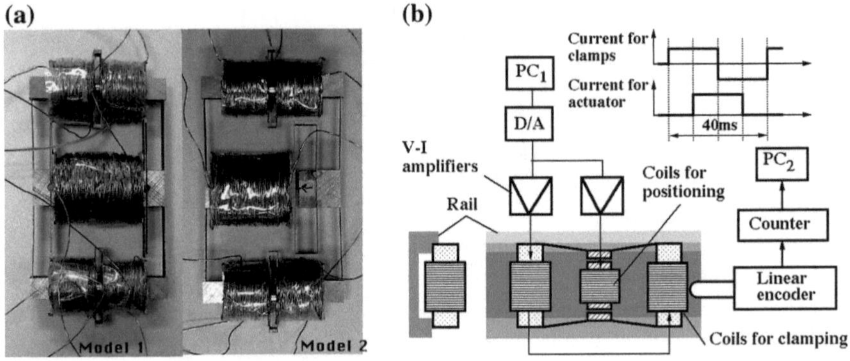

Fig. 4 **a** Two microsteppers, and **b** driving and measuring setup for microstepper

4 Experimental Results

Figure 5a shows the displacement measurement results for the model 1 microstepper. The relationship among the applied current, averaged positioning speed and averaged step is shown in Fig. 5b. This figure shows that a coil current of more than 1.1 A remarkably increased the speed and step. It was expected that a current more than 1.1 A closes both small and large gaps c_1 and c_2 of the positioning actuators, and then considerably increases the output displacement of the reducer. A 1.3 A current achieved the maximum positioning speed of 1.02 mm/s. This feed rate was 30 % faster than expected value of 0.78 mm/s described in Sect. 2.3. More current, however, gradually decreased the speed. It is likely that too much current generates an undesirable attractive force between the electromagnet for positioning and the rail to spoil the positioning speed.

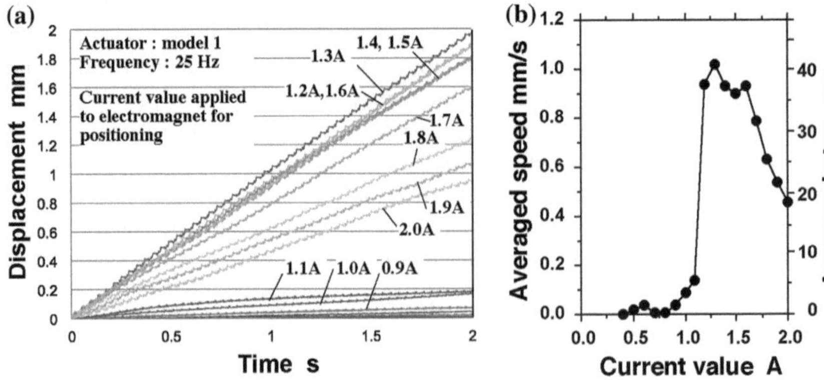

Fig. 5 **a** Measured displacement of model 1, and **b** effect of current on averaged speed

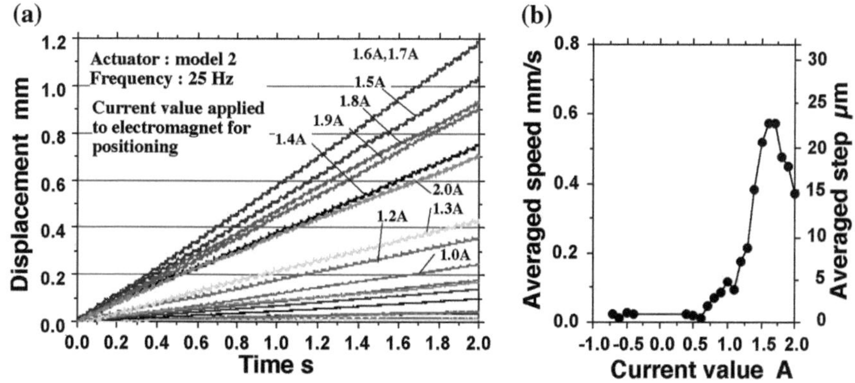

Fig. 6 a Measured displacement of model 2, and b effect of current on averaged speed

Figure 6a, b depict the measurement results for model 2. Currents of 1.6 and 1.7 A achieved the same maximum positioning speed of 0.573 mm/s. This was 32 % slower than 0.84 mm/s expected from the calculation in Sect. 2.3. The coil current within ±0.6 A kept the positioning speed very low. When the current was more than +0.6 A, the positioning speed and step increased with increases in current. However, a coil current more 1.6 A spoiled the positioning speed in the same manner as in model 1.

Figure 7 depicts the step responses of the two types of the microsteppers at the positioning actuators' currents of 0.4 A, which are shown in only five sequences. The average step or the positioning resolution of one sequence was approximately 50 and 80 nm, respectively. Although each displacement increases in a staircase pattern, at least in theory, the experimental result seems to indicate forward-and-backward motions of 0.2 or 0.4 μm. In this regard, it is speculated that some geometrical deviations of the contact surfaces between the microstepper and the rail, such as out-of-flatness, produces motion errors of the microstepper, especially angular motion errors.

Fig. 7 Positioning resolution

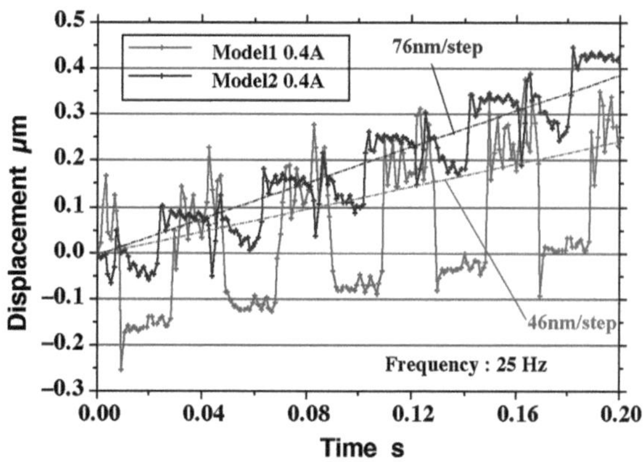

5 Conclusions

Two types of microsteppers equipped with displacement reducers and electromagnet actuators instead of a piezoelectric actuator have been developed to improve both high-speed and high-positioning resolution. The reducer can vary the step length in the positioning direction because of its non-linear characteristics between input and output displacements. The resulting microsteppers achieved approximately 1-mm/s maximum speed at a driving frequency of 25 Hz, and positioning resolutions better than 0.1 μm.

References

1. Slocum, A.H.: Precision machine design, 10.5 Piezoelectric actuators, p. 666. Society of Manufacturing Engineers, Dearborn (1992)
2. Li, J., Sedaghati, R., Dargahi, J., Waechter, D.: Design and development of a new piezoelectric linear Inchworm® actuator. Mechatronics **15**, 651–681 (2005)
3. Moon, C., Lee, S., Chung, J.K.: A new inchworm type actuator with I/Q heterodyne interferometer feedback. Mechatoronics **16**, 105–110 (2006)
4. Sclater, N., Chronis, N.P.: vMechanisms and Mechanical Devices Sourcebook, p. 153. McGraw-Hill, New York (1992)
5. Slocum, H.A.: Precision Machine Design, 8.6 Flexural bearings, p. 521. Society Manufacturing Engineers, Dearborn (1992)
6. Oiwa, T., Senjo, M., Nakajima, M., Masuda, T.: Hand-operated translation stage with subnanometer resolution. Rev. Sci. Instrum. **76**(076105), 1–4 (2005)

Author Index

B. Corves et al. (eds.), *Mechanisms, Transmissions and Applications*,
Mechanisms and Machine Science 31, DOI 10.1007/978-3-319-17067-1

MIX
Papier aus verantwortungsvollen Quellen
Paper from responsible sources
FSC® C105338

If you have any concerns about our products,
you can contact us on
ProductSafety@springernature.com

In case Publisher is established outside the EU,
the EU authorized representative is:
Springer Nature Customer Service Center GmbH
Europaplatz 3, 69115 Heidelberg, Germany

Printed by Libri Plureos GmbH
in Hamburg, Germany